The Open Mind

The Open Mind

Cold War Politics and the Sciences of Human Nature

JAMIE COHEN-COLE

THE UNIVERSITY OF CHICAGO PRESS   CHICAGO AND LONDON

The University of Chicago Press, Chicago 60637
The University of Chicago Press, Ltd., London
© 2014 by The University of Chicago
All rights reserved. Published 2014.
Paperback edition 2016
Printed in the United States of America

23 22 21 20 19 18 17 16   2 3 4 5 6

ISBN-13: 978-0-226-09216-4 (cloth)
ISBN-13: 978-0-226-36190-1 (paper)
ISBN-13: 978-0-226-09233-1 (e-book)
DOI: 10.7208/chicago/9780226092331.001.0001

Library of Congress Cataloging-in-Publication Data

Cohen-Cole, Jamie Nace, 1972– author.
  The open mind : cold war politics and the sciences of human nature / Jamie Cohen-Cole.
    pages ; cm
  Includes bibliographical references and index.
  ISBN 978-0-226-09216-4 (cloth : alkaline paper) — ISBN 978-0-226-09233-1 (e-book)
  1. Human behavior models—Political aspects—United States. 2. Cognitive science—Political aspects—United States. 3. Social sciences—Political aspects—United States. 4. Social sciences—United States—History—20th century. I. Title.
  BF39.3.C645 2014
  153—dc23

2013020551

♾ This paper meets the requirements of ANSI/NISO Z39.48-1992 (Permanence of Paper).

FOR EUGENIA AND NAOMI

Contents

Introduction

The Cold War was a time when psychology came into its own as a tool of social analysis. With marked rapidity the structural, institutional, and economic ways of understanding American society that had dominated academic and public discourse in preceding decades gave way to explanations framed in terms of the psyche. Historian Carl Schorske, recalling the intellectual currents of the immediate postwar period, found the "sudden blaze of interest in Sigmund Freud" particularly memorable. "Truly the premises for understanding man and society," he wrote, "seemed to be shifting from the social-historical to the psychological scene."[1] The sociologist Daniel Bell observed at the threshold of the 1960s that the previous decade "mark[ed] the difference" between "a Marxist analysis of America" and one cast in a "cultural anthropology *cum* a Jungian and nervous sociological idiom."[2] So warmly, it seems, had American intellectuals and social critics embraced the psychological idiom that eight years later the political writer Samuel Lubell could write, in the influential political journal *Public Interest*, "our society seems to have developed a predilection, even craze, for reading psychological explanations into anything and everything that happens, moving as far toward this extreme as Marxians once did in assigning an economic cause to anything and everything."[3]

If psychology could explain everything, there was one aspect of the self that held special importance to the intellectual and policy worlds: open-mindedness. Open-mindedness was a kind of mind characterized by autonomy, creativity, and the use of reason. To the scientific experts, intellectuals, and policy makers who developed and utilized the concept of the open mind, this type of self served simultaneously as model and ideal of national and intellectual character. They projected upon the open mind their aspirations for the American character and liberal pluralist democracy, for scientific thinking and true intellectual inquiry. Indeed, for some of these individuals the open mind transcended the academic and political, as its traits were even conscripted to serve as criteria for human nature itself.

Cold War intellectuals and policy makers saw in open-mindedness solutions to the most pressing problems faced by the nation. Those who defined American foreign policy believed that open-minded autonomy, a hallmark of American virtue, posed a threat to the communist system.[4] Traditional or authoritarian societies could not be sustained in the presence of a citizen body that thought autonomously, but for a modern democracy like America, open-mindedness would have the opposite effect, offering social cohesion. The open mind meant a respect for individuality, tolerance of difference, appreciation of pluralism, and appreciation of freedom of thought. If citizens were sufficiently equipped with these virtues, thought policy makers and social critics, the nation would flourish.

The various traits associated with open-mindedness reinforced one another and became organizing features of Cold War politics and intellectual life. Yet even when Americans agreed on the importance of freedom of thought, determining the specific characteristics of free thought was not easily done. It was only through concentrated attention to the pressing national problem that experts, educators, policy makers, and public intellectuals came to develop a common language through which they understood the cognitive virtues sibling to free thought. The concept of the open mind did not spring fully formed at the dawn of the Cold War era but was rather invented as a characterological umbrella that could unify the political and intellectual desiderata of the time. By studying the process of invention, that is, the efforts made in scientific and other contexts to understand, define, measure, and explain autonomous thinking, we can better understand the role of the open mind in shaping the intellectual, social, and political life of Cold War America. That role, as this book will demonstrate, was central, a success indicated by the way in which the virtues of the open mind became, for a time, nearly invisible norms of American culture.

*   *   *

The members of the community most responsible for defining, promulgating, and implementing the concept of open-mindedness were the founders of cognitive science, leading social scientists, the officers at grant-giving bodies, and the directors of science policy. Based in Cambridge at Harvard and MIT, this community also had a strong presence in New York, especially at Columbia University, the Ford Foundation, and the Carnegie Corporation. These scientists and patrons extended their reach in intellectual outposts at other universities that were restructured or otherwise shaped by their ideas. Institutions such as Berkeley, Stanford, the University of Michigan, and the University of North Carolina received advice from scholarly visiting committees, often with the promise of funds, to revise existing programs along the lines favored by the core intellectuals. Another major outpost was in the federal government, after John F. Kennedy staffed his administration with intellectuals from Harvard and MIT. McGeorge Bundy for instance, one of these Cambridge intellectuals, headed the National Security Council.

While members of this community were at different times based in specific institutions, it is important that we avoid identifying the community with fixed locations. Just as instrumental in nurturing their ideas as the places that served as their primary professional homes were the temporary spaces that tended their intellectual social lives—the conferences, cocktail parties, dinner clubs that they avidly attended, and the academic retreats that fostered intellectual socializing, such as the American Academy of Arts and Sciences, the Institute for Advanced Study, and the Center for Advanced Study in the Behavioral Sciences. Informal social occasions, not simply the daily grind of formal business and research, gave clarity and texture to the concept of the open mind. Members of this community themselves spoke and wrote voluminously about how much they valued such gatherings, and the informal discussions that occurred at these meetings, as a seedbed for their intellectual creativity.

It was especially in quotidian social experience that members of this community engaged a self that was multidimensional, simultaneously political, academic, and natural. Social experience carried over to intellectual work: notable in the transcriptions of many of these discussions, as well as in their academic and popular writings, is the way their treatment of the open mind reflected that intermingling of the three domains. As a result, scientific accounts of human nature echoed the concerns of national and academic politics. Conversely, political struggles on the national stage and within university departments employed the analytical tools and terms that social scientists and psychologists had defined.

We may distinguish, then, three primary roles fashioned for the open-minded self in the Cold War era. First, this self had a political role in which it served as an exemplary model of citizenship, engaging with others to make America a free and democratic society. Second, this self participated in academic society, partaking of the intellectual and social life of the university community and displaying characteristics of a model researcher, scientist, or thinker. And third, the open mind served as a universal model of human nature. In its first two roles, this self was a model in a normative sense—it was a role model *for* the proper modes of democratic or academic thinking. In its third role, the open mind was sublimated to become a model *of* normal human nature.

What did the civic open mind look like? Psychologists saw it in terms of a recurrent constellation of features: the open mind was tolerant, broad, flexible, realistic, unprejudiced. To sharpen its definition, psychologists also developed the tools to detect its antithesis: the closed mind was rigid, narrow, conformist, intolerant, ideological, and prejudiced. A closed-minded person rejected new ideas and people, and, because of compulsive adherence to ideology, lacked his or her own thoughts. Thus the closed mind characterized the subjects of totalitarian states, but it posed nearer threats as well. As Cold War intellectuals and policy makers saw it, in America the closed-minded citizen was responsible for two of the leading domestic concerns of the day, bigotry and mass society. It was imperative that the open mind be enlisted, on the one hand, to help keep the Communists without, and on the other, to eradicate the racists and conformist robots of the crowd within.

To realize their ideals, members of this intellectual circle instigated and oversaw the implementation of nationwide educational reform. Those among them who were academics undertook to redesign college distribution requirements so as to make students open-minded citizens of a very particular stripe, centrist liberals like themselves, who would populate a centrist pluralist society of the future. Nor did these individuals neglect the shaping of more tender minds: at the elementary and secondary levels, new science curricula were developed to fashion more open, indeed more *human*, minds.

While communism was undoubtedly a target for those who worked to promote the open mind, this political agenda had secondary, internal targets. Oriented toward making America more liberal, the open mind was also positioned against Joseph McCarthy, the archetypal anticommunist, and against members of the military. McCarthyism represented precisely the kind of extremism incompatible with the definition of a tolerant, rational mind. The military was a bastion of conformity and rigidity. Open-mindedness in the classroom, devoted to using active, discovery-based learning and designed to break narrow, "authoritarian" pedagogical techniques, was meant to enhance the very mental attributes that could resist McCarthyism and military structures—flexibility, autonomy, and creativity. This was how the nation could be defended from the rising tide of conformity, how American individualism could be preserved.

Such was the political role that the intellectuals and policy makers at the center of this study envisioned for the open mind. To turn now to the academic role, the actualization of open-minded ideals is best understood by looking at the rise of interdisciplinarity during this period. Interdisciplinary research is today widely regarded as a mode of knowledge-gathering that is inherently virtuous. This assumption is in fact a legacy of the persuasive rhetoric of Cold War intellectuals, who successfully valorized the research methods that they preferred. This they achieved by attaching to their practices the virtues of the open mind.

Taking advantage of a cultural climate that celebrated pluralism, these intellectuals cast themselves as good citizens of the academy, for to be interdisciplinary was to welcome and thrive on difference of ideas and viewpoints. Almost by definition they were broad and flexible, and as the characterological traits of "broad" and "flexible" tended to travel with "creative," "autonomous," and "rational," interdisciplinary scholars naturally possessed the latter attributes as well. Disciplinary researchers by contrast, bound to the precepts established by their own kind, were rigid, narrow, conformist, intolerant of difference, prejudiced against other fields, and ideological. They had all the deficits of the closed mind. Such volatile terms had their intended effect: the researchers who cast themselves as interdisciplinary were vastly more successful in drawing outside patronage and support from university administrators than their disciplinary counterparts.

Among the numerous fields that benefited from the preference for interdisciplinarity was cognitive science, the field that legitimized the scientific study of mind. The success of cognitive scientists in drawing financial support greatly contributed to the establishment of their fledgling discipline, the very survival of which depended on overturning the then prevailing view that mind, to all intents and purposes, did not exist. From the 1920s to the 1960s, behaviorist psychology held sway as the scientific approach to human nature. With the argument that science required ob-

jective observation of measurable phenomena, and that mental phenomena, being immeasurable, either should be left to the philosophers or,
like the soul, simply did not exist, behaviorists had managed to gain control of the scientific and rigorous end of psychology.[5] In essence their
scientific vision painted a picture of humans as Pavlov's dog, organisms
defined by no more than conditioned responses to environmental stimuli. Although mind was studied in some areas of early twentieth-century
psychology, behaviorists were largely successful in tarring as unscientific
any attempts to study mental phenomena on terms other than their own
deterministic or stochastic connection of stimuli to responses.[6]

With the behaviorist vision packing so much muscle within experimental psychology, the new branch of cognitive science had to be prepared to fight a vigorous civil war. At stake was recognition as serious
scientists. It was not sufficient for cognitive scientists to demonstrate how
they could study the mind—Freudians, social psychologists, school psychologists, and intelligence testers did so already. Respectability was to
be won by showing how the mind could be studied while not relinquish-
ing their hold on scientific rigor.

In this enterprise the cognitive scientists succeeded, and the means of
their success hinged in part on associating both behaviorist practitioners and methods, and behaviorist thought—their Pavlovian vision of human nature—with closed-mindedness. For what was the authoritarian
personality if not mindless and merely responsive to external stimuli?
And what was behaviorist commitment to disciplinary research if not
rigid and narrow? To the bleakness of this view the cognitive scientists
offered a bright alternative. They envisioned humans and their internal
psyches as independent of the environment, as autonomous and creative.
They presented themselves and their work as inclusive of diverse fields of
thought. Cognitive scientists not only epitomized the democratic character, but their account of humanity was more attractive. To accept their
scientific vision was to find that being quintessentially American was one
and the same as being human.

*   *   *

Understanding the political meanings of open-minded autonomy accordingly sheds new light on the political and cultural significance of
cognitive scientists' argument that humans are autonomous and not simply products of their environment. Existing histories either do not fo-

cus on the constitutive role of the politics of autonomy in the cognitive revolution,[7] or they deny it.[8] Where historians of science and scholars of science studies have considered the political implications of the sciences discussed in this book, they have centered their attention on military funding and the RAND Corporation. As a consequence these histories present a clear and direct connection between the forms of human reason analyzed in the sciences and Cold War military imperatives.[9] However much funding for social and cognitive science of the Cold War period came from military or civilian government sources, it is nevertheless also the case that the cognitive revolution and significant segments of social science operated on a much broader political register than those defined by military concerns.

Identifying the constitutive role of the open mind in Cold War America highlights the fact that the sciences had no simple relationship to the state and military. Far from being devoted primarily only to fighting communism, or to command and control technologies, the open mind was intended to make America more liberal. Recognition of the specific political values attached to the open mind also shows the ways in which Cold War culture was internally divided.

By giving attention to the development and application of the scientific techniques that defined open-mindedness, this book explains the establishment and subsequent unraveling of Cold War centrism. I offer an analysis of Cold War culture and the maintenance of its apparent consensus by tracking the tools of psychological analysis through which intellectuals produced the very conformity that they feared. It was not Zeitgeist, nor hegemonic ideology, but specific psychological technologies that assigned non-mainstream ideas to irrational ideology, thereby helping to produce consensus and conformity among the remainder of people—those who held "rational" ideas.

Psychological technologies deserve their own attention because they cannot be reduced to particular ideologies. Although scientific tools for describing human nature developed in and supported a very specific political order, they were not tied inexorably to Cold War centrism, a particular domestic social system, or even to a specific foreign policy. In fact, the very same scientific tools and modes of social analysis that had been used during the 1950s and early 1960s to police the boundaries of what counted as acceptable or reasonable political debate became potent weapons to crack the centrist political order. Members of the New Left and Second Wave feminist movements repurposed the scientific vision of rational selfhood to challenge rather than support the status quo. The Left's successful appropriation of centrist political tools led some prominent intellectuals to abandon their former affiliation with liberalism and identify as neoconservative. Conservatives, for their part, undermined one of the pillars of liberalism by pointing to the political and ideological significance of human nature as described by the cognitive and social sciences. With this attack, conservatives managed to break the connection between human nature and the values of Cold War liberalism.

In the last decade, historians of conservatism have aptly contended that these modes of psychological analysis, so prevalent during the 1950s and 1960s, obscured rather than illuminated much of what was important and significant about the rising right-wing political movement.[10] But however inappropriate 1950s psychological analysis may be for understanding the history of right-wing politics, we must go further than dismissing these psychological tools out of hand. For to do so is to misunderstand the mechanism and measure of modernity in centrist and liberal social thought. If these psychological tools are less than reliable for the analysis of the right, once we move past declaring them simply accurate, inaccurate, or mere reflections of hegemony, we can disassemble and inspect them. By doing so we can recover the mindset and social imagination of the people who made and used them to construct a central part of the intellectual and political culture of the Cold War.

*   *   *

The first part of the book focuses on the American mind. Chapter 1 examines the general education movement from the 1930s to the 1950s. It details the efforts of curriculum designers to protect democracy and to direct the future of modern America by making their students open-minded and flexible. Chapter 2 examines the way in which ideals of open-mindedness and creativity structured Cold War political and social thought.

The book's second section centers on the role of open-mindedness in the academy. Chapter 3 studies the professional and epistemic norms in the postwar social sciences. By showing the connections between interdisciplinarity, democracy, and open-mindedness, it explains why interdisciplinarity became and remains a valued mode of scientific research. Chapter 4 examines intellectual salons to provide a picture of the social and intellectual norms held by midcentury intellectuals. It highlights how in both casual conversation and formal work, social theorists conflated their own social world with that of America more broadly. In this social world, creative open-minded people were welcome and those deemed to have closed minds were excluded.

The book's third section shows how cognitive science propagated the epistemic and social values of the open mind. Chapter 5 focuses on how cognitive science turned the mental virtues valued by salons into features of normal human nature. Chapter 6 examines the first institutional home for cognitive science to show how the directors organized their center to foster a research culture that emphasized the very mental virtues of creativity and open-mindedness that they were studying. Chapter 7 details the production of open-mindedness on a national scale. Elementary and secondary science curricula sponsored by the National Science Foundation aimed to inoculate Americans against McCarthyism by making them more open-minded and therefore more "human," as defined by cognitive science.

The concluding section of the book shows that, although the virtues of the open mind had specific political meanings in the early Cold War, by the late 1960s the same virtues came to be associated with an entirely new form of politics. Chapter 8 details how feminists and the New Left managed to adopt the virtues of the open mind for themselves. In the early 1970s, a newly energized right wing organized against open-mindedness. Conservatives became enraged by the liberalism they saw in the cognitive-based approach to learning in the curricula supported by the National Science Foundation and attacked social science for its "anti-American" focus on human creativity, autonomy, and equality.

The American Mind

CHAPTER ONE

Democratic Minds for a Complex Society

When America entered the Second World War, concern over the nation's disunity deepened and spread among its citizen body. A perception that unity was critical to the sustainment of national morale led many to call for efforts to promote social cohesion. Policy makers and intellectuals responded by conducting studies on the causes of religious, racial, and ethnic divisions. Community groups, labor and business organizations, and government agencies sponsored advertising campaigns and established discussion groups to further intergroup tolerance.[1]

A distinctive feature of the thought and action on unity in this period was the emphasis placed on the role played by cultural and intellectual life. American social thinkers and policy makers, viewing culture as a primary determinant of social cohesiveness, worried about the corrosive effects of modern life. They feared that science and technology, expansion of knowledge, and resulting rapid change in daily life would fragment individual experience, tear social bonds, and dissolve the nation's coherence as a political entity.

The social and political impact of culture, particularly intellectual culture, was a theme that surfaced often and in wide-ranging arenas of discussion. In an address at the annual meeting of the Conference on Science, Philosophy and Religion (CSPR), David Lilienthal, director of the Tennessee Valley Authority (TVA), went so far as to attribute the cause of the current world war to disunity produced by modern culture. In blaming disunity Lilienthal meant something more specific than the rivaling ideologies of Allied and Axis powers. He meant that the modern world had been fragmented by its "high degree of specialization of function."[2] For illustration Lilienthal pointed to the variety of experts working for the TVA—specialists in fields ranging from electrical engineering to soil chemistry and dendrology. Expertise was a central fact of modernity, and in Lilienthal's view that expertise had become a barrier to meaningful communication, preventing people with different disciplinary training from understanding one another's work.

Remarkable as it may now seem that a high-ranking public official should locate the "root cause" of World War II not in Hitler's imperialistic ambitions but in modernity's excessive specialization, Lilienthal was addressing an audience who felt as he did. The CSPR had been formed out of the view that modernization, brought about by the growth of science and technology and resulting in ever-mounting forms of specialization, had produced a dangerous loss of common culture. As CSPR participants saw it, intellectual and spiritual disunity was chiefly responsible for the current "world crisis," and it was the largest threat from which America was to be guarded.

If intellectual and spiritual disunity was the greatest danger the world faced, then intellectuals and spiritual leaders themselves needed to recognize their agency both in producing and averting the crisis. CSPR participants, over one hundred fifty leading academics, theologians, politicians, and social commentators who gathered annually, made a point of discussing how they, with all the variety of philosophies and specialties they represented, embodied the problem they were trying to solve. They saw their own differences as undermining the stability of the world, and participants even charged others within the same conference with responsibility for the rise of fascism and abetting its totalitarianism.[3] It was thus critical, in their shared goal of developing a common culture for American society, to find a resolution to the differences they represented. Was a free spirit of inquiry, or spiritual and religious values, more directly tied to democracy? Was democracy based on Protestant, Catholic, and Jewish faith in the dignity of man, or on freedom of thought, including freedom from religious doctrine? Did secular or religious values hold more promise as a way to unify culture?

Lilienthal and the CSPR were representative of a national sentiment that since the root of the modern world's ills was its fragmented intellectual culture, the proper way to cure world and national problems was to change that culture. Debates over the intellectual directions that the country should take occurred perhaps in their strongest form over edu-

cational policy and curriculum design. If rooting out modernity's problems meant changing intellectual culture, then fashioning a proper mental landscape for American citizens in their formative years surely had to
be part of the cure.

A report put out by the Educational Policies Commission (EPC), a
policymaking group appointed by the National Education Association
and the American Association of School Administrators, expressed a
view similar to Lilienthal's, that the war had not been "chiefly caused by
the machinations of evil men," but was "largely a result of profound dislocations in the culture and social structure caused by the advances in
science and technology."[4] To address these dislocations the EPC recommended curricular innovations that would unify American students' educational experience and provide them with a foundation for their future
lives as citizens. Like the CSPR participants who sought to unify American culture by reconciling their own intellectual differences, members of
the EPC placed responsibility on intellectuals to produce unity in a diverse nation, which they would do by offering a properly designed educational experience.

The EPC played a large role in the world of education policy. It had
a distinguished membership of nationally recognized leaders of education such as Edmund Day, president of Cornell, and George Zook, president of the American Council on Education. After the war, members
included representatives from the U.S. Office of Education and the Carnegie Fund for the Advancement of Teaching; Dwight Eisenhower, then
president of Columbia University; and Harvard president James Bryant
Conant, one of most widely recognized experts on education and its relationship to social policy. In addition, the EPC was connected with several thousand well-placed consultants in education and media who advised the EPC on its reports and helped advertise and implement the
reports once they were completed.[5]

In calling for curricular innovation as a means to unify the nation,
the EPC lent authority to a sentiment that was already in wide circulation. Precisely what form that innovation should take, however, was a
contentious and long-debated question. To the intellectuals, educators,
and policy makers who were involved in this debate, the stakes were particularly high. Resolution would determine what students encountered
in the classroom, what kinds of people they would become, and the ultimate shape of society. It was not just the character of an abstract intellec-

tual culture that was under discussion, but also what kind of minds made for right-minded citizens—for individuals who would become guardians of American society and its democracy.

The debate over curricular innovation came down to two sets of proposed solutions, characterized either as "liberal education" or "general education." These programs were both aimed at unifying the secondary and collegiate curricula. The differences between liberal and general education turned on how each sought to manage the growth of knowledge, the expansion of science and technology, and the modern world's increased social complexity. Would educators follow the general education model, in which science and technology would be central to a unified curriculum, or would these be treated as peripheral to the concerns of a liberal educational program defined by the great works of Western literature?

This chapter examines the debate over pedagogy that roiled education and policymaking circles in the 1930s and first half of the 1940s. I show how discussions were imbued with politics and with midcentury anxiety about the fracturing of the modern world. Questions of pedagogy frequently became philosophical discussions about the meaning of proper citizenship, the definition of a good society, and the true meaning of democracy. In the end, neither liberal education nor general education succeeded in achieving dominance. Solution came in the form of a synthesis developed at Harvard in 1943–45. The work of a committee of professors and outside consultants was to provide a vision of the right kind of mind for America that came to have lasting influence.

Educating for Unity

In the 1930s a sense grew within the education community that the American curriculum was fracturing. Some felt that changes in the university that had been accumulating since the late nineteenth century had destroyed curricular unity and undergraduate education. As these critics saw it, the introduction of electives had broken up a curriculum once anchored by Greek, Latin, mathematics, and moral philosophy and by the goal of disciplining mental faculties through the study of these subjects.[6] Further, with growing emphasis on a research mission in universities, disciplines and departments proliferated as academics pursued increasingly focused and specialized work. Instead of offering broad instruction in their fields, professors opted to teach courses best suited to their majors and future graduate students. As a consequence undergraduates were faced with a collection of disparate, disconnected, narrow courses, each aimed at specialists.

Fracturing of the traditional curriculum was due not only to the introduction of electives but also to new subjects offered as a result of the Morrill Land Grant Acts of 1862 and 1890. These acts used funds from the sale of federal land to support the founding of colleges that would teach agriculture and mechanical arts.[7] Responsible for the creation of large-scale state university systems, the acts made study of practical fields available to a large portion of people who went to college. Nonetheless, despite the radical transformation to the American university curriculum effected by these laws, critics in the 1930s paid more attention to the elective system than to the Morrill Acts.[8]

Intensified calls for curricular change followed on this perceived fracturing of the curriculum. The general education movement emerged as a result, and interest in liberal education renewed itself. Leaders of the liberal education movement in the 1930s included Mark Van Doren of Columbia, Robert Hutchins, president of the University of Chicago, and philosopher Mortimer Adler. By contrast, early general education advocates were associated with less prestigious institutions and had a less public profile. They were able, however, to compensate for their disadvantage in standing by aligning themselves with John Dewey and Sidney Hook, who articulated positions that they often found compatible with their own.

Proponents of both pedagogical programs argued that a "cafeteria" style education served neither the student nor the nation. They contended that a new curriculum that would provide "unity" was needed. Such unity seemed to have several virtues. It would connect the various fields of knowledge to each other, it would connect academic study to the life of the individual student, and it would provide a means of forging stronger bonds between citizens and between individuals and society.[9]

Although advocates of liberal and of general education both agreed that the curriculum should be more unified, the two camps had different goals. Liberal education advocates argued that modernity, science, and technology had destroyed "culture," which they defined as learning in the Western canon. Their aim was to restore to American society a vanishing elite, a body of cultivated individuals who had undergone a humanistic program of study. On the other side general education advocates sought to produce capable citizens. They too worried about "culture," but not in the sense meant by liberal education. General education advocates thought of culture in an ethnographic sense—the democratic values and ways of life that formed the bedrock of American society. In their view modernity, science, and technology were destroying the unity of that culture, separating citizens from one another and from decision makers. Their desire for common culture was instead oriented toward sustaining egalitarian democracy and community.

Beyond having opposing visions of what kind of American to cultivate, the two camps differed in their views of appropriate educational content and method. Liberal education was committed to teaching the classics, especially religion, philosophy (metaphysics in particular), and literature. Education in science was felt to be of peripheral importance, if not actually inimical to study of the humanistic tradition.[10] Truth was viewed as universal and something to be taught, not discovered. In contrast, those in favor of general education either included or emphasized the teaching of science and technology, and espoused the relative or pragmatic nature of truth. They urged scientific teaching methods, and often secular, modern, progressive, and student-centered approaches. Knowledge was to be used for practical ends.[11]

Structural aspects of existing curricula dissatisfied members of both groups, but the reasons for their dissatisfaction were quite different. Liberal education advocates faulted academic disciplines as overly narrow, while those in favor of general education characterized them as impractical and disconnected from real-life student interest, abilities, and needs.

What each diagnosed as the root problem conditioned the nature of the cure they prescribed. To remedy what they saw as a problem of insularity produced by disciplinary knowledge, liberal education advocates hoped to build connection between citizens through a store of common knowledge provided by the great works of Western literature. Since impracticality and disconnectedness from real life was the problem for general education advocates, they called for curricula that would either be "unified" by the interests of the individual student or by their relevance to the contemporary world.[12]

The cures prescribed by one group ran counter to the problems diagnosed by the other. The kind of practical learning that the general education camp supported was precisely what was deplored in liberal education circles as materialistic and unintellectual. General education advocates, for their part, found it hardly "democratic" that the entire

American population would be interested in and prepared to read the great books of the Western world. The liberal education program was in their view as elitist as the traditional disciplinary teaching that liberal education advocates criticized.

In 1943 James Bryant Conant, president of Harvard and future head of the EPC, joined this debate by establishing a committee to develop a general education program that would serve the needs of postwar America. At the time he was serving as chairman of the National Defense Research Council, a part of the Office of Scientific Research and Development, which was devoted to developing weapons, including the atomic bomb, for prosecuting the war.

Between his trips to Washington, D.C., to oversee the Manhattan Project and other wartime scientific developments, Conant turned his thoughts to reshaping America through educational reform. Drawing an analogy to the wartime service of scientists like himself, Conant thought of putting faculty not so directly involved in the war effort to work securing the peace. These faculty would shape the postwar nation by defining what the objectives of general education in a free society should be.[13]

Over the course of the next two and a half years, the committee Conant formed often met several times a week for several hours a session.[14] Discussions and reports during these meetings drew on the views and testimony of over seventy-five consultants, who ranged from state and city commissioners of education to preparatory school headmasters, sociologists specializing in education, union representatives, and theologians. In their discussions and final report, the committee members and their consultants considered and sought to offer a final answer to the problem of national unity.

The meeting transcripts provide a unique picture of the social thought of mid-twentieth-century America. Because the committee preserved the day-to-day and even minute-to-minute record of their activities, we are able to witness a more candid, less carefully edited and measured view of education and democracy than what the committee put into published form. The transcripts also archive the social thought of individuals who, whether because of their field of specialty or because of their social position, left no other record of their ideas about what constituted citizenship and the good society.

In its meetings and final report, the committee articulated an ambitious pedagogical philosophy for the nation. *General Education in a Free Society* (often called the "Red Book" for the color of its cover), published in 1945, contained both plans for the country's future and a method for achieving them through the shaping of student minds. The report gave an account of democracy and a definition of right-minded citizenship, and it added chapters on curricula for all colleges and high schools and for Harvard in particular. It provided, in addition, a political vision for how America could remain a democracy even while fundamental social functions were removed from the sphere of politics and public deliberation to be managed by experts or state bureaucracies.

Committee meetings featured expositions of participants' views of democracy, the United States, social relations, the proper way to think, the best ways to teach and learn, and the good life. Their views on these topics drew on an eclectic range of sources, from personal beliefs to expertise in American history, normative epistemology, and classical history and philosophy. For instance, historians Arthur Schlesinger Sr. and Paul Buck offered their knowledge of American society past and present. Classicists John Finley and Raphael Demos offered background on Greek education, democracy, and the Aristotelian view of how education contributed to citizenship and the good life.

Committee members referred to themselves with pride as general intellectuals, not education experts. For knowledge of contemporary school and social conditions they were glad to rely on outside consultants. Robert Havighurst and Byron Hollinshead, for example, education experts from the University of Chicago, spoke on the relationship of curriculum to social structure. Other consultants informed them of the effects of economic conditions and demography on high schools and colleges, conditions in local schools in upstate New York and Ohio, and the particular needs of labor unions and engineering firms for well-rounded workers. The product of the committee's work would therefore represent the joint effort of intellectuals whose particular areas of knowledge lay outside of pedagogy per se.

Ultimately they were confident enough in their own intelligence, knowledge, judgment, and wisdom on matters of democracy, education, and citizenship to base their proposal for reshaping American society largely on their individual and collective views. Periodically they did rely on the thoughts and ideas of those outside the room, living as well as dead. Figures who appeared in discussion ranged from ancient Roman master of rhetorical education Quintilian, Renaissance humanist scholar and theologian Desiderius Erasmus, and Italian philosopher Benedetto Croce, to Thomas Jefferson, Karl Marx, and Karl Mannheim.[15] The most recurrent sources of authority the committee looked to, however, were the ancient Greeks and resident Harvard philosopher Alfred North Whitehead. These figures were held in so much respect that committee members, including Paul Buck and John Finley, respectively its chair and vice-chair, reverentially called on their ideas to support particular views that were under discussion.[16]

One prominent contemporary figure whose work concerned precisely the question of the relationship between democracy and education was given short shrift: John Dewey. In hundreds of hours of discussion and numerous reports and memos, the only time John Dewey made a significant appearance was in a report by Robert Ulich, a professor of education. Ulich examined the history of pedagogical philosophy since the ancient world. In that context Dewey not only figured as just one of numerous educational philosophers, his thought was also swiftly dismissed. Ulich commented, "Through basing education on merely instrumentalist concepts [Dewey] gives no philosophically satisfying answer to the problem of values and goals of both education and democracy."[17] This rather casual dismissal of Dewey, which was received with no objection, indicates the extent to which the committee and its consultants were largely content to approach social thought, philosophy, education, and democracy through their own knowledge.

By the time its discussions had been polished for public consumption in the Red Book, the committee had to some degree modulated its private dismissal of Dewey. This modulation was due in part to the committee's effort to present a measured synthesis of the existing polarized positions on education. At one end there was the scientific or pragmatic approach, best represented by Dewey and his followers, and at the other there was the educational philosophy that centered on the Western tradition and the great books. By way of synthesis the report argued that the empirical approach central to pragmatism was well rooted in the Western tradition. Nevertheless, the authors' efforts to give a fair account of pragmatism did not prevent them from calling into question the propriety of using the scientific approach in all domains of human affairs.

This questioning of the "pragmatist solution" occurred in the context of an argument that placed it as just one of the four existing ways to unite a curriculum. The other methods were the program (offered mostly by Catholic colleges) that sought to provide unity through Christianity; a liberal arts program based on the Western tradition; and a curriculum based on the practical problems of modern life. This final option

was one that many read as "Deweyan" or "progressive." The committee
deemed all four of these approaches insufficient as unifying principles
for American general education. They were listed merely to function as
preambles to and foils for the synthetic program advocated by Harvard's
committee.[18]

The Mind as Synthesis

The Harvard committee and its consultants sought to find a resolution
between the two primary approaches to education, the rationalistic and
religious on the one hand and the scientific and pragmatic on the other.
The committee and its consultants argued that the various competing
positions of general education could be synthesized because they all
shared a belief in the "human spirit" and "human dignity."[19] Because
this humanism was supposedly shared by all versions of educational theory, the committee suggested following the "American spirit" of compromise and proposed "cooperation on the level of action irrespective
of agreement on ultimates."[20] This meant working directly toward developing the minds of students regardless of the manner of justifying such
activity—whether religious and traditional, or modern and scientific.

Synthesis would be achieved by putting a certain kind of mentality,
rather than specific books, the scientific method, or particular aspects
of modern life, at the center of general education. The mentality to be
molded was not based on knowledge but on intellectual skills.[21] The report envisioned Americans unified through the shared skills of effective
thinking, judgment, communication, and the ability "to discriminate
among values."[22]

To understand this point about the preference for mental skills over
knowledge, it helps to see how the Harvard committee came to a consensus that common knowledge was *not* an essential part of general education. This was a hard-won decision. Members of the committee and
their consultants had begun from the commitment that general education necessarily involved unifying the curriculum by giving students
shared knowledge.[23] This view reflected a position long held by advocates of liberal education, notably Hutchins and Adler, who equated
unity of curriculum with cultural and political unity.[24] Unlike Hutchins
and Adler however, the committee found that they could not agree on

a set of canonical works that would form the core of their own curriculum. Repeated efforts at resolution over several months in late 1943 ended in failure.[25] Committee members were also unwilling to treat the work of modern science either as peripheral to general education or to reduce modern science to a set of timeless principles as Hutchins had advocated.[26]

In the end the committee and its consultants turned their own failure to agree on necessary common knowledge from a problem into a solution. Testimony on American social conditions had convinced them that the traditional liberal arts curriculum was inappropriate for many students.[27] For that reason a truly democratic general education could not have a set curriculum. As Alonso Grace, Connecticut commissioner of public education, put it during committee discussions, diversity of student abilities and interests demanded a "flexible" general education curriculum.[28] For the committee, flexibility came to mean a program that would unify Americans without requiring them all to read the same books.

This position developed out of attention to several features of modern American society. First, the committee recognized the reality of social stratification, especially along lines of class, and individual differences in scholastic ability. Second, it recognized that the diversity of modern American life meant attention to the increasing role of specialization of knowledge as a "centrifugal" force in modern society.[29] Third, it recognized the position of progressive educators who advocated general education as a process of instruction in the issues and affairs of modern life rather than as cultivation through exposure to the great works of the past.

Harry Gilson, the Maine commissioner of education, argued during deliberations that what was important in general education was the method of analysis and mode of thinking that students learned, not the particular works they studied.[30] In its final report the committee incorporated Gilson's views, noting that "general education must accordingly be conceived less as a specific set of books to be read or courses to be given, than as a concern for certain goals of knowledge."[31] By taking this position the committee opposed the educational philosophy that promoted liberal education through the great books. When it adopted this view, the Harvard committee tilted toward progressive educators at schools such as the University of Minnesota and Sarah Lawrence College, who called for general education to bring unity to students' lives through coursework that directly connected the classroom to students' individual interests or the practical affairs of modern life.

The committee's difficulties with knowledge as a goal of education included survey courses as well as programs featuring a core curriculum. For instance, Paul Buck and John Finley specifically criticized the survey courses at the University of Chicago for providing instruction in the form of thinly spread "inert information" instead of teaching students how to develop skills such as ratiocination.[32]

As a result the committee's curricular suggestions in its final report consistently expressed preference for mental skills over knowledge.[33] It suggested that high school English classes should be taught with fewer rather than more books so that more time could be spent on analysis. The committee recommended that no particular set of books form the curriculum. Both suggestions were aimed at helping students develop general skills in reading and analyzing literature rather than at acquiring knowledge of the literature's content.[34]

The view that a set curriculum was not of primary importance in education echoed the results of other important curricular studies of the period. In 1944, a year before the publication of *General Education in a Free Society*, the EPC published a report entitled *Education for All American Youth*. The report was an exercise in speculative futurism written in the form of a history of curricular change from 1945 to 1955. As the EPC imagined it, one of the most significant changes that people in 1955 would appreciate was a movement away from prescribed sequences in high school courses to focus on developing general mental abilities.[35]

The Progressive Educational Association's "Eight-Year Study" had advanced a similar perspective in 1942. The research program had enrolled thirty high schools that had committed to different levels of curricular reform. The study examined the students' progress through their four years of high school and four years of college. One significant result of the study was that there was no particular high school curriculum necessary for success in college. Indeed it turned out that students who had been educated in nontraditional high schools received slightly higher grades in college than their peers. The report argued that the important factor was the kind of mind that students developed, not the specifics of what they read. College admissions officers, including Harvard's, ratified this perspective by concluding that they could choose their entering classes even without a set high school curriculum.[36]

Envisioning Society

The position taken by the Harvard committee that it was impractical if not impossible to form a curriculum based on a common core of knowledge was an effort to meet what it saw as an undeniable feature of American society—its diversity—on its own terms. Further, the committee hoped that its plan to put mental development at the center of education would resolve the broader societal problem produced by modernity itself.

In keeping with the prevailing view of the time, the committee noted that the modern world was characterized by innumerable modes of expertise. "In this epoch," wrote the committee, "almost all of us must be experts in some field in order to make a living. . . . I must trust the advice of my doctor, my plumber, my lawyer, my radio repairman, and so on."[37] Given the quantity of specialized knowledge, it seemed clear that no one could master every kind. The consequence, as the committee observed, was that the simple act of negotiating everyday life required each person to rely on others.

By casting matters this way, the committee put the distribution of expertise in an egalitarian frame. But it opened up a new problem. If every American needed to rely on other people and their particular areas of expertise, then would not everyday living need to be conducted on a great deal of trust? In order for society to function well, there had to be some rational basis for that trust. In other words, the kind of society envisioned by the Harvard committee was one in which the ability to judge specialist competence from a nonspecialist perspective was an indispensable skill.

This ability became the core of Harvard's vision for general education. What was needed was an educational program that would give American youth the essential mental tools to evaluate expertise in fields not their own. As the committee put it, "there are standards and a style for every type of activity—manual, athletic, intellectual, or artistic, and the educated man should be one who can tell sound from shoddy work outside his own field."[38] It concluded: "The aim of general education may

be defined as that of providing the broad critical sense by which to recognize competence in any field."[39]

Through this solution the Harvard committee again turned a problem into an advantage. Rather than try to counteract the overwhelming quantity of knowledge in the modern world by determining a priv-
ileged subset for all Americans to be acquainted with, the committee
accepted the fact on the ground and instead proposed a method of managing the large quantity of knowledge. This proposal both resolved the
problem of disunity and advanced the committee's collective moral and
political values. A society in which everyone had the same set of mental
skills would cohere by way of a special national character of mind. Be-
sides achieving unity through the sharing of mental traits, such a society
would also be fundamentally egalitarian, because everyone was either
expert or layman according to context. Modernity's overspecialization,
instead of being merely a social problem as Lilienthal and others characterized it, was thus placed in a democratic frame and assimilated to positive political ideals.

The committee's vision of America as a plurality of expertise provided not only a way of relating individuals to each other but also showed
how individual citizens could relate to the state. As the classical philosopher Raphael Demos characterized it, the committee's approach to
thinking and education was no different from the system of representative democracy:

> Since a single man can not become an expert in the ever so many fields of hu-
> man endeavor, mastery consists in the ability to recognize and choose the ex-
> pert. What I call the common or liberal education is just this ability: it is like
> the ability in democracy to govern ourselves; a layman is not necessarily an
> expert in government, but he is (or we hope is) an expert in choosing experts
> in government; he can appoint them, and can dis-appoint them.[40]

In the final report, the committee adapted this point to argue that the
same set of mental skills that ensured social cohesion applied in the domain of electoral politics: just as people should be able to judge experts
from any field, so too should they be able to evaluate politicians in their
"field" of governance.[41] Harvard's version of general education, then,
would safeguard modern representative democracy by developing citizens who were fully capable of making wise electoral decisions.

The Harvard curriculum was revealing of committee members' faith in the efficacy of reason. Their vision of the impact that their educational program would have on voting processes assumed that the business of choosing a political leader was essentially and properly a matter of technical evaluation. There was no consideration of the role that personal interests, preferences, and loyalties played in deciding which candidate was best. In fact the report naturalized the committee's conception of democracy so that an electoral process was only well-functioning when citizens made rational choices about which candidate was most qualified or competent for the job. It did not take into account the possibility that a candidate who was "best" for one voter was not necessarily best for another. Personal interests were delegitimized, since voting on the basis of any reason other than a candidate's competence was implicitly assimilated to ignorance, illogic, or prejudice.

A Society of Experts

This model of cultivation and evaluative thinking provided a means for imagining the possibility of cohesion in a complex capitalist society. The culture envisioned by the committee was not one that would require an ultimate defense of principles beyond the essential qualities of human nature and human talents. The democratic society that the committee imagined would be held together by right-minded people who could speak to and judge one another according to universal rational standards.

The report envisioned Americans unified one with the other through shared and universal mental skills. Through these skills, the American people could be the kinds of citizens necessary for a complex modern democracy. While this social vision expressed the committee's image of the mental, social, and political future of the entire nation, the faith that America was a disparate community of experts held together by the rational abilities of citizens to communicate across the boundaries of expertise was grounded by intimate experience with a specific social environment: Harvard itself.

As Joel Isaac has demonstrated, Harvard had fostered an environment that located the centers of intellectual activity and even institutional power within interdisciplinary settings. In numerous places—from the junior and senior common rooms in the houses (the undergraduate dormitories modeled on the colleges at Cambridge and Oxford Universi-

ties), in the Society of Fellows, in the Faculty Club, in discussion groups
and dinner clubs—Harvard culture encouraged faculty and students to
engage in intellectual socializing that bridged the disciplines in which
they specialized.[42] In places like the "Shop Club" the elect faculty members shared food, drink, and fellowship.[43]

These interdisciplinary settings emphasized the ability of members to
speak across the boundaries of disciplinary expertise either by eschewing narrow disciplinary jargon and adopting a language and manner of
speech appropriate to a varied audience or by developing a set of theoretical tools that could be applied in several disciplinary contexts. Beyond what they required of speakers, these interdisciplinary environments demanded that other participants, the listeners, be able to evaluate
the quality of the speaker's ideas and intellect. Notably, that evaluation
would need to be accomplished by individuals who lacked the particular
form of disciplinary expertise possessed by the speaker.

This set of intellectual and social values was perhaps best represented
in two institutions that Harvard established in the 1930s: ad hoc committees to evaluate candidates for tenure and the Society of Fellows. After
becoming Harvard's president in 1933, Conant established an up-or-out
policy for junior professors. The new system required junior professors
to be evaluated not only by the senior members of their own departments
but also by an ad hoc committee of administrators and scholars from
outside the department, selected for the occasion. Members of that committee could include both disciplinary specialists as well as intellectuals
trusted for their discerning judgment but who were not in the field.[44] The
advice of the ad hoc committee was then passed on to the dean of the
faculty and president and fellows of the university, none of whom were
presumed to have any disciplinary knowledge about the tenure case, for
final determining decisions.

At the Society of Fellows, the ethos of nondisciplinary communication and the faith that nonexperts could judge the intellect of people outside their own field were institutionalized in its day-to-day activities and
even in modes of selecting members. George Homans and Crane Brinton recount a selection interview where the candidate was told: "This is
not an examination. *No one in this room is competent to examine you.*
The purpose is for us to get acquainted, and the best way to do that is to
talk. So talk!"[45] Of course, the selection committee *was* examining the
candidate and *did* feel itself competent to do so.

The Society had been established after Conant's predecessor, Abbott

Lawrence Lowell, and faculty members Lawrence J. Henderson and Alfred North Whitehead noted that there were not spaces at Harvard for the development of original thought outside what they saw as the narrow and even vocational constraints of disciplinary training within departments. The model of intellectual society they sought to emulate was Cambridge University's Trinity College, its system of Junior Prize Fellowships, and the pattern of social and intellectual life at Cambridge High Table. Whitehead contended that he had always learned most from cross-disciplinary conversation with people he knew well. And so the Society established a pattern of encouraging such conversations with required dinners and lunches at which the right kind of interchange would occur.[46]

Lowell, Henderson, and Whitehead hoped that the Society would play as central a role in Harvard's culture as the fellowships in Trinity's culture. The Society's ideal of combining convivial discussion, fellowship, dining, and intellectual life for elite postgraduates would be expanded to Harvard's entire undergraduate body with the implementation of the "house" plan for all Harvard undergraduates. The first houses opened in 1930 and the Society of Fellows opened in 1933.[47]

It was this conception of intellectual merit and socializing that formed the social vision of the committee on general education. Thus when general education committee member Raphael Demos reflected on the role communication plays as a glue for American society, he drew on his experience of communication in the community at Harvard:

> Since it is so concrete itself, conversation thrives when aided by concrete physical things: good food, drink, and smoke, pleasant rooms and comfortable chairs. Surely the opportunity of the Harvard houses, in providing the setting for education conversation, needs no stressing; I have in mind especially the dining rooms (and the common rooms).[48]

Demos's argument about the centrality of common rooms to the community at Harvard played a critical role in the development of the committee's general education proposal not just because it was compelling to the other member of the committee. This argument was also important because it occurred at a critical juncture in the committee's deliberations. Demos suggested the common rooms as a way of thinking about America just as the committee was developing its account of the nation as a collection of experts. Thus American democracy could be a plurality of experts that was unified by emulating Harvard's Shop Club, the Society of Fellows, or the common rooms, each a place that encouraged cross-disciplinary intellectual banter.

Although deeply rooted in Harvard's own culture, this perspective on the good society as arising from a learning environment with "pleasant rooms" and "comfortable chairs" was far from idiosyncratic. It also appeared in, for instance, the Educational Policies Commission's *Education for All American Youth*. This book pictured communities all across the country centered not just on schools, but on schools equipped with rooms designed to increase education though specific creature comforts. These rooms would have "panelled walls, built-in bookshelves, indirect lighting, and pleasing harmony of colors." The "beauty" of the room would be "enhanced by the furnishings—the large tables and comfortable library chairs, the draperies, rugs, and floor lamps, the vases of flowers from the school garden, and the half dozen prints and paintings which hang upon the walls."[49] Such spaces would promote learning of humanistic subjects, and the schools with such spaces would serve as a civic hub for the larger community. Thus, where the Harvard vision was that its own common rooms would be the model for American society, for the EPC the basis of American society would be a well-appointed room in a rural high school.

Seeing convivial conversation as the basis for a smoothly functioning society had important consequences for Harvard in the subsequent development of its own general education program after the war as well as in the way it structured its own society of scholars. Like the committee on general education he had formed, James Conant saw general education as being directed toward a conversation in a social setting—such as at a cocktail party.

In articulating his vision of how the public should be educated in science, Conant contended that what was important was *not* knowledge of scientific facts or of the leading edge of scientific research. Instead, "understanding science" meant having a "feel" for the "tactics and strategies" of science. As Conant saw it, scientists possessed understanding of strategy and tactics by virtue of training in one field. As a consequence of this understanding, a scientist could have a meaningful discussion about science "even in an area in which he is quite ignorant." Given the martial nature of Conant's metaphor for understanding, it is worth remembering that Conant had personally had just such an experience of interdisciplinary scientific discussions when he served as chairman of the National Defense Research Committee (NDRC) during World War II.[50] Ultimately what his view of science meant for general education was that the general population should take classes that would enable them to have a conversation about science by understanding its tactics and strategies.

Subsequently, Conant designed and taught a general education course at Harvard that sought to incorporate these social and pedagogical philosophies. In the introduction to *The Copernican Revolution* by Thomas Kuhn, a book that was written for the general education class Conant taught, Conant lauded the kind of knowledge that allowed facility in conversation across disciplines. As he put it, when seeking facility with communication, Conant was not concerned with "a scholarly command of the ancient and modern classics" or even "a sensitive critical judgment of style or form." Instead what he believed was important was the kind of "knowledge which can be readily worked into a conversation at a suitable social gathering."[51] But Conant was quick to point out that the desired knowledge and communication skills were lacking in America—particularly in the domain of the sciences. As he put it, "it is very hard indeed to keep a conversation going about physical science in which the majority of participants are not themselves scientists or engineers."[52] It was because *The Copernican Revolution* made possible casual banter about science that Conant offered his enthusiastic endorsement of the text.

And this was why Kuhn's book was so important to Conant. It was an example of what general education in the sciences could accomplish for its students and the nation. It would enable the operation of democracy and modern society as envisioned by the Harvard committee on general education. This book would strengthen American society and its democracy by equipping nonscientists with a "working knowledge" of science; that is, knowledge that could "be worked into conversation at a suitable social gathering." That possibility, in turn, protected modern America from disunity and from undemocratic rule by experts unaccountable to a public that did not understand them.

Conclusion

Although deeply informed by local intellectual and institutional culture and the process of its composition, the Red Book sought to and did play

a large role in the way that pedagogues envisioned the relationship between education and citizenship. Because it distilled the perspectives of
a wide range of external consultants, the Red Book also marked out a
social vision that crystallized several strains of liberal social thought that
would be central to the postwar period.

The Red Book was a significant contribution to the discussion on curricula because it was a synthetic work, had been produced by Harvard,
and because James Conant was already a national leader on the relationship of education to society. The report's impact was further magnified
by the many people who read or bought it. Harvard itself financed send-
ing copies of the book to educational leaders and policy makers. The
book drew reviews in the popular press, including the *New York Times*,
and sold over forty thousand copies in five years.[53]

Only a year later, when the *Journal of General Education* was
launched at the University of Iowa, the Harvard report on general education became a—if not the—primary framework for consideration of
general education in the pages of the journal. It was the most common
benchmark against which other general education programs would be
measured. The prominence of Harvard in the pages of the journal was
certainly helped by the fact that the editor, Earl McGrath, made Harvard's vision of general education his primary and often sole point of
reference.[54]

By 1949, McGrath had resigned as editor of the journal. But as he
moved to serve as U.S. commissioner of education, McGrath's views on
pedagogy and society would continue to be influential.[55] Into the 1950s,
other educators continued to treat Harvard's vision of general education
as a significant synthesis of existing approaches to general education.[56]

Beyond serving as a synthesis of existing pedagogical programs, the
Red Book also articulated a program for solving the critical problems
facing America's modern and complex society. The book underlined a
commitment to the view that all Americans had a claim on the mental
virtues of citizenship regardless of whether or not they had attended a
school that taught from the great books. This view echoed James Conant's long-held position that individual merit should be assessed not on
the basis of what people knew but on how well they could think. Because
of this commitment to ability over knowledge, Conant changed Harvard's own admissions policy to focus on intellectual ability, authored
a number of works on the importance of selecting students on the ba-

sis of merit not wealth, and, perhaps most importantly, was instrumental in supporting the development of the Scholastic Aptitude Test (SAT), a primary way that Harvard and other colleges and universities came to choose their student bodies. The SAT, modeled on IQ tests, was most specifically not a test of knowledge. Instead, it was designed, marketed, and used as a test of aptitude for future performance.[57]

The future imagined in *General Education in a Free Society* was a technocracy in which each citizen was educated in how to appreciate, judge, and defer to expertise and in which political questions and even voting became technical problems. This technocratic vision fit with the emerging national security state that James Conant, in his role in aiding the development and deployment of the atomic bomb, helped bring into being. The Red Book's vision of a free society also fit the social and political order that developed at the Tennessee Valley Authority (TVA). TVA director David Lilienthal had worried that specialization had been disruptive enough to have been the cause of the world war. However, he, like the authors of the Red Book, found unity and democracy in interdisciplinarity. Thus when he published his account of the TVA, Lilienthal saw unification of numerous kinds of expertise on a single farm as "grass roots" democracy.[58]

However, if the TVA was an interdisciplinary democracy, it was also, even by 1944 when Lilienthal published his account, becoming the arsenal of democracy. Among the projects the TVA took up during World War II was supplying power to the Oak Ridge Laboratory, the enormous industrial plant that enriched uranium for the Manhattan Project and the atomic bomb that was eventually dropped on Hiroshima.

The technocratic national security state that Lilienthal and Conant helped bring into being at the NDRC, TVA, and AEC was a kind of society in which graduates of Harvard's plan of general education would be excellent citizens. It would be a system in which democracy, freedom, and natural culture would be unified by the rationality of its citizens and their trained abilities to separate good from bad in fields ranging from art to science, electrical work, and politics.

What ultimately mattered most to Conant, to the committee he assembled, to its consultants, and to other advocates of general education was that it was education for freedom and democracy. On this point Conant noted that although education in the disciplines was the same on both sides of the Iron Curtain, it was only the free world that pro-

vided its students with the general education that equipped them with
the mindset to be citizens in a democracy.[59] This was a free society, a society which would flourish because its citizens would have cultivated universal standards of judgment as well as faith in and commitment to the
mental virtues of rationality, creativity, tolerance, communication, and
open-minded inquiry.

CHAPTER TWO

The Creative American

Modern society filled Cold War intellectuals with anxiety. Its bureaucracy, social fragmentation, homogeneity, and soulless consumerism represented alienation, isolation, and the evaporation of true community. It meant conformity and the erosion of the nation's frontier spirit and entrepreneurial nature. Without proper care, it could lead to mob rule or totalitarianism.

This chapter examines how Cold War social critics and policy makers found the answer to these seemingly pressing social and political problems in the systematic and scientific study of individual character. Their science provided tools for understanding the kind of person who threatened to make America into a mass or even an authoritarian society. Scientific tools also crystallized a form of the exemplary self that would inoculate America against the dangers of mass society.

The defining feature of that positive personality type was creativity, a trait taken to be interchangeable with autonomy, rationality, tolerance, and open-mindedness. John Gardner, president of the Carnegie Corporation and patron of much Cold War social science including creativity research, noted: "[Creativity] is a word of dizzying popularity. . . . It is more than a word today; it is an incantation. People think of it as a kind of wonder drug, powerful and presumably painless; and everybody wants a prescription. It is part of a growing resistance to the tyranny of formula, a new respect for individuality, a dawning recognition of the potentialities of the liberated mind."[1] Men such as Gardner believed that creative people would bring coherence to America's increasingly complex and diverse culture. At the same time, these creative people would mitigate the conformity that many social critics feared was a key characteristic of both traditional society and modern mass society.[2] Their creativity would be the critical ingredient in making possible the "unity in diversity" that Cold War social critics believed to be the defining feature of the liberal pluralism they desired for America.[3]

Here I examine an effort in the 1940s and 1950s by liberal intellectuals to politicize personhood and promote autonomous selfhood in the service of making a better nation through reform of the self. My analysis centers on two ways in which intellectuals understood and advocated the connection between creativity and liberalism. First, I examine the frameworks for understanding self and society developed by liberal American intellectuals. They focused on the threats to unity and diversity, such as McCarthyism and racism. Concerned by reactionary efforts to maintain the status quo or even revert to the past, they searched for a way to achieve social and cultural unity while also respecting individuality and America's pluralist character.

The first step in addressing this tension was to understand scientifically the kind of individual character that intellectuals thought to be most threatening to their ideal social and political order: the closed-minded authoritarian. According to social scientists, authoritarians endangered liberal, pluralist democracy because they were handicapped by a lack of psychic freedom and full selfhood. Freedom, autonomy, and creativity were defined as the inverse of these authoritarian traits. In addition to their currency within the academy and the world of social science, these models of good and bad selves provided the American public with techniques of self-inspection, tools for self-management, and benchmarks to which they could aspire.

Second, I explain how these models of the self were used for political ends. Liberal social scientists sought to mold America into a nation that fit their vision of the good society. They did so by managing the definitions of creativity and autonomy in such a way that those traits, once redefined, would describe their political allies. Conversely, the opposite traits—conformity, rigidity, and narrow-mindedness—were defined so as to apply to the liberals' McCarthyite opponents on the right and their Communist foes on the left. By marking as irrational the social and political views they disagreed with, liberal social scientists played an important role in marginalizing non-centrist political ideas as irrational and thereby helped generate the apparent consensus of the Cold War era.[4]

Pluralism and the Challenges of Modern Society

From World War II into the Cold War, social critics, intellectuals, and policy makers came to see change, variety, and, especially, complexity as defining features of modern America.[5] Commentators noted the proliferation of institutions, professions, occupations, and forms of knowledge, the multiplicity of religions, races, and ethnicities. Even though, in principle, liberals appreciated variety, they also felt that if it was not carefully controlled, American society and culture would fracture and undermine the nation's democracy. The question, then, was how to develop a society that would facilitate cultural unity and integration while maintaining healthy room for diversity and toleration of difference.

This focus on unity remained a constant fixture in the postwar period. For instance, it pervaded a 1957 conference devoted to understanding the "American style," sponsored by MIT's Center for International Studies (CENIS), a research arm of the CIA and the State Department. The conference attendees, luminaries of the academic and policy worlds, kept circling back to discuss how America had "fragmented into numberless small communities organized on varying social, geographical, professional, or intellectual principles; a society infinitely diverse."[6]

Likewise, in 1961, Seymour Martin Lipset and Leo Lowenthal warned: "Democracy itself is threatened [when] all discussion and decisions become segmentalized, each only understood by a tiny informed coterie."[7] Here Lipset and Lowenthal cited and rephrased a concern C. P. Snow had raised in 1959 in *The Two Cultures*, a book concerning divisions between scientists and humanists.[8] But, as the earlier comments by David Lilienthal, in the general education movement, and at the conference on the American Style indicate, Snow did not uncover a previously unnoticed phenomenon. Rather, his book arrived in a hospitable cultural environment and was a particularly powerful articulation of a set of social fears that had long preoccupied intellectuals.

However, the necessity that citizens manage modern America's complexity meant much more than coping with the proliferation of fields of learning or of specialized professions. More substantively, complexity also referred to social, political, and social diversity. Hence, what was needed were people who could balance two tasks: negotiating a world with increasing professional specialization and appreciating America's pluralism. Accomplishing both tasks ultimately required a certain kind of mindset. Both tasks were also threatened by another kind of mentality: narrow-minded conformity.

Analyzing Conformity in the Social Sciences

Although intellectuals were concerned about how the complexity of modern life threatened the unity of American culture, they also identified a second challenge for the nation: finding a way for citizens to maintain their autonomy. How could citizens retain their individuality in the face of the pressures for homogenization and conformity that permeated American culture? What connected these two challenges—achieving unity and the maintenance of individual autonomy—was a consistent view and set of values regarding the self and its rational independence.

Autonomy was a theme common to discourse about life in America as well as in analyses of international politics and culture. When considering the domestic side of these issues, Americans—from the most elite circles of intellectual discourse to the popular media—focused in particular on autonomy's inverse: conformity. From the works of such popular social critics as William Whyte and David Riesman to mass-market magazines like *Reader's Digest*, *Woman's Day*, and *Life* and novels like *The Man in the Grey Flannel Suit*, Americans expressed anxiety about the growing conformity in the nation.[9] They believed that the conditions of modern American life, including the corporatization of work and suburban homogeneity, produced conformity and therefore weakness in American culture and society. As Daniel Bell put it, "Everyone is against conformity, and probably everyone always was. Thirty-five years ago, you could easily rattle any middle class American by charging him with being a 'Babbitt.' Today you can do so by accusing him of conformity."[10]

These concerns about individuality at home were energized by the way social commentators and policy makers looked at international affairs. Some argued that America's inventive spirit depended on its diversity. The homogenization of thought implied by conformity thus threatened to weaken the nation. In 1959, a survey of American culture noted that variety and "heterogeneity" had become one of America's new values.[11] Moreover, domestic conformity suggested a worrisome lack of distance between the United States and the Soviet Union. For the group of intellectuals concerned with "mass society," there was a direct connection between conformity and authoritarianism.[12] The bland and homogeneous American suburb and the totalitarian machine that was the USSR shared a common feature: they were both populated by a similar sort of subject. In the imagination of liberal intellectuals, it was that kind of person who, devoid of a true self, could undermine American democracy.

Americans consistently framed the distinction between capitalism and communism as a conflict between a system that allowed freedom of thought and one that did not. These views can be found in the founding document of American Cold War strategy, George Kennan's policy analysis in the journal *Foreign Affairs*. "NSC-68," the National Security Council's analysis of the USSR's quest for world domination and recommendation for massive increases in military spending, put an official stamp on these views. It characterized American democracy as constituted by freedom of thought, reason, tolerance, diversity, and creativity. On the other hand, the USSR was driven by a quest for domination both of the world and of individual minds. More pedestrian literature, such as pamphlets circulated to high school and college administrators, articulated a similar perspective.[13]

One such pamphlet was circulated by the Educational Policies Commission (EPC), a group, as noted in the previous chapter, comprised of school superintendents, the presidents of Columbia and Harvard (Dwight Eisenhower and James B. Conant), and representatives of the U.S. Office of Education and of the Carnegie Fund for the Advancement of Teaching. Comparing the "Soviet and non-Soviet worlds," the pamphlet defined democracy as "dedicated to the proposition that intellectual freedom is essential to the worthwhile life and development of mankind." Communism, by contrast, was "implacably opposed" to this freedom. The pamphlet noted: "There are many other differences between the two societies, but the issue of intellectual liberty appears to be the most basic, clear-cut, difficult, and persistent."[14] Left by the wayside in this analysis was any discussion of private versus state ownership of property or free markets versus centralized control of the economy.

Academic social scientists played an important role in this particular discourse on self and society. Their work gave structure and authority to ideas about politics and personhood that circulated among educators, social critics, and policy makers. It gave Americans a system for conducting social critique in the language of the individual psyche. And it contributed to the emphasis on psychic autonomy at home and abroad, while at the same time providing a set of formal tools for understanding persons that could be deployed by more widely read social critics.

By World War II, social scientists had devoted significant attention to political questions couched in psychological terms.[15] They combined psychology and anthropology in national character studies and in the culture and personality movement, as well as in psychological explanations for the political views held by individuals. The most important work in this last genre of social science was *The Authoritarian Personality* (*TAP*), a thousand-page, twenty-six-chapter study coauthored by Theodor Adorno, Else Frenkel-Brunswik, Daniel Levinson, and R. Nevitt Sanford.[16]

The main task of the project that led to *TAP* was to construct a full character profile for the authoritarian personality and a set of tools by which to identify him or her. This work consequently held that prejudiced and antidemocratic beliefs were only two symptoms of a character "syndrome" that had numerous pathological manifestations. To demonstrate this pathology, the book's research program involved constructing, administering, and statistically analyzing survey tests of ethnocentrism (the "E scale"), authoritarianism or fascism (the "F scale"), and anti-Semitism (the "AS scale"). In addition to these techniques, *TAP* included clinical interviews and use of the Thematic Apperception Test (TAT), a projective instrument similar to the Rorschach test. *TAP* repeatedly stressed that the scores produced with each of these tests were highly correlated with scores on the others.

What, then, were the characteristics of the authoritarian personality? First, authoritarians were remarkably similar to one another. *TAP* noted that, while those "extremely low" in authoritarianism were a diverse group, those who scored high on the authoritarian scale were quite uniform.[17] The book thereby made a social scientific argument for the relationship of individualism to democracy and posited the nondemocratic nature of social homogeneity. It also made the identification of people afflicted with authoritarianism simple, since one was just like another.

The beauty of *TAP* was that it identified psychic traits that would appear both in relationship to other people and also intrapsychically. For instance, authoritarian people exhibited prejudiced and stereotyped thinking. On the social side, this meant "generalized ethnocentrism"—a reactionary rejection of different social groups of all kinds. The authoritarian person's almost compulsive prejudice came from the "need for an outgroup" and inability to identify with "humanity as a whole."[18]

On the personal, intrapsychic side, such prejudice meant "stereotyped thinking," "rigidity," "narrow-mindedness," and "intolerance of ambiguity." All of these terms had technical definitions and indicated forms of cognitive deficiency that would occur even in contexts stripped of social cues. Consider, for instance, one of the more widely discussed cognitive disorders associated with authoritarianism: "rigidity." Earlier psychologists had linked rigidity to lower and simpler organisms, feeble-mindedness, mental disorders, lack of creativity, lack of intelligence, and ethnocentrism. "Rigidity" had much the same connotation in *The Authoritarian Personality* as in the earlier psychological studies. However, *TAP* also emphasized the irrational nature of those afflicted with this trait.[19] The stated reason for this irrationality was that authoritarians operated by "taking over conventional clichés and values. . . . There is no place for ambivalence or ambiguities. . . . Every attempt is made to eliminate them. In the course of these attempts a subtle but profound distortion of reality has taken place, precipitated by the fact that stereotypical categorizations can never do justice to all the aspects of reality."[20] The authoritarian's distortions of reality occurred not only in connection with his or her social judgment, but also under conditions of pure sensory stimulation. For instance, Frenkel-Brunswik found that ethnocentric children dealt particularly poorly with ambiguous perceptual stimuli.[21]

Although widely regarded as one of the important and influential works of the postwar period, *TAP* received sustained criticism for its reliance on neo-Freudian models to explain authoritarianism. For instance, noting the failure of *TAP* to control for social variables, Herbert Hyman and Paul Sheatsley were unconvinced that the cause of an authoritarian personality could be found in an individual's early family experiences and their ultimate effects on the psyche.[22] The community of social psychologists seemingly agreed with this point. Ultimately, the psychodynamic theory in *TAP* was the portion of the work that was least often explored in follow-up studies, where attention was primarily focused on statistical studies using the F scale.[23]

Though psychodynamic explanations were largely dropped by social scientists, the focus on the authoritarian's crippled cognitive processes remained at full strength. For instance, Jerome Fisher conducted a set

of experiments that compared the memory abilities of the ethnocentric
and the non-ethnocentric. Like the authors of *TAP*, Fisher found "rigidity" and "intolerance of ambiguity" in ethnocentric subjects. He tested a
group of college students for ethnocentrism using the questionnaire developed by the authors of *TAP*. He also showed them figure 2.1. After

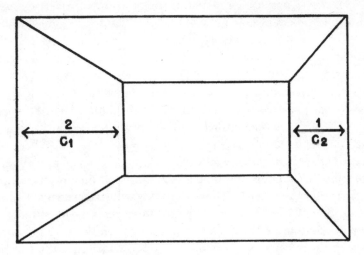

FIGURE 2.1. Stimulus image for memory tests. From Jerome Fisher, "The Memory Process and Certain Psychosocial Attitudes, with Special Reference to the Law of Prägnanz,"
Journal of Personality 19 (1951): 406–20.

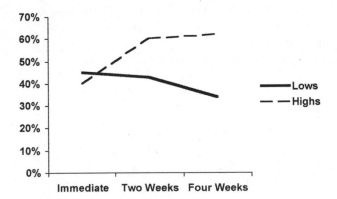

FIGURE 2.2. Percentage of individuals with High or Low authoritarianism scores who mis-
recalled figure 2.1. Chart generated from data given in Fisher, "The Memory Process and
Certain Psychosocial Attitudes."

removing the figure from view, he asked the students to reproduce it. He subsequently asked them to draw the figure after two weeks and again after another two weeks. Noting that the figure was asymmetrical, such that C_1 is twice the size of C_2, Fisher then graphically compared those high in ethnocentrism (the "Highs") against the "Lows" on the basis of their reproduction of figure 2.1 as more symmetrical than it actually was. His graph showed that those high in ethnocentrism displayed an increasing tendency to reproduce the figure symmetrically. By the fourth week, over 60 percent of the "Highs" were erroneously recalling the figure as symmetrical.[24]

This demonstration of the cognitive deficits associated with ethnocentrism would become a touchstone of Cold War social science. They were, for instance, part of Gordon Allport's seminal book, *The Nature of Prejudice*, the most important work on racism in the postwar period apart from Gunnar Myrdal's *An American Dilemma*.[25] Allport, like Fisher and the authors of *TAP*, explained racism by pointing to the irrationality of racist views.

A second and significant attribute distinguished the democratic character from the authoritarian one: autonomy. The authoritarian's social and cognitive deficits extended well beyond the prejudice that closed him or her off from experience of the world. Ultimately, the root of the authoritarian's illness was the lack of a true self. He or she was consequently dominated by other people, by experience, or by society. The authoritarian's attachment to "conventional clichés and values" was a "crutch" that substituted for the absent self. *TAP* noted the ethnocentrist's "conformity to externally imposed values," "blind submission to the ingroup," and "uncritical obedience" to authority figures.[26] The authors also observed that people with democratic minds possessed "greater autonomy," "an internalized conscience . . . oriented toward genuine, intrinsic values and standards rather than toward external authorities,"[27] and an inner core that defined the autonomous self.

TAP concluded that autonomy allowed individuals not only to be true to themselves but also to maintain a connection with truth and reality. Conformity, on the other hand, produced only lies and errors in vision, memory, or logic. Ultimately, this account of conformity was embedded in the very tools that psychologists developed to diagnose it. In the work of Solomon Asch and Richard Crutchfield, the measure of conformity was defined as the percentage of times that a subject yielded to community consensus when that community was in error. By contrast, the inherent value of individuality could be seen in the fact that truth could be achieved only through difference from the group.[28]

Fashioning Tools to Measure Creativity

As they had with other character traits, social scientists looked to develop systematic measures of creativity and autonomy. Even though these were, in principle, democratic traits that everyone but authoritarians possessed, in practice it was important to be able to rank people according to their level of creativity. To accomplish this task, intellectuals relied on well-conditioned sensibilities as well as elaborate systems of popular and expert characterological analysis to identify and measure autonomy and creative thinking. Ultimately, scientific and lay psychological knowledge reinforced each other and provided psychologists, social critics, policy makers, and their audiences with consistent and reliable techniques for molding themselves and judging others on the basis of their autonomy and creativity or lack thereof.

Major work on the psychology of creativity began after World War II and grew out of military and defense concerns. The Atomic Energy Commission (AEC) and the National Science Foundation (NSF) supported psychological research on creativity as a part of their graduate research fellowship programs.[29] For grant officers, the problem was one of predicting the future. They needed to identify individuals who would, at a later date, be the most productive architects of the next generation of atomic weaponry. Ultimately, the trait of creativity was one of the factors the officers settled on to make these predictions.

With the support of the Carnegie Corporation and the Rockefeller Foundation, a second major research project was initiated in 1949 at Berkeley's Institute of Personality Assessment and Research (IPAR). IPAR united members of the *Authoritarian Personality* project with alumni of the personnel assessment project of the Office of Strategic Services (OSS), predecessor of the CIA, to develop standards for evaluating creativity and "personal effectiveness."[30]

Three critical assumptions linked these projects for identifying the creative kind of person. First, creativity was assumed to be a useful, productive, social trait. It was not to be understood simply as a mental process. It would consequently be measured in terms of the novel creation of actual products, whether they were poems, patents, buildings, or bombs. Second, the procedures for identifying creative people were developed out of techniques used to study potential OSS officers, airplane mechanics, or atomic scientists. Third, psychologists devised instruments for measuring and understanding creativity that were calibrated by preexisting nonscientific notions of what it was and who possessed it. That is, they found exemplary individuals who were already known for their creativity and then built tools that could distinguish these people from everyone else. Although the methods of defining, measuring, and predicting creativity varied from psychologist to psychologist, the one thing that they shared was an effort to standardize and ratify *preexisting* views and understandings of who was creative and who was not.[31] Psychologists thereby constructed psychological theory directly on top of a foundation of popular wisdom about creativity.

As a consequence of using preexisting criteria for creativity, psychology's measures of autonomy, creativity, and conformity aligned closely with cultural and socioeconomic divisions in American culture. Thus, for example, the IPAR researcher Richard Crutchfield found that scientists were creative and nonconformist, while the opposite held true for military officers.[32] He also found that creativity and conformity were opposite traits. Thus, one could expect to find the least creativity among the segments of society that were most conformist, rigid, or authoritarian.

Crutchfield and other psychologists noted that women were more conformist than men. Presumably because he was so convinced by his own work and because it fit with prevailing stereotypes, Crutchfield subsequently went on to publish a widely used textbook in which he erroneously cited the work of other psychologists as confirming his own finding of gender differences in conformity.[33] The majority of researchers on creativity joined Crutchfield in finding women both more conformist and less creative than men. Frank Barron, another IPAR researcher, articulated a typical view when he wondered whether men created ideas and women created children.[34]

Other social scientists confirmed Barron's findings that related creativity and its opposite, authoritarianism, to class and cosmopolitan cultivation. They found that authoritarianism was inversely correlated with both education and measured intelligence.[35] In fact, education was so closely tied to liberal democratic thinking that college seniors were less authoritarian than first-year students.[36] Further, adults who had attained only a grammar school education were over twice as likely as those with a college education to agree with statements that were indicators of authoritarianism. For instance, 80 percent of those with grammar school education agreed "the most important thing to teach children is absolute obedience to their parents." In contrast, only 35 percent of college educated respondents agreed with this statement.[37]

While questions such as these managed to divide people along educational lines, other tools for identifying authoritarianism were similarly effective in making social distinctions. A reliable way of distinguishing the Highs (in authoritarianism) from the Lows was to ask them to name people they admired. Those low in authoritarianism responded with such names as "Whitman, Pushkin, Beethoven, Voltaire, Bertrand Russell, Comte, Maimonides, Confucius, Sir William Osler, Freud, [and] Pestallozzi." On the other hand, the Highs responded with "Marshall, MacArthur, Lindbergh, the Pope, Henry Ford, Washington, Teddy Roosevelt, Kate Smith, [and] Bing Crosby." The authors of *TAP* read these two groups of names as an indication that democratically minded people valued "intellectual, aesthetic, and scientific achievement, social contribution, and democratic social change," while authoritarians valued "power and control, conservative Americana, etc." On the other hand, critics of *TAP* noted another difference: the individuals admired by democratically minded Americans were "largely unknown to nonintellectuals."[38]

The difference between these two classes of responses is underlined by research that made direct connections between class and authoritarianism. *TAP* found that a person's ethnocentrism score reflected his or her parents' and grandparents' professions. On the one hand, people who appreciated ethnic and racial difference had parents and grandparents in professional fields. On the other hand, it noted "salesmen, policemen, firemen and their families seem to be more frequently among the prejudiced."[39] Given the oft-repeated opposition of authoritarianism and creativity, it would have been easy in the 1950s to conclude that the highly educated children of professionals were the most likely to have democratic, creative, and autonomous characters.

Measures of creativity and authoritarianism were incredibly effective in dividing American society at its joints. They were tools so well attuned to their context that one psychologist recalls that as an untrained student without ever having seen the psychological instruments, he could predict with 80 percent accuracy whether a person was authoritarian or not simply by seeing him or her.[40] These psychological instruments and the traits they reified thus crystallized the Cold War culture that connected class, cultural sensibility, religion, and political orientation.

Social scientists were quite explicit about the cultural foundations of authoritarianism. In his early work on the subject, for instance, the sociologist Seymour Martin Lipset repeatedly connected the tolerant, democratic mindset with "sophistication," education, and cosmopolitan experience.[41] At the same time, he and the other contributors to the leading social scientific examination of the radical right found authoritarianism and support of McCarthy concentrated among the working class, farmers, and small businessmen.

However, subsequent research would question the assertion that support of McCarthy was based in the working class. Even Lipset would begin to see concentrations of McCarthy defenders among middle- and upper-class Republicans. This transition in perception highlights the elitist and historically rooted nature of the 1950s research that linked democratic thinking with cultural sophistication.[42]

Applying the Closed Mind in Politics

Having framed their vision of society in terms of the individual character and psyche, midcentury Americans turned to managing political discourse by shaping the very meanings of the open, autonomous mind and the closed, conformist mind. Social scientists thereby prepared a technology for conducting politics in psychological terms. Their tools enabled those who adopted them to wear a mantle of apolitical, nonideological science while at the same time labeling certain political positions as objectively irrational.

According to some Cold War centrists, political views that fell outside the mainstream had all the attributes of conformist authoritarianism—rigidity, closed-mindedness, and intolerance—and were therefore unworthy of consideration. Although the specific political views criticized as rigid and closed-minded varied, through the 1950s and into the 1960s it was liberal centrists who most often applied such epithets to their opponents—whether Communists or others on the left or racists, Joseph McCarthy and his supporters, or members of the John Birch Society on the right.[43]

The centrist character of this psychological/political technology is evident in the publishing history of *TAP*, in its reception, and in work that followed in the same genre of political psychology. One of the works that inspired *TAP* was developed using a Marxist framework. In this form, the scale of political personality ranged from bad to good as the political spectrum moved from authoritarian to neutral to revolutionary. However, this tripartite structure of political personality that celebrated revolutionary attitudes was dropped in *TAP* and replaced with a model that had two poles: authoritarian and democratic.[44]

The new frame opened up the possibility of seeing authoritarianism on both the right and the left. While *TAP* helped initiate a pattern in Cold War culture of linking communism and fascism, the University of Chicago sociologist Edward Shils thought that *TAP* had not gone far enough. In one of the more famous critiques of *TAP*, Shils argued that its opinion scales and analyses missed the obvious fact that Communists displayed all of the mental deficiencies of authoritarians on the right.[45]

Following Shils's criticism, social scientists largely looked for ways to frame authoritarianism as a trait that could characterize individuals on either the right or the left. This move had the advantage of casting the study of authoritarianism as apolitical and, consequently, more scientific. The social psychologist Milton Rokeach continued this trend by substituting a nominally politically neutral term, "closed-minded," for the loaded term "authoritarian." For psychologists, Rokeach's new instrument possessed two virtues. It could be used to label anyone, regardless of his or her politics. In addition, because of the neutrality of the tool, the act of labeling itself could be innocent of politics since it was determined not by sentiment but by purely rational expert judgment.

The use of psychology as a means of pursuing centrist politics can be readily seen in the interview forms developed by Rokeach to diagnose closed-mindedness. A primary characteristic of the "closed mind" was its tendency to engage in inappropriate social critique. Rokeach marked as closed-minded those who agreed with any of the following statements: "Most people don't give a 'damn' for others," "Unfortunately, a good many people with whom I have discussed important social and moral problems don't really understand what's going on," or "In times like these, a person must be pretty selfish if he considers primarily his own happiness."[46]

Pessimism and ambition were also markers of closed-mindedness. On Rokeach's scale, respondents who agreed that "it is only natural for a person to be rather fearful of the future" or that "fundamentally, the world we live in is a pretty lonesome place" would be scored as closed-minded. The same held true for those who hoped to make a significant difference in the world. In this case, the closed-minded person would agree with statements such as "The main thing in life is for a person to want to do something important," "If given the chance I would do something of great benefit to the world," or "While I don't like to admit this even to myself, my secret ambition is to become a great man, like Einstein, Beethoven, or Shakespeare." To Rokeach, all of these statements indicated closed-mindedness because they displayed classic symptoms of authoritarianism, including "concern with power and status" and "self-aggrandizement as a defense against self-inadequacy."[47]

Of course, there are numerous ways to read criticism of contemporary events or a desire to be of benefit to the world. One might interpret these sentiments in a neutral or even positive light. One could even reverse Rokeach's moral calculus and criticize not the desire for change but, rather, complacency. As we will see in chapter 8, that is precisely what happened. Thus, Rokeach's particular way of measuring closed-mindedness need not be seen simply as a psychological instrument; it also serves as a characteristic display of the limited range of action and aspiration that 1950s experts like Rokeach deemed acceptable.

In addition to crafting statements to elicit replies that would uncover respondents' social critiques, pessimism, or hopes to have a personal impact on the world—all of which centered on discontent with the current state of affairs—Rokeach added others that were specifically designed to reveal forms of right- or left-wing ideological commitment or "opinionation." Opinionation signaled essentially the same cognitive characteristics as closed-mindedness, including the inability to think logically. Agreeing with such statements, then, would mark the subject as being a closed-minded person of a particular sort. He offered the following examples:

Left Opinionation:

A person must be pretty stupid if he still believes in differences between the races.

A person must be pretty short-sighted if he believes that college professors should be forced to take loyalty oaths.

Only a simple-minded fool would think that Senator McCarthy is a defender of American democracy.

Thoughtful persons know that the American Legion is not really interested in democracy.

It is all too true that the rich are getting richer and the poor are getting poorer.

Right Opinionation:
It's the fellow travelers or Reds who keep yelling all the time about Civil Rights.
Any intelligent person can plainly see that the real reason that America is rearming is to stop aggression.[48]

Rokeach argued that he was not measuring specific beliefs but, instead, the form in which they were expressed. What made these sample statements particular markers of closed-mindedness was the tone of assurance they expressed and the way they characterized opposing views. Rokeach implicitly demanded that—on pain of being labeled irrational —people remain unemotional on significant issues. Thus, for instance, an unwillingness to discuss differences between the races calmly was a marker of irrational ideology.

Rokeach's method of character analysis largely mirrored the techniques in *The Authoritarian Personality*. Both interpreted the opinions and beliefs expressed by subjects as symptoms of fundamental character structure. As a system for diagnosing irrationality and lack of connection with reality, Rokeach's work provided the means to dismiss social criticism from either the right or the left as unworthy of consideration.

In its very aspirations to political objectivity and neutrality, the social psychology of Rokeach and his colleagues bears the mark of its time. Although they did not always agree about where the boundaries of proper belief were located, social scientists marked certain social and political forms as sacrosanct and developed scientific tools that demonstrated the irrationality of those who dissented. At the same time, they often demanded that certain areas remain open to debate. For instance, Gordon Allport's *The Nature of Prejudice* (1954) characterized those who believed in the inequality of the races as having much the same cognitive handicaps as the authoritarians and ethnocentrists described in *TAP*. In taking such a definitive position, Allport might have been marked by Rokeach as closed-minded. On the other hand, Allport took it as entirely rational for people to be viscerally opposed to interracial marriage.

Rokeach's work, *TAP*, and the psychological research it inspired provided a foundation for much postwar cultural criticism. For instance, as was the case in *TAP*, Seymour Martin Lipset explained fascism by its irrationality.[49] Another opinion survey conducted during the 1950s furthered this genre of social criticism by concluding that people who deviated from centrist political views did so because they had crippled cognitive functions, "personality malintegration," and "social maladaptation."[50] For the purposes of this survey, unlike in Rokeach's work, the vehemence with which people spoke was not at issue. All that mattered was the extent to which they held unpopular views.

Such analysis was not simply a form of independent social thought. Indeed, two organizations backed by the CIA, the Fund for the Republic and the American Committee for Cultural Freedom, suggested and financially supported research on the "radical right" by Lipset and the historian Richard Hofstadter. This work ultimately led to contributions to Daniel Bell's volume *The Radical Right*. In his chapter, Hofstadter relied heavily on the mode of psychological analysis used in *The Authoritarian Personality*.[51]

David Riesman and Nathan Glazer, two contributors to *The Radical Right*, borrowed a page straight from Else Frenkel-Brunswik by explaining the behavior of the "discontented" classes in terms of their "intolerance of ambiguity." This analysis ultimately held sway in the social sciences and led to the view that the concerns of the right wing were largely an emotional "status anxiety," rather than legitimate or serious. It also likely helped lead liberal intellectuals like Lionel Trilling and Richard Hofstadter to conclude that there was no serious argument to be made for conservative politics.[52] As one of Hofstadter's graduate students, the historian Dorothy Ross, recalls, the prevailing sentiment was that conservatives "had no mind." Such treatment of the right as irrational "pseudoconservatives" may have helped steer intellectuals away from serious consideration of conservatism until the 1980s.[53] It was not only liberal intellectuals who used psychology to constrain political debate, however.

Like *TAP*, a report by the Educational Policies Commission equated proper politics with mental health. The EPC warned of the "impetuous" who might call for preventive war and the "escapists" who might call for national isolationism. Such escapists, the EPC warned, were the sort of people who sought "personal isolation by trying to cut their individual lives loose from the social context." But, the EPC noted hopefully, "It is probable, however, that the great majority will come to terms with reality. Difficult and unpleasant though that will be, it will in the long run be psychologically wholesome." In making these arguments, the EPC thoroughly mixed psyche and politics. It used politics as a gauge for mental health, conflated the politics of isolationism with the "antisocial" personality, and even promoted the adoption of the politics of containment as a cure for individual emotional distress.[54]

Psycho-political critique frequently involved labeling unacceptable politics as "ideology." Social critics often contrasted free (and hence democratic) thought with ideological thought by mobilizing social psychology's conceptual apparatus for understanding deviant politics. In this model, ideology meant conformity to a system of dogmatic ideas. As psychologists had argued, such conformity meant the loss of individual autonomy and, therefore, loss of connection with the real world. Accordingly, politicians, social critics, intellectuals, and academics suggested that the highest form of thought and political engagement was nonideological.[55]

This argument was a refrain for the members of the Congress of Cultural Freedom (CCF), an international group of intellectuals covertly funded by the CIA. One member, the historian Arthur Schlesinger Jr., echoed the analysis of authoritarianism in *TAP* while serving as special assistant to President John F. Kennedy. He explained the dangers of communism to the Indian people and argued that ideology was rigid, theological dogma that "obscured reality" and operated contrary to democracy, pragmatism, and empiricism.[56]

This veneration of lack of ideology operated on much the same terms as the psychological discourse on conformity. The ideologue, much like the conformist, lacked autonomy because ideological commitment was synonymous with the surrender of freedom of thought. This was a primary reason why liberal intellectuals ranging from Sidney Hook, a leading figure in the American Committee for Cultural Freedom, the U.S. affiliate of the CCF, to Arthur Lovejoy, as well as the EPC, argued that Communists should be barred from teaching positions.[57] As the EPC argued, the facts were quite clear: "The whole spirit of free American education will be subverted unless teachers are free to think for themselves. It is because members of the Communist Party are required to surrender this right . . . that they should be excluded from employment as teachers."[58] The point here was *not* that Communist teachers would corrupt their pupils. Rather, Communists were unsuitable as teachers because, owing to their ideological commitments and party membership, their thoughts were not their own.

While serving as dean of Harvard's faculty (1953–60), McGeorge Bundy delivered testimony to Congress in which he took almost precisely the same position. Although Bundy contended that "the real scientific strength of the country is in its free minds" and insisted that Harvard applied no political tests for employing faculty, it was nevertheless the case that the university excluded "Americans who still surrender to Communist discipline."[59] However, even those who were not Communists but who, in Bundy's eyes, leaned suspiciously far to the left were suspect as well.[60]

Ultimately, Bundy demanded "complete candor" about past associations and activities from individuals under suspicion. This was, perhaps, because Communists were usually taken to be constitutionally dishonest. Hence a person's ability to be truthful indicated that he or she was free from one of the primary disabilities of Communism. A complete airing of the person's past therefore also indicated a break from it. Finally, the requirement of candor had the advantage that it put Bundy in a position to retain a nominal commitment to the freedom of ideas and yet to dismiss staff, faculty, or graduate students not for their beliefs but because they had failed to tell the whole truth.

Raymond Allen, president of the University of Washington, Seattle, built his justification for having fired tenured members of his faculty on claims similar to those of Bundy and the EPC:

> A member of the Communist Party is not a free man. Freedom, I believe, is the most essential ingredient of American civilization and democracy. In the American scheme educational institutions are the foundation stones upon which real freedom rests. . . . [A] member of the Communist Party is not a free man, . . . he is instead a slave to immutable dogma. . . . His lack of freedom disqualifies him from professional service as a teacher. Because he is not free, I hold that he is incompetent to be a teacher. . . . Because he has failed to be a free agent, because he is intolerant of the beliefs of others and because education cannot tolerate organized intolerance, I hold that he is in neglect of his most essential duty as a teacher.[61]

Allen, Hook, Bundy, and the EPC thus reached the paradoxical position of calling for free thinking while excluding certain people from the classroom because their unconventional ideas proved that they did not exercise free thought. In fact, one of the faulty fired by Raymond Allen was not even a Communist but simply uncooperative.[62] Operating within the same system as that advanced by *TAP*, these educators and administrators believed that the improper nature of particular political views could be reduced to and understood in terms of the specific kinds of mentality and deficits associated with them.

Society and Well-Mannered Creativity

If improper politics could be explained by a certain form of mentality, narrow-minded conformity, intellectuals of the Cold War period offered a contrasting, positive model of individual cognition that would advance the values they believed constituted America at its best. That positive character type possessed inner autonomy, the ultimate form of which was creativity. Like conformity, creativity was more than a personal attribute. It had social ramifications. The creative person was uniquely capable of coping with the rapid pace of change that characterized modern life.[63] It was to be the very foundation of the modern pluralist society that social critics hoped to build.

In creativity could be found the inverse of all of the personal, emotional, cognitive, social, and political deficits of the conformist: health, flexibility, openness, tolerance, and democratic character.[64] According to the authors of *The Authoritarian Personality*, the democratic person's autonomy produced greater "creativity," "spontaneity," "imagination," and "self-actualization" than the authoritarian was capable of.[65] This meant that creativity was essential to democracy. Indeed, it was often linked to autonomy and "inner-direction."[66] Consequently, praise of creativity and criticism of authoritarianism often traveled together, as in the work of Arthur Koestler, one of the founders of the CCF.[67]

As one Defense Department official indicated, many Americans believed in the opposition of authoritarianism and creative insight: "In practically every discussion of creative effort, great emphasis is placed on the point that new ideas constitute departures from conventionalized views. On this account, it is argued that any form of rigorous indoctrination tends to limit intellectual freedom and, therefore, to reduce creative capabilities."[68]

For many psychologists, social scientists, public intellectuals, and their readers in the 1950s, creativity and autonomy were unalloyed aids to building a bourgeois society.[69] The social nature of creativity stands out particularly because of how carefully it was contrasted with genius, a trait that traditionally carried antisocial connotations. In opposition to creativity, genius was seen more as a product of heredity than of environment. While one *Reader's Digest* article proclaimed, "You Can Learn to Think Creatively,"[70] scientists from Francis Galton to Lewis Terman and Cyril Burt described genius as an organic property.[71] Articles on genius appeared in eugenics journals, while articles on creativity appeared in the *Journal of Engineering Education*. Genius had a history of being associated with physical illness such as tuberculosis and forms of mental disorder ranging from insanity to neurosis. In contrast, experts insisted that creativity was a sign of health and that neuroses hindered creativity.[72] A genius, unlike a creative person, could operate—perhaps even operated best—in social isolation.[73]

Spurred by social concerns and, as I detail below, funding from the Atomic Energy Commission, research on creativity exploded after 1950. The importance of creativity to postwar Americans was significant enough that there were as many creativity studies published between 1950 and 1965 as there had been in the previous two hundred years.[74] A comprehensive bibliography on the subject indicates that although research on creativity expanded exponentially after World War II, studies on genius declined. Unlike the work on creativity, which dates predominantly to after 1955, the literature on genius dates predominantly from the nineteenth century and the early twentieth century. Studies of genius are also more European, while the creativity literature is mostly American.[75]

What we see in the way postwar Americans consistently marginalized genius is how very important it was to them that positive mental traits be adaptable to society. They consequently pictured creativity as a social affair. While numerous publications examined the kinds of environments that would foster creativity, there was almost no literature on its genetic basis. Many psychologists, cultural critics, and educators agreed that creativity was much more a product of nurture than of nature. Accordingly, educational experts filled their journals with articles on how to promote student creativity.[76] Critics of education charged that schools, in fact, bled creativity out of students. But both camps believed that social environment played a vital role.

The importance of society in constituting creativity was relevant not just to everyday schoolchildren but to the best scientists and artists as well. For them, as for children, creativity demanded training, knowledge, work, and practice.[77] Industry and business publications lauded the technique of brainstorming to solve problems at factories or make marketing programs more productive.[78] Vocational guidance experts noted that creativity made for a more effective sales force.[79] Business leaders and engineers sought ways to speed up the rate of product innovation and development by improving the work environment.[80] Complementing this trend, psychologists examined how group processes affected creative thought.[81]

While society affected individual creativity, so, too, did creativity impact society. As portrayed in most literature of the 1950s, individual creativity was a productive and positive force in society. The word did not then have the connotation of critical and oppositional character that it would later acquire. Thus, although the opposite of creativity was conformity, anti-conformity was by no means equivalent to creativity.

Commentators on creativity in the 1950s consistently emphasized that the rejection of social norms did not imply creative thought or character. Consider the 1959 book *The Uncommon Man: The Individual in the Organization*, by Crawford Greenwalt. In this work Greenwalt, the CEO of DuPont, proposed that his corporation should be the model for American society, since most Americans were wage earners working in similar organizations. Greenwalt suggested that America's historic strength and its best hope for future progress lay in promoting the creative potential of the common man. If organizations like DuPont would reward creative ideas, he suggested, American society would prosper.

Greenwalt called for people to practice what he called "good manners" and to follow social norms while at the same time retaining their independence of thought. His imagined society, where people could maintain their individuality while exhibiting good behavior, rested on such conservative political views as the belief that a progressive income tax hindered individual motivation.

While it might not be surprising that someone like Greenwalt would laud independence of mind as long as it was safely secured within the frame of "good manners," in fact very similar sorts of analyses came from the liberal end of the political spectrum as well. For instance, the preface to *The Creative Mind and Method*, a 1960 volume on creativity, argued that it was a mistake to identify creativity with the "prevailing picture of the . . . artist as a starving, isolated, and rather crazy individual, a social and sensual Bohemian who works in fits of inspiration—when he isn't too drunk or hopped up to work at all." Moreover, this work noted that the "tragedy" of the Beat Generation lay in its "frantic noises over nothing, its desperate rebellion without purpose, and, even more, in its urgent attempts to put equal marks between abstraction and aberration, between freedom and carelessness, between the creative and the deviant."[82]

Although this might seem to be a conservative position, the volume's organizers were by no means cultural reactionaries. One of them, Lyman Bryson, had championed scientific humanism and social engineering, both hallmarks of left-wing politics, in his 1948 work, *Science and Freedom*.[83] Bryson argued that the strongest society was the one that was the most diverse and creative. He accordingly hosted and provided commentary for a radio series for WGBH that presented creative artists, "scientists, philosophers, anthropologists, psychologists, theologians, and educators" to the general public. These interviews provided the raw material for the *Creative Mind and Method* volume.[84] Given Bryson's commitment to creativity as a productive social force, the volume's prefatory critique of the Beats should be understood not so much as opposition to their unconventional modes of behavior but, rather, as an argument that there is no necessary connection between creativity and anti-conformity.

Social scientists made precisely the same point. The liberal sociologist David Riesman, for instance, wrote that "today, whole groups are matter-of-factly Bohemian; but the individuals who compose them are not necessarily free. On the contrary, they are often zealously tuned in to the signals of a group that finds the meaning of life, quite unproblematically, in an illusion of attacking an allegedly dominant and punishing majority of Babbitts." Riesman was joined in this diagnosis by many others, ranging from Paul Goodman, a left-wing poet and social critic, to Richard Crutchfield, a psychologist who specialized in the study of creativity, to Betty Friedan, who, before publishing *The Feminine Mystique*, had spent the late 1950s combating conformity among high school students. They all held that, in being unconventional, Beats and bohemians were merely slavishly following their unconventional peers.[85]

Most of the discussion on the relationship between creativity and conformity was based on positions roughly similar to those staked out by Riesman and Greenwalt. Liberal and conservative social critics imagined well-mannered creativity as a solution to several different kinds of problems in postwar America, and they saw conformity as inhibiting the creativity they so desired.

Although such interpretations of the polar opposition of conformity and creative autonomy were widely shared, what was not so universal were the cultural and political values placed upon such mental traits or how to identify them. For instance, both liberals and conservatives supported modern art and jazz as creative expressions that were emblematic of the freedom of the American spirit. Fostering the appreciation of the right kinds of culture, then, would be a way to shape individuals and politics. Intellectuals and the cultural and political elite accordingly determined that specific forms of cultural production were weapons in the Cold War struggle with the Soviet Union.

Frank Barron discovered congruence between the kinds of art that individuals preferred and where they fell on the continuum of psychological simplicity/complexity. "Complex" individuals were more creative and flexible and consistently preferred modern art, whether of the primitivist, expressionist, impressionist, or cubist variety. The "simple" person, on the other hand, evinced authoritarian personality traits (conformity, stereotyped thinking, rigid and compulsive morality, dogma, repression) and expressed a preference for more traditional representational works such as Botticelli's *Virgin and Child*, Fra Filippo Lippi's *The Adoration*, and Gainsborough's *Blue Boy*.[86]

Barron's findings on art preference fit well within the cultural system of 1950s America. Consequently, the CIA supported the Congress of Cultural Freedom, which, in turn, was instrumental in arranging exhibits of abstract expressionist art in Europe.[87] The pure forms of creativity displayed by Jackson Pollock and demanded of his audience seemed to highlight the freedoms of the American way of life. While Pollock's expressionism suggested the liberation of the individual from politics and society, the realism of Soviet art—as well as that of Pollock's early work and the work of his mentor, Thomas Hart Benton—was directly engaged with the politics of class.[88]

The question of creativity's value demonstrates divisions within the United States over what was most genuinely American. Many intellectuals, the CIA, and the organizations it funded sought to promote modern art in the service of fighting communism and advancing autonomy, creativity, and independent thought. And psychologists produced research that demonstrated the un-American character of the right wing and people who disliked modern art. On the other hand, reactionaries ranging from real estate developers in Los Angeles to art critics for the *National Review* linked modern art and jazz with communism.[89]

The conservative critique of modern art was not just about the art itself. Just as liberal social scientists used their measures of creativity and authoritarianism to criticize people who maintained a preference for traditional art, conservative opinion makers drew on their own criteria of proper citizenship to criticize those who produced and consumed modern art. This critique was continuous with McCarthy's attack on the

Eastern establishment and the elites who populated the State Department. In both cases, waging the Cold War involved fighting those who were perceived to be eroding America from within.

Conservatives would fight the Cold War by attacking precisely those parts of culture that liberals saw as both most central to America and most antithetical to Communist values. Thus, while liberals advocated tolerance and saw conformity as characteristic of Communist totalitarianism, the sociologist Samuel Stouffer found that McCarthy drew support from people who were both ignorant and intolerant of nonconformity.[90]

This association between conformity and McCarthyism appeared not only in the survey work of social scientists but also in the words of some of McCarthy's most famous defenders. In the context of a 1954 book defending Joseph McCarthy's goals and methods, William Buckley and Brent Bozell, rather than critiquing conformity as antithetical to American values, argued that it was only proper to support America by demanding conformity to its central values. Buckley contended that liberals' advertised support of academic freedom was hollow. In an earlier book, *God and Man at Yale* (1951), he had pointed out that liberals at his alma mater did not extend academic freedom to Communists and argued that the homage paid to it was merely a cover for indoctrinating students in liberalism, undermining their proper religious faith, and promoting egalitarianism and collectivism.[91]

McGeorge Bundy, then a professor of government at Harvard, wrote what would be used as one of Yale's semiofficial responses to Buckley's book on Yale. His scathing review rebutted Buckley's accusations, charged him with dishonesty, and attacked him as a "twisted and ignorant" man.[92] In this review and the interchange that followed, several of the problems that Bundy and other liberals found with Communists would appear in connection with Buckley. Although Buckley's published riposte argued that it was Bundy who was dishonest, Bundy's reply maintained that even Buckley's self-defense was mendacious.

One significant issue that Bundy highlighted was Buckley's Catholic faith. He found it "strange" and "remarkable" that Buckley was "an ardent Catholic," hid that fact, and yet called for Yale to return to its religious tradition—one that, Bundy pointed out, was Protestant. Bundy was particularly bothered by Buckley's effort to conceal his "special allegiance."[93] Here Bundy raised the same questions about Buckley's character that he used to justify excluding Communists from Harvard and other universities: they were constitutionally dishonest and they "surrendered" their own freedom of thought in the name of political or religious dogma.[94]

Bundy asked Buckley for the same candor that he would later, as dean, demand from ex-Communists and other left-wing faculty at Harvard. In the view of Bundy and other liberal anti-Communists, both communism and Catholicism involved "special allegiance" to an institution that dictated truth rather than seeing it as something to be discovered by the individual or though a process of inquiry. Further, although Bundy did not highlight this specific point, *God and Man at Yale* advocated precisely the aspect of the Catholic educational system—that is, teaching that truth is a given rather than something to be discovered—that liberals like John Dewey and Sidney Hook had critiqued as authoritarian and that *General Education in a Free Society* had rejected as inappropriate for most American schools.[95]

Bundy's treatment of Buckley as akin to a Communist in his mindset was far from idiosyncratic. Despite the fact that Catholics from Buckley to Joseph McCarthy made careers of rooting communism out of American life, they were often read as sharing mental attributes with these enemies. As David Hollinger and John McGreevy have noted, during the 1940s and early 1950s liberal scholars, including Dewey, Hook, Rokeach, and Lipset, saw considerable overlap between the attitudes of the Catholic Church and the authoritarian mindset. This view hinged on the support of Catholics for Francisco Franco as well as on the anti-Semitism of such famous Catholics as Father Coughlin. Hook, for instance, linked the Catholic intellectual Jacques Maritain to communism, opining that "Catholicism is the oldest and greatest totalitarian movement in history." Hook supported this claim by noting the ability of Catholics to make peace with Fascists like Mussolini; he also suggested that Maritain was comfortable with communism because, as he put it, communism and Catholicism shared "dogmatic metaphysics."[96]

* * *

The lauding of mental attributes such as a penchant for open-minded inquiry or flexibility was a partisan endeavor, even if it did not always travel under an explicit banner of political activity. When authoritarianism could be successfully measured by a person's admiration of the pope and democratic character assessed by a person's appreciation of

Bertrand Russell, these traits ultimately marked more than political affiliation and cognitive capabilities. They simultaneously served as effective tools for measuring and marking an individual's social position.

Conclusion

Ultimately, psychological tests of creativity and open-mindedness, much like other robust psychological instruments, such as IQ tests and the SAT, were tuned in favor of certain social, cultural, religious, economic, and political groups in America.[97] Just as psychologists such as Lewis Terman used psychometric tests to crystallize particular norms of gender and intelligence, the psychological measurements of creativity and authoritarianism reified American political, social, class, and gender divisions of the 1950s so well because they were cast in the mold of preexisting social judgments.

Despite, or perhaps because of, the partisan nature of psychological measures of the open and creative or closed and authoritarian mind, these measures spread widely though American society. By the mid-1990s, over two thousand studies on authoritarianism had been published; they most likely involved screening hundreds of thousands of Americans. Further, anecdotal evidence indicates that unpublished examinations dwarf published studies, as academic psychologists in almost every university and college screened students in their classes for authoritarianism.[98] Consequently, despite the substantive methodological criticism *TAP* received, especially for its Freudian aspects, the social practices for identifying authoritarianism and related deficits of mind were widely propagated by social scientists and their readers. And, as with the SAT, despite the specific origins of these measures of creativity and authoritarianism, they became useful general tools for evaluating and shaping both individuals and society.

"Creativity" worked well as a measure of individual merit and as a metric for social inclusion or exclusion. In so doing, it joined other systematic means of ranking individuals such as "intelligence," "merit," and "character." The latter two terms functioned as devices for elite colleges that found intelligence and academic ability insufficient means for keeping their populations wealthy, white, and Protestant.[99] Like other mental traits, creativity was found in highest measure among the people who framed the term. As psychologists eventually recognized, the prototype of the creative person they described was none other than the psychologists themselves or other intellectuals who were defining creativity.[100]

The values, sensibilities, preferences, and forms of thinking that were prevalent within the community of psychologists helped to form the definition of rationality, creativity, and right thinking in mid-twentieth-century America. The creative self they saw as the solution to the problems of American life was ultimately based on a form of selfhood desired in leading research universities.

The lines of influence ran in more than one direction, however. Where intellectual values and ideals encoded in ideals of autonomy functioned to police the boundaries of acceptable politics and social thought in America, it was also the case that the values of liberal pluralism helped to structure the daily lives of intellectuals. The next chapter takes up the influence of liberal democratic thinking within the microculture of the elite academy.

The Academic Mind

CHAPTER THREE

Interdisciplinarity as a Virtue

Interdisciplinary social science was hot in Cold War America. American academics, administrators, and foundation officials saw it as paving the road both to better theory and also to the production of practical and useful results. A 1959 survey noted, "by now, of course, the popular maxim is that 'all good social science is interdisciplinary.'"[1] Social scientists and their patrons agreed that the critical topics facing their fields and those of central theoretical importance for the development of the social sciences were ones that fell between disciplines—small group interaction, "reference group," "role," "mobility," "status," "self," "personality," "values," "culture," "character," "national character," "action," "information," and "decision making."[2] As one review suggested, "avant-garde" social scientists were those who shaped their research with an interdisciplinary "problem focus" and who had interest in "broader theoretical questions common to all the social sciences." On the other hand, this review characterized "traditionalism" in pejorative terms by suggesting that it meant "clinging" to "standardized fields."[3] An even more significant development in the social sciences in the postwar years was the invention of the "behavioral sciences," which were often taken as inherently interdisciplinary. Since the behavioral sciences subsumed much of the social sciences' intellectual territory to the extent that fields such as psychology, sociology, or anthropology became behavioral sciences, they also became interdisciplinary.

Excitement about interdisciplinarity in the social sciences ran so broad and so deep in mid-twentieth-century America that one of the most frequent criticisms leveled against interdisciplinary research and education programs was that they were insufficiently interdisciplinary.[4] The cachet of interdisciplinarity is indicated by the fact that interaction of several disciplines was almost always discussed in positive terms such as "cross-fertilization" and essentially never in terms such as "cross-sterilization."[5] This perspective appears in a self-survey the University of Chicago conducted of its behavioral sciences. Chicago social scientists contrasted the active intellectual life of interdisciplinary endeavors with the "more *vegetative* existence" of remaining within disciplinary boundaries.[6] Echoing the Chicago view, one commentator remarked, "the assumption is frequently made that interdisciplinary research has more potential power for broad and deep insight and for change than unidisciplinary research."[7] To the anthropologist George Murdock, interdisciplinarity was itself a goal to be achieved. For Murdock, interdisciplinarity ranked with both "pure science" and "practical results" as a target toward which social scientists should strive.[8] Richard Wohl noted that the largest objection to interdisciplinarity was that some people were overexcited by it.[9] As he put it, some social scientists esteemed interdisciplinary research merely because "interdisciplinarity" itself was good.[10]

If Wohl himself thought that interdisciplinarity was overvalued, many social scientists in the 1950s would have disagreed. While there were differences over the inherent value of interdisciplinarity, social scientists did agree on one thing—that it was highly regarded. Even those who had reservations about the intrinsic value of interdisciplinary work observed the prevalence of the view that interdisciplinary research was better than work that was disciplinary.[11]

Interdisciplinarity became so important that social scientists and their patrons devoted considerable effort, time, and resources merely to understanding how to make it function better. Leading social scientists participated in numerous interdisciplinary conferences devoted to integrating their several fields of study. Seeking to capture the fluid social and intellectual dynamics of these conferences, the participants transcribed and published the contents of their discussions.[12] One such project, an NIMH survey completed in 1958, examined interdisciplinary research projects that focused on the subject of mental health. This survey involved organizing five conferences at which the participants could discuss their experiences with interdisciplinary research. One hundred and seven people from nine disciplines attended these conferences.[13] In addition, the survey included sessions at meetings of the professional societies. Eight hundred people attended these sessions.[14]

Explanations of such excitement over interdisciplinarity have taken several canonical forms. Some scholars have analyzed the phenomenon by arguing that disciplines do not divide the world at its joints. For instance, Robert K. Merton argued that the social sciences became increasingly interdisciplinary as a matter of natural and pragmatic adjustment to the inevitable process of disciplinary specialization.[15] According to this kind of argument, important areas of the social world fall between the disciplines' areas of specialization. Others have seen in interdisciplinarity a strategy for creative innovation. And still others perceived excitement over interdisciplinarity to be mere superstition.[16]

This chapter takes a different route by examining the popularity of interdisciplinarity as an expression of historically and culturally specific values. These values linked politics, cultural sensibilities, definitions of science, visions of the proper ordering of the academy and American society, as well what it meant to be a right-minded person. These values defined which sorts of inquiry would count as science, which questions were "interesting," what it meant to be a creative person, as well as the most plausible strategies for learning about society.

Historians have noted how, in different historical contexts, scientists have sought to improve both their work and themselves by moving toward quantitative methods, rigor, precision, or objectivity.[17] Similarly, in midcentury, American social scientists could improve their own standing or the status of their work by incorporating the perspectives and methods of several disciplines. Adopting an interdisciplinary perspective had several implications from the philosophical to the political. It was attached to a specific view of how scientific knowledge and human thinking operated. By making their work interdisciplinary, social scientists aligned themselves with a rationalist and deductive philosophy. In so doing they often portrayed themselves as true scientists and attacked those attached to empiricism as nothing more than partisan social reformers or mere pedantic collectors of social facts. Interdisciplinarity was more than philosophy. In the community of the most influential social scientists, interdisciplinary work also marked an individual as creative, practical, open minded, tolerant, and scientific. Interdisciplinarity operated on a register that linked the unification of knowledge to the operation of America's pluralist society. It did so because proper interdisciplinarity, as then understood, was a social matter. It did not require individuals know more than one field but rather valued their capacity to cooperate with researchers in different fields.

These themes—the nature of society, the question of whether the social sciences were scientific, the optimal form of social scientific thinking, the institutional structures of the Cold War academy, and the nature of American society—were intertwined with one another. Political culture, institutions, individuals, curricular reforms of general education, and disciplinary practices interacted to make it a foregone conclusion that the interdisciplinary mode of research was the best way of pursuing (and thinking about) answers to social scientific problems and questions.

Making Disciplines

The seeds of the Cold War interdisciplinarity boom were planted well before World War II. To understand the particular arguments for interdisciplinarity that held sway in the Cold War period, it helps to first look briefly at the emergence of disciplines in America from the late nineteenth century through the interwar era as well as the counter reactions they provoked. This section pays special attention to psychology because the field's intensive disciplinary focus in the first half of the twentieth century would prove to be critical to the subsequent counter argument for interdisciplinarity during the Cold War period.

After the Civil War, American social sciences professionalized in parallel with the emergence of the research university and increasing organization along disciplinary and departmental lines.[18] Academics developed increasingly clear boundaries around their own fields, mechanisms for distinguishing proper from improper modes of inquiry, new requirements for credentials, and modes of professional conduct. They founded new journals and professional societies in economics, history, sociology, psychology, and political science. Despite continuing debates over where precisely to draw disciplinary boundaries and over which methods would characterize proper forms of inquiry in each field, there were nevertheless increasing numbers of disciplines that were easier and easier to separate one from another as well as from the nonspecialized knowledge held by members of the public. These developments meant that the newly emergent social sciences captured from religion, moral philosophy, and political economy a measure of authority over defining the problems of social order and the solutions for resolving them.[19]

The rise and differentiation of the social sciences drew on a commitment to an empiricist vision of knowledge. Those who supported a pragmatic and scientific vision of humanity believed that all humans acquire valid knowledge through experience rather than rational deduction or intuition.[20] Philosophers such as Horace Kallen, Sidney Hook, and Philip Frank, sociologists such as Robert Park and Ernest Burgess, and economists such as Wesley Mitchell framed their own work as empirical and therefore modern and scientific. Their opponents, on the other hand, sought to understand knowledge as depending on deduction from rational and a priori truths. By the 1930s this orientation was associated with a religious, often Catholic and "medieval," viewpoint. The rationalist position was often seen by both supporters such as Robert Hutchins and Mortimer Adler, and opponents such as Sidney Hook, as having been articulated by Thomas Aquinas—hence its association with the Middle Ages. Because rationalism's supporters often claimed the supremacy of metaphysics over science, both its supporters and opponents saw the rationalist vision of humanity as opposing the empiricist and modern scientific vision.[21]

Thus, two clusters of terms developed. There was, on the one hand, an alliance among modernity, science, and empiricism and disciplinary development. Opposed to these views were values, rationality, deductive reason, metaphysics, and medievalism.[22] This construction held sway in the social sciences, popular culture, and among granting agencies such as the Laura Spelman Rockefeller Memorial.[23]

The conjunction of disciplinary specialization, secularism, empiricism, and science found no stronger faction than among experimental psychologists. Indeed, psychologists made empiricism a basis of their claims to disciplinary autonomy and authority.[24] For psychologists, following empiricism had a special consequence. Not only a norm for professional conduct, it also defined how they understood their object of study: human nature. From the first decade of the twentieth century, psychologists sought to place their field on a firmer scientific footing by centering its study on directly observable phenomena. In the service of "objectivity of the facts of science" experimental psychology consequently eschewed discussion of soul, mind, or consciousness and focused increasingly on behavior.[25] As John Watson put it, since thoughts are neither reproducible nor public, they could not serve as data in a psychology that took itself to be engaged in science.[26] As he saw it, those who attributed consciousness to humans were nothing better than self-serving medicine men who engaged in "religious philosophy."[27]

According to the view increasingly adopted by the field, science's goal was to explain (or explain away) interior or mental experience through the exterior, physical world of behavior. Drawing on the system of val-

ues that gave preference to the modern and scientific, many behaviorists made a practice of branding the study of mind as an atavistic return
to philosophy or religion.[28] This program set up scientific psychology as
an empirical process that opposed itself to the rationalizing, theoretical,
and therefore unscientific versions of psychology held by the lay public
and by obsolete psychology. As psychologists cast themselves as objective empiricists, they were increasingly committed to the view that their
understanding of the world derived from experience alone. In so doing,
psychologists downplayed the significance of the activity of their own
minds and sought to turn their work into one of recording directly visible and measurable behavior.

Having adopted these practices for policing their own scientific research, psychologists ascribed similar processes to the people and animals they studied. Where the psychologists determined that the only
legitimate scientific activity was empirical investigation, they also concluded that all organisms' psychological processes operated on strictly
empirical principles as well. Where psychologists conceived of their own
work as one of making readings of pointers on their instruments, they
focused on organisms learning by rote or trial-and-error and came to
see behaviors as a series of stimulus-response reflexes chained one after
the other. Consequently, those psychologists who sought to rely strictly
on observation and refrain from engaging in speculation or intuition in
their own mental lives also pulled back from studying the topics of intuition, insight, will, mind, reasoning, and thinking in the mental lives
of others. For instance, behaviorist Clark Hull studied trial-and-error
learning in rats and argued that scientific research is also a conditioned
reflex acquired by trial-and-error learning.[29]

In a 1939 survey of the previous fifty years of psychological literature,
Gordon Allport, then president of the American Psychological Association, and his graduate student, Jerome Bruner, noted how deeply behaviorism had transformed psychology. Over this period the treatment of
mental processes as real in psychology journals had declined by over two
thirds as psychologists began to consider mental processes as mere scientific constructs, which, like Ptolemy's epicycles, were supposed to be
devices for calculating observed phenomena.[30] Consequently, between
1888 and 1938, studies of "higher mental processes" of "learning, reasoning, concept-formation, reverie, or creative thinking" declined from
over 22.1 percent of the literature to 6.1 percent (i.e., a 72 percent drop).[31]
Together, these trends indicated psychology's increasing movement away

from the study of active, intelligent, and self-motivated aspects of human nature.

At the time that behaviorism was gaining strength, psychologists engaged in a related project of separating scientific experimental psychology from that which failed to reach the standard of true science. In the 1930s, this boundary drawing between real, scientific psychology and other studies found no stronger and more effective advocate than the Harvard experimentalist S. S. Stevens. In addition to developing a scientific philosophy that purified psychology, Stevens also served in a critical role as the organizer of an important institution that both propagated his disciplinary vision and, by design, excluded psychologists who failed to fit Stevens's own view of what counted as scientific psychology.

Stevens specialized in psychophysics, the branch of experimental psychology that sought to determine mathematical relationships between the physical world and the senses. Stevens's own research focused on the psychophysics of hearing. He sought, for instance, to find the relationship between the psychological experience of sound and sound's physical properties.[32] One particular line of work involved determining the mathematical relationship between the physical properties of a pure tone (i.e., its frequency and its intensity), and how loud it was experienced to be by a listener.[33]

Stevens developed a philosophy to match his brand of experimental psychology. In the 1930s he authored a series of articles on operationism.[34] Drawing on his own experimental experience buttressed by the authority of the physicist Percy Bridgman, who had first introduced operationism, and on philosophers of the Vienna Circle such as Rudolph Carnap and Otto Neurath, Stevens staked out a philosophy for his science.[35]

According to Stevens's "rigorous" formulation of the philosophy, "science is a set of empirical propositions agreed upon by members of society." Stevens added that scientific terms derive their meaning from a set of physical operations and, consequently, "a proposition has empirical meaning only when the criteria of its truth or falsity consist of concrete operations which can be performed upon demand." Arguing in a similar vein as Watson had, Stevens wrote that "only those operations which are public and repeatable are admitted to the body of science."[36] Thus what could not be measured would not be considered a proper subject matter for science.

In his depiction of science, Stevens tacked between two kinds of arguments. On the one hand, Stevens claimed that he, Bridgman, and the logical empiricists delivered nothing other than "empirical" observations on the nature of scientific research. For instance, he wrote, "operational principles are induced generalizations rather than *a priori* fiats," and "operationism seeks only to discover how scientists do what they do."[37] In fact, he called this philosophy "the science of science." On the other hand, Stevens turned these purported empirical observations into normative judgments. In this vein, he divided "tough-minded" scientists (who, by his definition of science, work in the operationist mode) from "tender-minded" philosophers of the rationalist variety.[38] He also equated the operationist brand of tough-mindedness in a science with its "maturity."[39]

Following the logical empiricists, Stevens argued that those terms that were not operationally defined were either metaphysical issues or pseudoproblems and were therefore not a part of proper science.[40] Stevens also blurred the lines between normative and descriptive accounts of science in his laudatory discussion of logical positivism. He described the movement and operationism as "an empirical study of the actual doings of science-makers,"[41] even though several pages earlier he had remarked that logical positivism describes how the "household" of science "*should be run.*"[42] Thus, according to Stevens, science both *is* and *should be* conducted according to the rules of operationism and logical positivism.

This philosophy of science fit nicely with Stevens's brand of psychology. Just as he argued that the meaning of ideas, including scientific ones, needed to be referred to a set of physical operations, Stevens's own research was an effort to give rigorous and scientific meaning to the common and everyday discourse of perception and sensation. Correspondingly, he was interested in the ways that sensory processes were determined by events in the physical environment. For instance, Stevens produced a curve by plotting decibels against psychological loudness, or, as he termed it, "sones." He noted that a second curve of remarkably similar shape could be produced by plotting decibels against the electrical potentials measured in the cochleae of guinea pigs when they were exposed to sounds of varying intensity. Remarkably, though they matched one another, both the sone-decibel and the decibel-cochlea output curves were *not* of the same shape as the decibel loudness scale.

Stevens, then, had found a lawful relationship between events in the physical world (decibels) and the events of psychological experience (sones). It was an interesting finding because the relationship to the decibel scale was not of a simple, linear form.[43] In addition, Stevens's experiments on guinea pigs suggested that the difference between the human experience of loudness and the physical properties of sound was produced by the anatomy of the ear. If the human ear was sufficiently similar to a guinea pig's, then the cochlea could be understood as a "measuring instrument" and the psychological experience of loudness would be like the brain "reading off" the cochlea's electrical output.

Stevens thus envisioned the people taking part in his experiments (i.e., the psychological subjects) as much like the scientific empiricists he described in his philosophical papers. Humans, like scientists, made regular and systematic judgments based on empirical experience. Stevens treated this empirical process of perception just as philosophers of science treated the empirical process of science. Just as physical scientists were to ground concepts such as distance through use of a meter stick, so too was the psychological experience of loudness to be grounded by the operation of measuring the electrical output of the cochlea.

Stevens extended his normative view of science in general to a more focused argument about the proper way to conduct the discipline of psychology. Having noted the affinity between operationism and logical positivism, he noted that these two philosophies were in accordance with the behavioristic brand of psychology. For instance, he concluded his article "Psychology and the Science of Science" (1939) by asking: "Does it not appear that the Science of Science must go directly to psychology for many of its answers? Is it not also plain that a behavioristic psychology is the only one that can be of much use in this enterprise?"[44]

Stevens's article appeared within several supporting intellectual and institutional contexts. First, at the time that Stevens wrote it, behaviorism had become the dominant strand of experimental psychology.[45] Second, as Stevens's article indicates, operationism was quickly becoming an important way of distinguishing good from bad work in psychology. Among the various contributors to this trend were Carol Pratt and E. G. Boring.[46] Third, in sociology George Lunberg used operationism just as Stevens did in psychology: to mark off the proper forms of science.[47]

At the more local level, Harvard had resident or visiting proponents of both operationism and logical positivism such as Percy Bridgman, Philip Frank, and Herbert Feigl. If both philosophies made sharper distinctions between the valid and invalid forms of science, such a division had practical consequences outside the domain of intellectual discourse. In fact, the distinction between science and non-science was instrumental in the creation of Harvard's Department of Psychology. In 1934, psychology at Harvard separated from the Department of Philosophy on the basis of psychology's experimental focus.[48]

Stevens himself played a critical role in giving institutional backing to his own methodological prescriptions separating the true from the impure forms of science. He did so setting up an exclusive community designed to separate the good from the bad psychologists. In 1936, Stevens and the Swarthmore psychologist Edwin B. Newman founded the Psychological Roundtable.[49] Newman and Stevens had been disgusted with the level of scientific work presented at the previous meeting of the APA. Moreover, they had felt that the group founded by Titchner, the Society of Experimental Psychologists (SEP)—itself an exclusive group—was insufficiently scientific because it included psychologists who were no longer actively conducting experimental research.

As a consequence, the Psychological Roundtable, reserved for men under the age of forty, was initially named "The Society of Experiment-*ing* Psychologists." Its rules specified that, in contrast with the SEP, there would be no life members and that participation was renewed annually and limited to active researchers. E. G. Boring took the name "The Society of Experiment*ing* Psychologists" as it was intended: an attack on the SEP. As a member of the SEP and as Stevens's advisor and department chair, Boring was sufficiently disturbed to get Stevens to change the organization's name to the Psychological Roundtable (PRT).

As a founder of the PRT and a multiyear member of the "Secret Six" who directed the organization, Stevens played an important gatekeeping role in an exclusive and secretive club. Meetings of the PRT gave participants the opportunity to form important social and professional contacts and to have early access to unpublished research of the other attendees. Individuals could not become members of the PRT, but instead were invited to meetings at the discretion of the Secret Six, who were also known as the "Autocratic Minority." Invitation to a meeting one year did not guarantee invitation in following years. Thus the Secret Six played an extremely important role in setting the boundaries of experimental psychology.

Additional rules of the PRT that Stevens had a hand in drafting did further work dividing insiders from outsiders. Meetings themselves were kept as secret as possible. Participants were not to report on events of the meetings and were to keep silent about their own participation; their travel expenses were not to be reported on tax forms and their CVs were not to mention the PRT.[50] By the late 1950s membership in the PRT had become so significant that when Harvard's Psychology Department sought to make a senior appointment, it began shaping its list of candidates from members of the SEP and from the pool of people it guessed were members of the PRT.[51] This way of managing the PRT—as an exclusive and secret organization with a secret membership and leadership hidden even from the members—continued through the 1960s.[52]

Aside from drawing heavily on only a few schools, the PRT—like its predecessor, the SEP—was also exclusive in that its membership was restricted to men. These two organizations thus played a role in restricting the membership of the "elect" in psychology to men. The reason that the SEP offered for such sexist exclusion was that it improved the opportunity for homosocial fellowship that might be inhibited by the presence of women.[53]

Indeed, as Charles Gross recalls, the distinguished lectureship that capped the PRT was punctuated with scatological humor and pornographic slides reminiscent of *Hustler*. Such occasions could simply not have occurred with women present. With the second-wave feminist movement, however, mores changed; women were included in the PRT, and the memorial lecture added images of male genitalia to the repertoire that had once included only images female genitalia.[54] If the PRT lecture had become "inclusive," it was certainly far from so in the preceding years.

Thus, S. S. Stevens played a critical role in shaping the institutional and intellectual structures of academic psychology in the late 1930s. With the growth of the PRT's significance, with its policies of secrecy and of separation of insiders from outsiders, with the control vested in the Secret Six, and as a member of this executive committee and a founding member of the organization, Stevens held substantial authority to determine which psychologists and what kinds of work would count as either scientific or experimental.

Beyond his personal influence in defining operationism, in founding the PRT, and in choosing which men were sufficiently scientific enough to be invited to participate, Stevens played an important additional role in shaping the definition of scientific psychology in America. Specifically, he was the editor of the *Handbook of Experimental Psychology*, a 1,400-page tome for educating and training graduate students on the state of the field. Although each chapter of the *Handbook* was written by a separate author, Stevens was nevertheless in a position to shape the outcome of the volume by choice of the topics that it would cover as well as of the authors who would write on those topics.[55]

Psychology's disciplinary power as a science rested on linked processes of exclusion. Under the prescription of operationism and behaviorism, it would exclude from consideration statements and data not acquired in a thoroughly empirical manner. Under the social system erected by the SEP and PRT, groups of psychologists would establish groups of insiders according to their status as proper, empirical, male scientists. Excluded from this group were those deemed to threaten community feeling and the scientific status of insiders: all women, and male psychologists who were insufficiently focused on the experimental or "hard" end of psychology. This process of exclusion drove the study of emotions, personality, and, indeed, all higher mental processes out of psychology's inner sanctum.

Interdisciplinarity

Running against the development of disciplinary boundaries whether in psychology or the other social sciences, scholars, intellectuals, and policy makers regularly and periodically bemoaned such specialization as fragmentation and made efforts to foster integration in the social sciences. This set of views motivated the American Council of Learned Societies (ACLS), a coalition of social scientific and humanistic disciplines founded in 1919. The ACLS sought to foster international scholarly cooperation and to develop relationships among its eleven constituent scholarly organizations. A similar impulse served as the motivation for the establishment of the Social Science Research Council in 1923 and of the conferences and research it supported. The SSRC aimed to serve as an umbrella for economics, sociology, political science, statistics, psychology, anthropology, and history. By 1925 the council was organized with representatives from the professional organizations of each field and had committed itself to "ordinarily" dealing with problems that "involve two or more disciplines."[56]

Social scientists offered multiple reasons why social science in America needed to be interdisciplinary. For the SSRC members and patrons at the Rockefeller Foundation and Laura Spellman Rockefeller Memorial, commitment to cross-disciplinary interaction sprang from the view that this cross-disciplinarity would make the social sciences both scientific and more realistic.[57] As SSRC leaders Beardsley Ruml and Robert Crane saw it, integrated social science would be better at solving public or real world problems.[58] By the end of 1939, the SSRC and social scientists it supported were absolutely convinced that the best way to proceed in social science was to frame a problem and then attack it from multiple directions using the techniques of several disciplines.[59]

In taking the position that a multi-pronged attack on a problem would produce results, SSRC members adopted a vision of knowledge production that was consistent with that developed in the National Research Council (from which it sprung) and which was promoted by the major private patrons of both the natural and social sciences. By the 1930s the Rockefeller Foundation was supporting science on a large scale that was consistently structured on interdisciplinary terms.[60]

At the same time, other advocates of interdisciplinarity linked methodological pluralism to democracy. According to this view the pursuit of social science in an interdisciplinary way was equivalent to conducting intellectual life in a democratic fashion. For instance, John Dewey used the pages of his contribution to the *Encyclopedia of Unified Science* to make this point and implicitly attack the effort of the editors to "mechanically" unify the sciences.[61] Similarly Horace Kallen took the *Encyclopedia* to be an "authoritarian" endeavor.[62] For his part, Alain Locke contended that when the empiricist movement had developed into behaviorism and positivism it had made the fight for "scientific objectivity" into a "dogmatic," "fanatical cult."[63]

In a presidential address to the American Psychological Association in September 1939, Gordon Allport made similar points. In an especially pointed attack on S. S. Stevens and E. G. Boring, who were members of his own department at Harvard, Allport called operationism a "magic concept" that demanded of its practitioners a "ritual of method" which stood in the way of obtaining useful results.[64] For Allport, the methodological practices of operationism had several connected social implications. Calling on the trope that linked science to democracy, Allport suggested that operationism was a narrow, Muslim faith, that it was unconnected with the practical affairs of everyday life, and that it was, as a consequence, unscientific. Opposing this narrow and exclusionary form of methodological religiosity with true science, and perhaps true religion, Allport remarked that "psychology, as a science—may I repeat?—can be justified only by giving mankind practical control over its destinies, not by squatting happily on a carpet of prayer."[65] But if methodological fetishism was not bad enough, perhaps even worse to Allport was that committed operationists systematically trimmed the prayer carpet on which they kneeled so as to exclude everyone who strayed from the one true faith.[66]

To these points about operationism's narrow, religious character, Allport added remarks on the philosophy's undemocratic nature. He envisioned methodological plurality as the means both to achieve the best science and to further the cause of democracy in general.[67] He called on American psychologists to "avoid authoritarianism" and to "keep psychology from becoming a cult" that rules out "original" and "daring" inquiry with "one-sided" methodological tests.[68] Referencing the Nazi efforts to expunge the work of Jewish psychologists, the recent German invasion of Poland, and the "heavy darkness [that] has descended over the European continent,"[69] Allport argued that "now, if ever" psychologists needed to learn and "apply . . . to ourselves"[70] the lesson of the importance of diversity in democracy.

Thus, in addition to condemning operationism as methodologically and philosophically narrow (and therefore short-sighted and undemocratic), Allport also opposed it on the grounds that it excluded from psychology much of the discipline's most important subject matter. In doing so, Allport questioned the entire basis of Stevens's operationist and positivist philosophy.

The difference between these two psychologists stemmed from their opposed conceptions of the fact-finding enterprise. Stevens, like the logical positivists he cited, divided empirical statements and observations from theories and logical arguments. According to this philosophy, it is possible to make neutral and objective observations; theory comes afterward as an explanation of those observations. Allport advanced the opposite view. Rather than operations defining the meaning of concepts, as Stevens had argued, Allport held that operations acquire meaning from concepts.[71] Choosing his own philosopher, Alfred North Whitehead, for authority, Allport argued that facts only become facts when placed within a conceptual framework. He added that dynamic psychology itself had demonstrated that what counts as a fact depends on the evaluation of specific individuals.[72] Having noted that observation, theory, and method are inseparable, Allport continued by pointing out that operationism's methodological narrowness implied a consequently narrow subject matter.[73] Allport thus argued that operationism served not only to delimit *modes* of fact-finding in psychology but also managed to exclude entire classes of facts as well. In so doing Allport here questioned the same fundamental premise that undergirded the empiricist project: that there are such things as neutral facts or observations.

The position staked out by Allport thus inverted both the political and the epistemological claims of empiricists like John Watson and S. S. Stevens. As Allport saw it, the antiscientific and religious position was found not in the study of mind but rather among followers of empiricism, behaviorism, and operationism whose search for methodological purity had denied the possibility of multiple approaches to the study of human nature.

Beyond such frequent analogies between methodological and national pluralism, proponents of interdisciplinarity also offered a methodological argument for why interdisciplinary social science was the only way to properly explain culture and social life in a democracy. As with his call for pluralism, Allport's argument was grounded by the view that the best kind of scientist subordinated facts to theoretical frameworks. On this view, the scientific self was defined not by commitment to data-gathering but to creating systems of thought for making sense of the world.

This particular relationship between the scientific self, democracy, and social science was articulated in "Toward a Common Language for the Area of the Social Sciences," an influential manifesto for unification of the social sciences coauthored in 1941 by an interdisciplinary group of Harvard social scientists including the anthropologist Clyde Kluckhohn and the sociologist Talcott Parsons. They contended that interdisciplinary approaches would promote theoretical integration and unify the social sciences, thereby making them more rigorous and theoretically sound. This text, used as a teaching document for a newly formed undergraduate major in "The Area of the Social Sciences," sought to integrate methods in anthropology, sociology, psychology, economics, and government.[74] It is worth examining in some depth because it indicates the philosophical position that would mature into one of the central texts of Cold War social science, *Toward a General Theory of Action*.[75] This position on theoretical integration would also play a major role at the national level in subsequent years.[76]

The authors of "Toward a Common Language" began by pointing out the need for organizing the various findings of their respective fields. A single framework or language, they argued, would allow them not only to communicate better with one another but also to increase the "rigor" of their fields.[77] Significantly, the authors were confident that they had selected a proper framework because interdisciplinary collaboration had enabled participating social scientists to abstract from raw data of behavior the universal "foci" around which the human social world was organized.[78] These foci were universal because they were conditioned by human nature, the nature of social systems, and "social action."[79] By using these foci, the social sciences could study nature and be as objective as the physical and biological sciences that took natural phenomena as their subject.

Although it put the authors in conflict with the empiricist program of the experimental psychologists, this argument was not novel. Indeed, the argument for treating social phenomena rather than individual behavior as both natural and as the proper focus of investigation echoed methodological precepts that Emile Durkheim had articulated in *The Division of Labor in Society*. Although they did not make reference to Durkheim, the authors of "Toward a Common Language" would have been familiar with this aspect of his work. Talcott Parsons had made Durkheim a primary point of analysis in own his first book, *The Structure of Social Action* (1937). Moreover, a biography on Durkheim published two years before the report on the "Common Language" had treated social facts as natural entities.[80]

For the committee that wrote "Toward a Common Language," identification of natural foci did more than provide an epistemology for countering the empiricist program. It also had two other benefits. Abstraction from individual behavior demonstrated the benefits of developing an interdisciplinary language to bridge their several fields. Even more, a second abstraction ("second order" abstraction) showed why unification and basic theoretical development of the social sciences were absolutely necessary. Making the suggested second-order abstraction involved finding regularities in a society's patterns or foci. The other authors of "Toward a Common Language" termed these regularities the "ethos" of the society.

As they put it, ethos penetrated all aspects of social and cultural life, whether public or private. It was not merely a property of individual behavior, or even of behavior patterns; instead, ethos *determined* behavior and its patterns. This is to say that socialization practices, morals, or aesthetic values did not influence ethos; instead, the causal arrow ran in the other direction.[81] Ethos was thus like a *Zeitgeist* that both characterized a society and shaped it.[82] Consequently "Toward a Common Language" proposed a method that did more than allow communication among the several social science disciplines. Its interdisciplinary method revealed and explained facts of individual behavior that so-called "narrow" specialization missed. Thus proper collaboration would enable the true scientists to develop a unifying theory and explain to empiricists working in a disciplinary mode the data that they themselves gathered.

The authors of "Toward a Common Language" also argued that their form of social science had political valence. By respecting and integrating the plurality of contributions and perspectives of the several social sciences, their form of unified social science could operate in a liberal, democratic fashion. But the political significance of interdisciplinary unification extended beyond how the community of researchers organized themselves. As the authors of "Toward a Common Language" saw it, interdisciplinary unity literally allowed social scientists to understand democracy better than disciplinary-minded researchers did. The reason was that ethos, which could only be discovered by interdisciplinary methods, was most significant in liberal societies because in these societies their citizens' behavior was structured by a shared ethos. On the other hand, in a tyranny social order would be maintained by the direct physical expression of power.[83] However, since ethos would be invisible to empiricists and those working within a single discipline, only those interdisciplinary people who adopted the pluralistic and rationalist epistemology advocated in "Toward a Common Language" could possibly explain democratic culture. If the study of ethos was tinged with and could promote the democratic spirit, then social science produced from a single discipline was biased, undemocratic, and best suited to the study of authoritarian social systems.

Interdisciplinarity as Weapon in Academic Politics

These arguments served as more than theoretical positions on what made for practically relevant and theoretically powerful social science in a democratic country. They were also political statements about how the academy should function. Just as S. S. Stevens had used his operationist vision to distinguish the good from the bad scientists and to cultivate the Psychological Roundtable, so too did advocates of interdisciplinary social science use their own epistemological values as tools for building their own institutions. A close look at how the authors of "Toward a Common Language" promoted their own vision of science indicates how seemingly abstract discussions about method were simultaneously techniques for arguing about the proper forms of scientific selfhood and weapons in academic political struggles.

By 1943, Harvard social scientists Talcott Parsons, Gordon Allport, Clyde Kluckhohn, and Henry Murray began arguing for replacing the informal committee structure involved in the Area of the Social Sciences major with a departmental restructuring at the administrative, faculty, and graduate levels. On June 10, 1943, they sent a letter to Paul Buck, the dean of the faculty, pointing out changes in academic geography that had made traditional lines dividing anthropology, sociology, and psychology outdated and an impediment to "progress."[84]

This letter adopted two tactics in its call for the creation of a new interdisciplinary Department of Basic Social Science. On the one hand, it used rhetoric in line with the SSRC's long-held position: disciplines were reactionary "anachronisms" unprepared for the postwar world.[85] On the other hand, interdisciplinary social science better represented the facts of modern life and could better address the needs of the community.[86]

Extending this case, Parsons and his collaborators claimed that their department would be the foundation of all other forms of social inquiry. The basic social sciences covered in their department would "deal with things which *ought to be presupposed by the other disciplines.*"[87] The "other disciplines" which were to rely on the basic social sciences were history, government, and economics.[88] As such a focus on the "basic" indicates, these social scientists framed their mutual affiliation as an issue not only of the convenience of collaboration and mutual stimulation but also of making their work more rigorous.

This attachment to breaking down disciplinary barriers came from the sense that not to do so would not only prevent the advancement of knowledge and its useful application but also damage their students. On their account, departments produced students who were narrow and both "ignorant" and "intolerant" toward "related branches of knowledge."[89]

Just as students' mental horizons could be narrowed by it, disciplinary specialization could also make students lose their freedom of thought and personal autonomy. The authors of the "Confidential Memorandum" warned of the over-focused student becoming "primarily a disciple of magnetic personalities, hypnotized by their brilliance," and they thereby called on the trope of authoritarian leaders mesmerizing the masses. Adding that interdisciplinary instruction would hopefully have "the desirable effect of destroying the cultism that has grown up in the

social sciences whereby a student became an exclusive follower of one favored teacher,"[90] the authors of this report called for requiring graduate
students to take classes across disciplines in the social sciences.

This memorandum, like Allport's presidential address in 1939, raised
issues of academic organization in the context of forms of national political culture. In both, the authors slipped back and forth between the
academy and national politics, eliding distinctions between desirable
academic thought and desirable and politically proper thought. They
stitched arguments about departmental structure to discussion of freedom of thought, and they linked plurality of academic method to a social
structure that respected a plurality of political views. This report linked
basic science, interdisciplinary methods, open-mindedness, democratic
social organization in the academy and the nation, and the achievement
of practical effects such as advancing democracy and improving social
affairs. But while this memorandum overtly pointed to intellectual and
social concerns at the national level, it did not mention the stormy mi-
cro-politics in Harvard's social sciences that had made its authors so ea-
ger to restructure their respective departments.

Between 1935 and 1945, members of the Psychology and Sociology
Departments could not agree on what did and did not count as belonging
to their fields. In psychology, the faculty were barely on speaking terms.[91]
They divided on every possible issue that bore on the nature of the discipline and the department. So deeply ran the disagreements that psychologists could not agree on what constituted fact, experiment, or the "general" core of their field.[92] They could not agree on content and structure
of graduate exams and they fought bitterly about faculty appointments.[93]
For instance, E. G. Boring and Karl Lashley opposed the appointment
of psychoanalytically concerned Henry Murray as untenured associate
professor in 1938, and Lashley threatened to resign when Murray was
promoted over his objections.[94] Subsequently Gordon Allport sought to
prevent S. S. Stevens from appointment to a tenure track position on the
grounds that Stevens was more of a physicist than a psychologist.[95]

When the "Confidential Memorandum" was submitted in 1943, the
Harvard administration was not prepared for the wholesale reorganization of the social sciences entailed by the creation of a new interdisciplinary department. Yet the authors of the memorandum deployed the
social scientific framework they had developed in continuing institutional struggles. Accordingly, where the "Confidential Memorandum"
had warned of the intolerance that disciplinary pedagogy created in stu-

dents, Parsons would shortly turn to attacking other sociology faculty for their narrow and defensive "departmentalism" and to lauding his own interdisciplinary approach as "pragmatic" concern with carrying out the interests of the university as a whole.[96]

The authors of the "Confidential Memorandum" also extended this language of interdisciplinary breadth versus rigidity and narrowness in an effort to appoint Robert K. Merton to a tenured position within the Sociology Department. Even in the context of a disciplinary appointment like this one, Talcott Parsons praised Merton because he had "broad" interests, was not "narrowly departmental," and avoided the "rigidity of departmental lines."[97] While Parsons noted the intellectual virtues of Merton's cross-disciplinary interests, he framed his discussion in terms of social values. Thus, he argued that Merton would be "an unusually 'good citizen' of the Department and the Faculty, and not merely an effective scholar and teacher as an individual." This meant that Merton got "along extremely well with a wide variety of people." Particularly attractive was the sense that "his influence" seemed "very generally to be in the direction of helping bring conflicting interests and personalities together, but without sacrificing his own integrity or independence." For his part, Gordon Allport sought Merton's appointment because his interdisciplinary interests meant he had the ability to get along with other people as well as that he was "realistic, vigorous, cooperative of others, and focused upon actual social problems of the present day."[98]

But if claims of interdisciplinarity could be used in lauding Merton, they could also be used to disparage those who remained within disciplinary frameworks. For instance, in writing Paul Buck, Parsons claimed that Pitrim Sorokin and Carle Zimmerman had opposed the appointments he wanted because of what they saw as the candidates' insufficient commitment to sociology itself.[99] Whether claiming that his own position was the "pragmatic" way to run a university or arguing that the disciplinary view advanced by Sorokin and Zimmerman promoted "intolerance," Parsons inserted epistemological arguments into departmental politics.

Although Parsons and Allport failed in their effort to appoint Merton, these arguments in favor of interdisciplinarity soon found enough purchase with the administration that Paul Buck eventually joined and supported their effort to join the social sciences in an interdisciplinary department in the fall of 1945. Buck, I would suggest, supported Parsons and his allies because he already agreed with their epistemological values.[100] Buck displayed his own values through two years of directing the committee that produced *General Education in a Free Society*. Both the general education project and Parsons's vision for the social sciences placed high value on intellectual breadth achieved through communication. Even more, the final product of the committee's work appeared in the summer of 1945, essentially the same time as Buck joined Parsons in the project of social science unification.[101]

Buck and the Harvard administration were so firmly on the side of interdisciplinary unification that they were prepared to steamroll all opposition. Ultimately, the administration made any serious opposition impossible.[102] Buck's priorities were so evident that even those such as the psychologist E. G. Boring who were viscerally opposed to the transition were effectively silenced.[103] Not only was serious opposition impossible, but even indication of reservation was squelched by the Harvard administration.[104] Ultimately, Boring echoed the party line even while resisting the formation of the interdisciplinary department by pointing to the "breadth" of his own teaching and research. He even defended himself against the charge that he was narrow while others were broad by suggesting that it was not he but his opponent, Gordon Allport, who was narrow.[105]

With the administration committed to speed and its willingness to squelch any opposition, Allport, Parsons, and Kluckhohn's dream of integration was realized in January 1946 when the faculty voted for the establishment of the Department of Social Relations. The department combined sociology, social anthropology, social psychology, and clinical psychology. The Department of Psychology would retain experimental psychology as represented by Boring and Stevens. Consequently, the branch of psychology identified by rigorous operationism, commitment to scientific empiricism, and a deterministic, behavioristic perspective on human nature was divided institutionally from the psychology identified by its interests in personality, social issues, and a commitment to the view that humans are naturally autonomous.

Working across Departmental and Disciplinary Lines: Tooling-Up the Social Sciences

The organization of the Social Relations Department signaled a broad movement in the social sciences. The interdisciplinary methods of research that had been developed in the 1920s and 1930s under the sponsorship of the SSRC and the Rockefeller Foundation were expanded dramatically during the war. This emphasis on interdisciplinarity appeared across the social, biological, and physical sciences. In the 1930s it was, for instance, central to work on cyclotrons and the study of radiation at the interface between medicine and physics.[106] Foundation officials like Warren Weaver spent the 1930s funding interdisciplinary research on the belief that practical problems could best be solved by integrating approaches from several fields. Ultimately Weaver transferred this method and philosophy of scientific management directly from his work at Rockefeller to his work in directing Section D-2 of the National Defense Research Council.[107] Yet it is important to recognize that it was not the war that produced the belief that interdisciplinary methods were more practical or more scientific. This commitment was in place even before the war had been won or successful research had been conducted. Even before the United States entered the war, attendees at a conference celebrating the SSRC's support of interdisciplinary social science at the University of Chicago noted again and again that it was the best way to produce practical results.[108]

Accordingly, when planning for the social scientific contribution to the war, administrators again and again organized in interdisciplinary fashion. Whether it was the Office of War Information, the Research and Analysis Division of the Army, or the personnel assessment project of the Office of Strategic Services, social scientists were put into interdisciplinary teams.[109] They ended up in teams during the war not because the war had already shown the effectiveness of interdisciplinary work but because of the preexisting view that interdisciplinary work was practical.

At the end of World War II they noted how the war had demonstrated the productivity of solving specific, practical problems by assembling a set of research techniques from a range of disciplines. Many drew the conclusion that a similar set of strategies would be productive for social scientific research conducted during the following years. Reflecting on his experience with the postwar bombing survey in Japan, the sociologist William Sewell noted the "most innovative insightful ideas were generated as a result of group discussions in which little regard was paid to the disciplinary origin of the idea."[110] Sewell continued, "my colleagues on the Bombing Survey, as well as those who had participated in other interdisciplinary social psychology research projects, were equally impressed with their experiences and were determined to promote interdisciplinary training and research programs on return to academic life."[111] Recalling his experiences in the Information and Education Division of the Research Branch of the Army, Robin Williams remarked that the war had led him to conclude "team research is feasible and productive to a degree which would not have been generally acknowledged as possible in many academic circles a few yeas ago."[112] For his part Jerome Bruner thought the research organization during the war provided lessons for the postwar period. His experiences were shaped in the context of morale and public opinion surveys conducted for the Office of War Information. By the war's end, Bruner was convinced that his wartime experiences pointed the way to conduct future research. In 1945 he wrote to Gordon Allport, noting that the war had made social scientists "less monosymptomatic." Bruner noted that they found that "problems like morale, readiness to buy bonds, fear in battle" were "too rich to be tackled with single instruments." He added: "This is all so utterly *obvious*, I feel ashamed to labor it. But in practice it is not obvious." Whether they forgot or were unaware of prewar emphasis on social scientific interdisciplinarity, the recollections of researchers like Sewell and Williams helped establish the view that wartime experience proved how important interdisciplinary work was.

If the war showed the efficacy of interdisciplinary work, it also pointed to a specific method for postwar improvement of the social sciences: development of and training in multiple social scientific tools. From his own experiences Bruner drew a clear lesson about how researchers should be trained: a researcher should "to be forced to think of all the possible conceptualizations of a problem in all fields of social science, all possible research tools which might be used, and then decide very self-consciously how [he or she] wants to conceptualize and limit the problem by choice of method."[113] So convinced was Bruner of the importance of spanning disciplines that he sought appointment in both the Psychology and the Sociology Departments at Harvard.[114] While this effort at interdisciplinary departmental membership failed and Bruner was appointed to the Psychology Department in the fall of 1945, he was moved to the Social Relations Department when it was founded later that academic year.

After the war's end, Bruner's own focus on tools became a standard means of improving the social sciences. Such a perspective appeared at a conference analyzing the state of the social sciences funded by the Carnegie Corporation. There the sociologist Samuel Stouffer argued that the national strategy for improving the social sciences should revolve around the invention of research tools.[115] Leonard Cottrell agreed. He noted that the "progress of science is determined in large part by the effectiveness of its tools of research." Cottrell added that tools were the critical criteria of true science. As structured information-gathering devices, tools prevented mindless "random and raw empiricism." At the same time, well-constructed research tools also kept scientific research grounded and protected against "futile theory building divorced from the invigorating experience of empirical testing." Properly crafted tools therefore promoted real science, namely, "skillful use of theoretical formulations which will guide the systematic and additive accumulation of knowledge."[116] To Charles Dollard, a friend and patron of Stouffer's and president of the Carnegie Corporation, the distinction between the creative and "first-rate research men" and those who merely had the ability to "perform useful routine functions" was that the former, the first-rate social scientists, possessed the imagination to invent new methods and techniques for furthering social science.[117] It should not be surprising that Dollard would be interested in how to distinguish between good and bad researchers, for as president of Carnegie, the decisions about the distribution of much of the available private funds for social scientific research lay in his hands. For our purposes, it is enough that Dollard signaled the role of tool and instrument creation as a criterion for distinguishing the excellent from the less so.

While the creation of new tools marked the route to developing the social sciences, the question remained: how best to acquire or produce those tools? For midcentury academics and their patrons, the answer to this question frequently lay in assembling an interdisciplinary community of researchers. Such a community would itself be a tool, more precisely a machine tool, for creating the new instruments that would help bring the social sciences into their maturity as sciences.

The Ford Foundation took just this approach in its early years of support of the behavioral sciences. Of the $300,000 the Foundation's Behavioral Sciences Division had scheduled in 1952 for "Improving Content and Methods," $150,000 was directed toward the "Interdisciplinary Research and Study Program." Moreover, the Behavioral Sciences Division devoted the majority of its support to the Center for Advanced Studies in the Behavioral Sciences (CASBS). A significant part of the Center's mission was exposing its members to disciplinary cross-fertilization that was supposedly impossible at their home institutions.[118]

In 1952 the CASBS was expected to use $1,150,00 (or 61 percent) of the $1,900,000 the Foundation's Behavioral Sciences Division would spend that year. On the other hand, programs not specifically marked as interdisciplinary, such as "mathematical training for behavioral scientists" and a mathematical seminar at Michigan, were to receive only $60,000 and $25,800, respectively.[119] Since strengthening the methodology of the behavioral sciences remained a major concern of the Foundation, that cross-disciplinary interaction at the CASBS and in the research program each dwarfed mathematical training indicates the significance that interdisciplinarity held for the patrons of the social sciences.

If at all possible, the Carnegie Corporation pushed interdisciplinarity even harder than other patrons of the social sciences. In 1948, John Gardner, a Carnegie officer and founder of its area studies program, gave voice to the foundation's preference for interdisciplinary research when he proclaimed, "every thoughtful person who has taken a serious look at the whole range of the social sciences is convinced that great gains will come from what has been called the 'integration of the social sciences.'"[120] So deep was Carnegie's commitment that it identified interdisciplinarity with making social science both more useful and more rigorous.[121]

This preference for interdisciplinary research on the part of patrons was patently obvious to their clients. One university self-study report commented on this bias in patronage by noting that "there is the belief that the large research project, especially if dressed up as 'interdisciplinary' gets first call on foundation funds."[122] And regardless of whether they approved of or were bothered by such bias toward interdisciplinarity, social scientists had by then learned well enough how to frame their research proposals in order to secure foundation support.

The focus of external patrons on interdisciplinarity allowed enterprising social scientists not only to capture funds from those sources but also to acquire more resources from university administrators. For instance, the pattern of funding from the Carnegie Corporation helped shape the structure, size, and intellectual program of Harvard's Department of Social Relations. The Carnegie Corporation ended up giving over $250,000 to the Laboratory of Social Relations.[123] Not incidentally, the laboratory was headed by Samuel Stouffer, who leveraged his personal connection with Carnegie president Charles Dollard in securing funding for it.[124] Carnegie both funded *and even suggested* individual projects such as the one that resulted in *Toward a General Theory of Action*.[125] Even though this project was an effort in theoretical social science, it, like the more empirical studies of the period, was seen as a project for advancing the field by the development of appropriate tools. In this case, it was tools for theory instead of tools for measurement, observation, or computation.[126]

Samuel Stouffer, in mediating between the Carnegie Corporation and Harvard, made sure to emphasize to Harvard administrators that Charles Dollard and the other Carnegie officials did not want their funds to be used for faculty salaries but only to supplement those expenses that Harvard would provide as a part of its regular budget.[127] Stouffer added that, in the long run, Harvard would likely receive more money from the Carnegie Corporation if it asked for support of research projects rather than salaries.[128] This interaction elicited a response from James Conant and Paul Buck that emphasized their substantial and long-term commitment to the Social Relations Department.[129]

By dividing the support of personnel from support of research projects, the Carnegie Corporation sought to make interdisciplinarity a normal feature of the social sciences. As Dollard explained to Stouffer, the foundation aimed "to commit Harvard funds so heavily to [the Department of Social Relations] enterprise as to cut off any line of retreat when foundation support comes to an end."[130] When Stouffer explained to Paul Buck that the Carnegie Corporation would not "put strings" on its funds,[131] it was true to the extent that no particular line of research would be demanded or precluded. On the other hand, to receive the grant, Harvard had to apply "strings" to itself. The Carnegie grant thus tied Harvard into supporting social science in an interdisciplinary form.

Although Carnegie was instrumental in making Harvard as an institution more interdisciplinary, it did not by itself create interest in interdisciplinary studies. Instead, Carnegie provided ammunition to the leaders of the Social Relations Department that allowed them to realize their goals. Indeed, although certainly encouraged by the Carnegie Corporation, the interest in interdisciplinarity within the department was a preexisting concern for Parsons, Allport, and their allies. Parsons's own *Structure of Social Action* (1937) as well as the collaborative efforts in "Toward a Common Language" (1941) all preceded direct monetary support from foundations.

Like Carnegie, the Ford Foundation's interest in interdisciplinarity helped to reshape the behavioral sciences through its preferential treatment of certain fields, certain methods, and certain scholars. In 1953–54, the Ford Foundation funded self-study projects in the behavioral sciences at the University of Chicago, the University of North Carolina, the University of Michigan, Stanford, and Harvard. Harvard's study concluded that it needed two new senior positions in the Department of Social Relations. The response by the Ford Foundation was to give $400,000 to endow a chair in the psychology of personality.[132] Its treatment of the University of Chicago followed a similar pattern.

Ford, then, like Carnegie, gave money to the Social Relations Department in a form that the social scientists themselves desired. However, individual universities did not always lead in the move to interdisciplinarity. For instance, Ford gave a $300,000 grant for the behavioral sciences to Berkeley that the university, apparently because the grant did not mesh with local priorities, did not spend for years.[133]

Scientists did not merely adapt to external force exerted by the foundations; instead the lines of influence flowed in both directions. For instance, the Ford Foundation carefully consulted individual social scientists to advise it on how to set up its behavioral sciences division. Subsequently, it assembled a committee to advise it on how to best address the goal of improving advanced training in the social sciences.[134] This committee unanimously suggested the establishment of an independent research center to address the need for training. After running the idea past supportive university administrators, the Ford Foundation had another committee of social scientists write a formal proposal. After the Foundation's trustees had approved the idea, yet a third committee was assembled to advise the Foundation on the more detailed aspects of the idea.[135]

The depth of the social scientists' ability to shape the Foundation's program is perhaps best indicated by the case of economics. This one discipline received its own program area outside the umbrella of the behavioral sciences. Despite the interests of Foundation officials and the advisory committee to bring economics into interdisciplinary conversation with the other social sciences, neoclassical economists successfully resisted this pressure by contending, as Milton Friedman did, that the "interdisciplinary fad" stood in the way of economics becoming a "cumulative" (a term typically interchangeable with "mature") science.[136]

For its part the Carnegie Corporation made a practice of surveying academics in order to determine whom to support. This activity was not the same as conducting peer review of grant proposals. Instead, Carnegie officials often conducted surveys *before* receiving requests for funding. Such surveys helped define the broad areas that should be funded as well as the individuals and institutions that should receive support.[137]

The pattern of patrons designing their funding and programs by depending on the advice of social scientists resulted in concentration of support in the areas favored by academic consultants. Those social scientists who had best access to foundations and to the ears of grant officers were able to play a role in directing patronage funds in ways that affected departmental battles both within their own institutions as well as other universities. In sociology, anthropology, and psychology, generally recognized as the central "behavioral sciences," funds from patrons went to those who had interdisciplinary interests and thereby strengthened the position of these scholars. On the other hand, in economics, those who took a neoclassical approach were able to use foundation funds to help maintain an upper hand against those who desired wider connections with other fields.[138]

At Carnegie Mellon and Harvard, social scientists connected with the Ford Foundation, Social Science Research Council, and Carnegie Corporation were able to use foundation funds to establish new departments at their own universities that matched their own vision of proper social science. The actions of the Carnegie Corporation played an important role in the drama that unfolded in Harvard's departmental squabbles in the social sciences. It enabled the formation of the Department of Social Relations and the separate existence of the Laboratory of Social Relations. For the formation of this department, the interest of the Carnegie Corporation in interdisciplinarity operated much like the rhetoric of unified science—it was a resource that Parsons, Stouffer, Allport, and their colleagues used to promote their professional interests at the expense of people like Boring and Stevens. At Carnegie Mellon, Herbert Simon used connections with patrons to subvert the Psychology Department and establish his vision of proper interdisciplinary behavioral science in the Graduate School of Industrial Administration.[139]

Social scientists who had the ears of patrons were not only in a position to alter their own institutions but were also able to use their leverage with foundations to change both the intellectual content of their fields and even the institutional arrangements at other universities.[140] Well-placed social scientists like Robert K. Merton, who advised the Ford Foundation on how to structure its behavioral sciences program, also served on the visiting committees for the "Self-Study" programs conducted by Harvard, Stanford, and the Universities of North Carolina, Chicago, and Michigan that the Ford Foundation funded.[141] Recommendations resulted in targeted Foundation grants that, in some cases, operated against the wishes of local faculty.[142] Thus, Foundation funds were a resource that scholars like Robert K. Merton, Samuel Stouffer, and Herbert Simon used to win battles at their own universities and to reshape the social sciences elsewhere.

The enthusiasm for interdisciplinary social science and social scientists' success in drawing external patronage changed older institutions and produced new ones that included, among others, the CASBS in Palo Alto and the Bureau of Applied Social Research at Columbia, and other survey research centers at Harvard, Yale, Princeton, Berkeley, UCLA, Illinois, Minnesota, and Washington. By the mid-1950s there were interdisciplinary social science research and training programs at Harvard, Yale, Cornell, Columbia, Chicago, Stanford, Berkeley, Minnesota, Wisconsin, Michigan, and North Carolina, among others.[143] As figure 3.1 indicates, the increase in interdisciplinarity occurred not only in terms of the number of universities that added interdisciplinary programs but also in terms of growing interdisciplinarity *within* individual universities such as Chicago.

At Harvard, the Social Relations Department aimed to make the social sciences more rigorous on a theoretical level by making them more interdisciplinary. The department's associated Laboratory of Social Relations was organized to facilitate the creation of experimental techniques. The philosophy of science that underpinned its program was one that favored the invention, creation, or application of new or untried research tools over the continued use of existing and well-used methods.[144] The laboratory's funds were therefore devoted to small grants for preliminary or test studies.

In practice this funding structure meant a real choice in the kinds of knowledge acquisition and research atmosphere that the laboratory fostered. In particular, the laboratory hoped to foster the casual exchange of ideas between researchers of different disciplines, and it favored the "long shot," untidy, or untried research programs. Giving voice to this attitude, Jerome Bruner and Leo Postman argued that interdisciplinary integration often required "tolerance toward chaos and disorder." They continued by remarking, "we can tolerate disorder if we remember that it is often a condition *sine qua non* of organic growth in science."[145] The view that interdisciplinarity, creativity, chaos, and tolerance of disorder

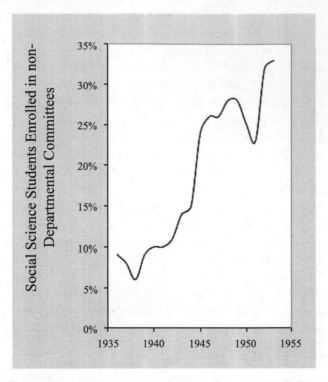

FIGURE 3.1. Interdisciplinarity at the University of Chicago. Chart generated from data given in Self-Study Committee, *A Report on the Behavioral Sciences at the University of Chicago* (Chicago: University of Chicago, 1954), 71.

often went together was a concept that played well with the directors of the laboratory. Accordingly, in its first few years, Bruner received more support from laboratory funds than any other of its researchers (Postman departed Harvard before the tabulation was undertaken).[146]

Treatment of interdisciplinary skills as desired and necessary characteristics came to be second nature in the Department of Social Relations. It helped to determine how senior faculty evaluated their junior peers. In supporting Jerome Bruner's tenure case, Talcott Parsons remarked in 1947, "he seems to us to be the best type of stimulating man who is individually immersed in creative research, continually experimenting with new ideas and new methods, and unusually successful in kindling the interests of his students in collaboration with him in these activities." For Parsons, Bruner's breadth of interests and ability to integrate ideas from across the social sciences far outweighed the criticisms that Bruner "spread himself too thinly" or that he was "not likely to be a creative contributor to the most fundamental theoretical development of his field."[147] Thus Parsons supported Bruner's promotion on grounds that gave more weight to experimenting with new ideas and methods than to the possibility that he had spread himself thinly.

In supporting the tenure of Frederick Mosteller, Samuel Stouffer wrote that he was looking for someone who exhibited a "versatile interest in the scientific problems of social psychology and related disciplines." In addition, Stouffer sought a person who had "the selflessness and *savoir faire* to be a flexible member of an interdisciplinary team." This person, Stouffer hoped, would be one who would not be "so occupied with a single track problem of his own that he lacks time or interest to help his colleagues wherever his skills will be useful." While Stouffer wanted each of these characteristics, he also remarked of one person he did not want to see appointed that "he suffers from a certain rigidity and arrogance which keeps him from being an ideal team man in the long pull." Of another person he did not want to see appointed, Stouffer wrote, "he is distinctly an individualist, with a good deal of rigidity. [He is a] colorful personality, he would be a decided asset to Harvard and I am one of his admirers, but do not see his fitness for our departmental spot." In recommending Mosteller above the other possible people, Stouffer was willing to overlook the fact that Mosteller had "yet to prove his ability to stay year after year if necessary with a single large problem and worry it until he solves it." To Stouffer it was acceptable to "sacrifice unusual staying power" because Mosteller possessed "other qualities," namely the flexibility that allowed him to contribute to interdisciplinary research.[148]

For both Parsons and Stouffer, then, the most desired characteristics of faculty they hoped to tenure were flexibility, breadth, and creativity. These characteristics would enable the teamwork that was deemed central to the department. They were considered to far outweigh their candidates' potential thinness or lack of depth or sustained interest in any one field. By 1951, the department was so committed to seeking out broadmindedness and flexibility that it would divide on an appointment, with Parsons and Allport on opposing sides, by arguing over whose preferred candidate had a broader or more interdisciplinary mind.[149]

Interdisciplinarity as a Route to Scientific Status

Much like the private foundations, the National Science Foundation (NSF) gave preferential support to interdisciplinary social science. In this case, interdisciplinarity would help the social sciences achieve the status of true sciences.

As it was originally organized, the NSF did not include a specific mandate to support the social sciences. This was because the public, Congress, and many natural scientists either equated the social sciences with socialism or did not find them to be sciences at all.[150] Others found social science invasive and not useful. Some disliked the way that social scientists violated social norms. Speaking for the "average American," Congressman Clarence Brown (R-Oh.) noted that "[he did] not want some expert running around and prying into his life and his personal affairs and deciding for him how he should live." Brown continued:

> If the impression becomes prevalent in Congress that this legislation (for the National Science Foundation) is to establish some sort of organization in which there would be a lot of short-haired women and long-haired men messing into everybody's personal affairs and lives, inquiring whether they love their wives or do not love them and so forth, you are not going to get this legislation.[151]

Brown offered a prescient warning about the methods of social science: two years later, Alfred Kinsey published his bestselling *Sexual Behavior in the Human Male* (1948), which was to be followed by *Sexual Behavior in the Human Female* (1953), both of which made use of "prying" questions.[152] Because of the social sciences' debatable status as real sciences and the distaste of "average" Americans for these disciplines, members of the science establishment such as Vannevar Bush came to the conclusion that including the social sciences in the NSF's charter could endanger the entire foundation.[153]

Social science's proponents responded to these criticisms by seeking to highlight the methodological similarities between their fields and those in the natural sciences.[154] As sociologist Bernard Barber put it, "Science is a unity, whatever the class of empirical materials to which it is applied, and therefore natural and social science belong together in principle."[155] In 1945 testimony before Congress, psychologist Robert

Yerkes contended that "demarcations" between the sciences were "entirely artificial."[156] Charles Dollard concurred:

> The goal is acceptance, at least by the literate public, including scholars in other fields, of the fact that the behavior of men, like the behavior of materials, is characterized by certain uniformities and patterns which can be studied systematically, and further that the discovery of these uniformities and patterns is a matter of importance to society at large.[157]

By highlighting the unity of the sciences, social scientists and their most important patrons emphasized that what made a field scientific was its method rather than its topic of study.

Such belief in and valuing of the unity of science supported a move toward interdisciplinarity. Moreover, the argument for the unity of the social and natural sciences suggested that the division between these two areas of human knowledge was a human artifact rather than one given in the natural state of things.[158] It was this effort to find links between the social and natural sciences that justified the NSF's initial move to fund social science by supporting the convergence of the social and natural sciences.

The strategy of identifying social science with natural science involved settling on a vision of what science actually was and what it meant to do scientific work. Since they had little authority or credibility on this matter, social scientists often made use of definitions of science written by physical scientists. They bolstered their claims by relying on the popular writing of James Bryant Conant.[159] He held that the advancement of science involved a "reduction in empiricism."[160] By this Conant meant that, in an advanced science, numerous empirical facts could be derived from a single theoretical or general proposition.

This connection between Conant's understanding of science on the one hand, and the search for interdisciplinarity on the other, was a matter of epistemological values. For instance, Conant's account of science made the development of conceptual schemes the criterion of a science's quality. To social scientists, this often meant modeling themselves after the physical sciences more than the "less adequate" biological sciences.[161]

Just as epistemological commitments could lead to judging biology as "less adequate" than the physical sciences, so too could adhering to Conant's epistemological values lead to preference for certain forms of social science. This kind of preference appears in Margaret Luszki's summary of the NIMH survey of interdisciplinarity. She noted, "there is a philosophical distinction which separates two groups who probably cannot work together: (1) those who believe that data are just there and that all we have to do is develop techniques to observe the data, and (2) those who believe that the data are a function of the theory." Luszki's interdisciplinary informants fell into the second category and felt that it would be impossible to include those with a positivist orientation in their projects.[162] Among other things, this particular philosophy of science meant that most experimental psychologists like S. S. Stevens were not good interdisciplinary citizens due to their commitment to behaviorism, operationism, and positivism.

A related philosophical orientation was critical to the fate of the social sciences within the NSF. Because the social sciences had no place of their own within the NSF, for several years all funding for the social sciences came through divisions designed to support the biological, medical, physical, mathematical, or engineering sciences. The Division of Biological and Medical Sciences, for instance, supported "basic research of an *interdisciplinary* nature involving convergence of the biological and social sciences." This division also funded the "interdisciplinary journal" *Human Biology* and an "interdisciplinary conference" sponsored by the Wenner-Gren Foundation in 1955 on "Man's Role in Changing the Face of the Earth." For its part, the Mathematical, Physical, and Engineering Sciences Division supported "basic research of an *interdisciplinary* nature involving convergence of the social sciences and the physical, mathematical, and engineering sciences."[163]

In 1954 Harry Alpert, the NSF's bureau chief of the social sciences, analyzed a collection of surveys of the social sciences in order to begin making the case that the social sciences deserved NSF funding. Alpert noted that the sources he read had argued that to become real sciences, "the social sciences must progress toward firmer theoretical formulation and conceptualization." This meant following Conant's vision of science and "lowering the degree of empiricism involved in solving problems" and "fabricating a web of interconnected concepts and conceptual schemes."[164] In order to be recognized as true sciences, the social sciences needed to "overcome their tendency towards diffuseness, unorganized specialization, insufficient integration, ineffective interdisciplinary communication."[165] Ultimately, the importance of interdisciplinarity to the NSF is indicated by the fact that, until 1958, adopting an disciplinary approach was essentially the only way that social scientists could receive its support.[166]

Creativity and the Persona of the Interdisciplinary Researcher

While "interdisciplinary" was most obviously a way of characterizing the organization of research projects, institutions, or educational curricula, it was more significantly a means of separating one type of person from another. In the mid-twentieth century patrons, scientists, administrators, and intellectuals saw interdisciplinarity as a characteristic, a trait of mind, or a way of thinking that a single person could possess. Being an interdisciplinary person meant *not* knowing several fields, but being tolerant, having the patience necessary for getting along with different types of people, having a proficiency in the languages (but not the factual content) of foreign disciplines, and possessing facility in translating one's own perspective and disciplinary ways of thinking into terms accessible to those steeped in a different tradition. Such an interdisciplinary individual was a person who could negotiate interdisciplinary contexts. Rarely was it necessary that these people have knowledge of more than one discipline. In fact, interdisciplinarity was not an issue of knowledge of several fields. Instead, as programmatic studies of interdisciplinarity by NIMH and the RAND Corporation indicated, what was needed was the ability, cognitive skills, and personality to get along with people of other disciplines.

According to RAND, aside from an interest in collaborative work, the primary requirement of an interdisciplinary researcher was "ease of communication, and the ability to properly disseminate necessary information to enable others to work expeditiously."[167] The secondary traits of the interdisciplinary individual were "training in the tools of science," "training in logic and philosophy of science," "knowledge of mathematics," and a "command of modern statistical theory and method." Significantly, these characteristics were depicted *not* as forms of knowledge but as means of improving communication. That knowledge of mathematics or statistics was taken not as domain knowledge but rather as a communicative or problem-solving skill was highlighted by a warning against team members' having too much disciplinary or content knowledge.[168]

The NIMH survey of interdisciplinary research concurred with the idea that interdisciplinarity was not about acquiring knowledge of several disciplines. In fact, it recommended that individuals who participated in interdisciplinary projects *not* acquire too much connection to foreign disciplines. The perceived value of interdisciplinary research came from its pluralist mixing of different kinds of people. By that logic, if team members on an interdisciplinary project acquired too much familiarity with one another's fields, then interdisciplinarity would end and creative heterogeneity would dissolve into uniformity and homogeneity.[169]

Recognizing how "interdisciplinary" could function as a character trait helps explain why it was so highly valued. What it meant to be an interdisciplinary person was practically synonymous with what it meant to be a creative person.[170] Thus creative people could be identified by their inner-direction, nonconformity, high tolerance of (or even preference for) disorder, flexibility, and open-mindedness.[171] These same mental virtues were regularly ascribed to the truly interdisciplinary person.

As we have seen, these were the traits that the interdisciplinary Department of Social Relations rewarded in its junior faculty seeking tenure. The similar self-presentation of the RAND Corporation's game theorists indicates how broadly such a view of creativity circulated. As Sharon Ghamari-Tabrizi has noted, RAND analysts marketed themselves to policy planners, the military, and opinion elites by emphasizing how their interdisciplinary techniques could stimulate creativity, "instill tolerance for ambiguity and uncertainty," and help avoid rigid thinking.[172] The NIMH survey of interdisciplinary projects noted repeatedly the importance of filling interdisciplinary research teams with "flexible" people.[173] According to the survey's informants, a good interdisciplinary researcher "should be *open-minded* and *creative* regarding *conceptual schemes* and the need for new approaches, *open* to the re-educative processes of the group, and able to translate his training from his original specialty to the interdisciplinary problem under consideration."[174] In an article on how to manage interdisciplinary projects, Gordon Blackwell suggested staffing interdisciplinary research teams with people who had "*flexibility* in theoretical orientation [and were] *not doctrinaire*."[175]

Ross Mooney, an educational psychologist and expert on creative thinking, exemplified this perspective in an article that treated traits of creativity and interdisciplinarity as if they were interchangeable. As Mooney put it, real understanding of creativity could not be accomplished unless it was studied by psychologists who were themselves creative thinkers. He went on to say that creative psychologists were the interdisciplinary ones. Mooney's reasoning was that the study of creativity would require an interdisciplinary endeavor and flexibility on the part of the researchers—and therefore creativity would best be studied by creative (i.e., interdisciplinary) scientists.[176]

The equation of creativity and interdisciplinarity becomes more evident on consideration of their opposites. In marked contrast to interdisciplinary people, disciplinary individuals were often represented as the inverse of creative. They were thus understood as closed, rigid, or narrow-minded conformists. This diagnosis involved several sorts of attacks on disciplinarity; closed-mindedness, rigidity, narrowness, and conformity each implied significant cognitive disorders. As we have seen, having a closed, rigid, or narrow mind meant an inability to cope with or "tolerate" ambiguous or complex information and situations.[177] Further, these character traits were primary symptoms of the ethnocentric, authoritarian mind—a fact that testified to the inability of the disciplinary-minded to work with different kinds of people. Psychologists also commonly noted the inverse correlation of intelligence with narrowness and rigidity. In 1947, for instance, a couple of psychologists suggested mental "rigidity" as an explanation for a person's failure to solve a conceptual problem. "Flexibility" was their explanation for successful solution of the same problem.[178] Significantly, according to these tests, thinking itself was defined by mental flexibility rather than any other number of options, whether skilled performance, complex calculations, or feats of memory.[179] Narrowness of mind thus implied more than a simple epistemic orientation but rather a serious handicap. As intellectuals habitually insisted, the natural world and the social world were complex. Consequently, those with narrow, disciplinary minds would be at a loss to accurately perceive or understand their surroundings. Confined to a single track, the disciplinary-minded would have trouble generating appropriate research strategies when faced with novel problems.

Conclusion

This chapter has examined how it came to be that interdisciplinarity seemed an unqualified good in postwar America. In the era of the Cold War, interdisciplinarity meant not only creativity but also democracy, rigor, and practicality. Interdisciplinarity's meanings came from the history of its emergence as well as the success of its advocates in pursuing a specific set of intellectual and even political values. In the interwar period patrons like the SSRC and Rockefeller Foundation decided that interdisciplinarity was the shortest route to solving practical problems. During World War II, administrators like Warren Weaver gambled that this prewar value system would produce effective results. Subsequently drawing on their experience in World War II, social scientists came to believe that "problem orientation" would be most productive in generating both practical results and advancing the theoretical concerns of their fields. By the 1950s, the best way for social science to be seen as truly scientific was to be interdisciplinary. University administrators, foundations, and government agencies contributed to this trend by giving preferential support to these interdisciplinary research topics, projects, and programs. Consequently, many of the hottest and best-supported topics of social scientific research, such as area studies, small group processes, and culture and personality, were interdisciplinary.

However much interdisciplinarity seemed universally attractive, it was not a universal panacea that meant all things good to all people. Instead, interdisciplinarity had a quite specific set of meanings that was a product of the history of the twentieth-century academy. Much of the way that interdisciplinarity came to be understood was a response to the specific modes of scientific work and discipline-building that had been adopted by psychologists working in the behaviorist and operationist tradition such as John Watson and S. S. Stevens. These psychologists relied on an empiricist method and drew sharp lines between their own modes of work and others that they designated as unscientific. They used such distinctions to establish disciplinary gatekeeping mechanisms like the Psychological Roundtable.

In contrast, a broad range of social scientists and their patrons, whether in private philanthropies or government, concluded that the best way to improve the social sciences would be to use interdisciplinary approaches to generate new and powerful research tools. These tools would allow social scientists to see otherwise invisible forces such as "ethos" that determined the course of social life. This methodological focus on tools avoided both a pure form of empiricist data-gathering and a form of theorizing disconnected from reality. Indeed advocates of interdisciplinarity contended that attachment to empiricism was itself a religious, unscientific dogma that prevented collaboration between people in different fields. Even more, because of the way in which social scientists saw America—as a pluralist society—they often identified interdisciplinary research as a pluralist endeavor which was optimally suited to the study of democratic societies.[180] Ultimately, then, social scientists fashioned a new form of social character, the interdisciplinary researcher whose open-minded personality was at once creative and tolerant of methodological variety. This person would be an optimal citizen of the social world of the midcentury academy.

CHAPTER FOUR

The Academy as Model of America

In a series of lectures delivered at Harvard in the fall of 1963, University of California president Clark Kerr analyzed the modern university and its historical development. To his eyes, universities were no longer cloistered ivory towers or even small towns. They were like modern cities.[1] Kerr was far from alone in making this connection between modern social forms and the world of the academy. To the Columbia sociologist Daniel Bell, the university was a central driving mechanism of the transformation of the nation from a series of disparate regional, rural cultures into a modern, cosmopolitan country with a complex yet "national culture."[2] This point, articulated in a study of the general education curriculum at Columbia, became a keystone of Bell's influential account of post-industrial society. According to Bell, in the post-industrial society the university would replace business as the "primary institution" of society. Economic and political decisions, "social prestige," and "social status" would all originate from "intellectual and scientific communities" rather than from "business firms."[3]

On this analysis, universities and the world of the academy faced many of the same challenges as modern America. Both America and the academy could be characterized by faceless anonymity as well as by competing sub-communities that did not understand each other. Both the university and America might be characterized by the conditions described in the title of David Riesman's book, *The Lonely Crowd*.

Unity of spirit and purpose was threatened, in the eyes of some critics, as apathetic suburbanites and even academics came to be more concerned with salaries, creature comforts, and conforming to social expectations than starting the next business, breaking ground on a farm on the American frontier, or making significant advances in knowledge. The purpose of a corporate job might be reduced to no more than paper-pushing, and the goal of an academic life might be no more than publishing for its own sake.[4] Because of its dramatic growth during and after World War II, even science seemed afflicted with the malaise of a middlebrow, corporate mentality.[5] What had once seemed to be a lofty investigation of the higher truths of nature now threatened to be simply a job that provided the material means for a comfortable lifestyle.[6]

As both the world of the mind and American society generally became more complex and diversified, individuals would be increasingly unfamiliar with other people, their culture, and their ideas. Perhaps even worse, if intellect were truly the center of modern society and culture, then the growth and fracturing of the academy and of knowledge would threaten the social system and even endanger democracy itself.[7]

One answer to these changes in the nature of the academy and within individual scientific disciplines was the application of techniques of industrial or military organization. In this mode of work, the techniques of operations research, pioneered to coordinate the work of the military during the war, would be applied to the research fields as well as to the conduct and organization of scientific research.[8] However, academics sought strategies to counteract the increasing regimentation and organization of intellectual work and life.

This chapter examines how even as the leaders of the intellectual world benefited from the influx of money and from the rapid institutional expansion of the academy, they worried that the organizational requirements produced by these changes could increase intellectual mediocrity and alienation. Critics, most especially those at the top of the academic ladder, also looked for ways to maintain the virtues of intimate, face-to-face community, the means to ensure communication across disciplinary boundaries, and the will to keep the academy from becoming like a corporation in which intellectuals became mere bureaucratic functionaries.

The answers they sought for maintaining the cherished aspects of their community reflected the treatments they proposed for the nation as a whole. When contemplating a similar set of symptoms plaguing the academy, intellectuals prescribed the same measure they had for America: creativity. Through creativity, the problems of the world of the intellect could be cured. By arriving at this particular analysis and proposals for reform, intellectuals and policy makers linked the academy to America. While cultural and political expectations as well as scientific models established intellectuals as not only the most creative, but as the very model of the creative person, it was also the case that intellectuals' own communities would serve as either casual or formal models for American society.

Why was it that intellectuals found that the same set of measures were necessary for America and for their own community? In looking at this question. I examine how social thinkers used their own day-to-day experiences to think about America, and how their political visions of the nation informed the kind intellectual community they wanted.

In their own social world, intellectuals developed standards of comportment, self-presentation, intelligence, creativity, and right-mindedness. They enacted and articulated these standards in their writing and face-to-face in modern-day interdisciplinary "salons." These salons were instantiated in dinner clubs and discussion groups, as well as in interdisciplinary study groups and conferences. With regular meetings and overlapping memberships, salons helped leading intellectuals of the postwar era form a community as well as a shared sense of normal and proper forms of self and of thinking. That community and shared sensibility offered a set of values for regulating both America and the academy.

The salons of postwar America served to mediate and connect intellectuals to one another, to the world of politics, and to reconfigure the public sphere. Like their French predecessors, the salons of midcentury America were social occasions and places for displaying and conducting intellectual and artistic work outside of preexisting institutions, disciplines, and departments. They emphasized expressions of creativity and innovation and valued certain forms of intellectual work above others. Salons were also places for displaying, cultivating, and thinking about certain forms of selfhood. They welcomed people who cultivated intellectual breadth and the ability to communicate in the proper fashion with others, and they discouraged narrow pedantry.[9]

Beyond the exclusion of Communists and others labeled as dogmatic or ideological, the norms of selfhood as defined by liberal social thinkers were useful as a positive guide for achieving specific social ends. One such end was regulation of the academy. Just as America needed creative, open-minded, autonomous citizens to flourish, so too did the social world of the academy call for the same types of people. Such individuals (and they were almost always male) were those who could maintain a sense of themselves in the context of the complex and rapidly growing and changing ecology of the academy. At the same time, they also needed to be flexible, open to change, be able to communicate with others and to form and maintain connections with people who were and thought differently.

By looking at the relationship between daily personal experience and social thought, this chapter examines how the solutions academics sought for reforming their own community reflected the political and social visions they held for America. Political sentiments provided a language for discussing and regulating the functions of intellectual social life. At the same time, the visceral experiences that intellectuals had day-to-day functioned as touchstones as they developed their accounts of America—especially when they offered off-the-cuff, informal evaluations of the nation. Their social vision was grounded in the kinds of interpersonal interactions that they practiced and valued in their own professional lives.

Modeling America

To begin understanding how intellectuals could see their own subculture as a model for or analogous to American life more broadly, we need to focus on the leading forms of social thought in the post World War II period.[10] The most important innovations in social science in the postwar years were designed to erase distinctions between different forms and scales of social life. These included game theory and small group studies.

Game theory spanned the study of economic behavior, childrearing, and negotiating traffic as well as the examination of strategic conflict between nations. For instance, in 1960, Thomas Schelling compared keeping a child from hurting a dog to "constraining" Mohammed Mossadeq, the democratically elected prime minister of Iran, who had been deposed in a CIA-led coup in 1953 that had been initiated to protect the profits of British Petroleum.[11] Whether this comparison reflects an attitude of cultural superiority that personified non-Western countries as children, an easy comfort with antidemocratic American imperialism, or a call for certain modes of childrearing is unclear. What mattered (enough for Schelling to win a Nobel Prize in economics) was that each situation was comparable, and, as he put it, could be used to provide useful information about the other. Game theory also provided a platform for work in political theory and international relations.[12]

The study of small groups was similarly attractive to social scientists and their patrons. At a time where fields across the social sciences were considered to be scientific, creative, rigorous, and even practical simply by being interdisciplinary, the study of small groups offered a bridge between psychology, sociology, and political science, the possibility of close observation as well as experimental control, and compatibility with one of the leading modes of social science theory at the time, the structural functionalism of Talcott Parsons.

As with much of early postwar experimental social psychology, experimental studies of small groups looked to offer explanations of human nature and behavior that were independent of context as well as content. Such stripping of context and content was a means for social psychology to approach a universal, even natural, science.[13] Knowledge about the nature of one kind of small group was translated for use in an entirely different context. For instance, knowledge gathered in the context of experimental research on the interaction of children in playgroups was understood to bear on the practice of group therapy, child guidance, the organization of conferences of diplomats, the implementation of meetings designed to foster religious tolerance, and the training for the OSS.[14]

Research cast in a context-free mode enabled social scientists to use their knowledge about one part of society to think about the remainder of society. The wave of research on small groups therefore enabled social scientists to use the social world they knew best, their own, as a heuristic device for understanding social interaction in general. Accordingly, a substantive area of work within small group study was directed toward the study of academics, policy makers, and other experts.[15] As with the rats, pigeons, and college students who provided experimental psychologists a readily available model of human nature, groups of intellectuals and policy makers were a useful and accessible population for those who studied small groups.

Small groups, composed of people who knew one another, whether in neighborhood communities, the National Security Council, or the Supreme Court, appeared to some social scientists to be *the* forum where most, or even all, social, cultural, and political activity occurred.[16] For intellectuals and politicians, one of the most significant forms of small groups was the conference.

The anthropologist Margaret Mead saw conferences as a fundamental "new social invention" that was a characteristic feature of the modern world.[17] Conferences were important enough that she eventually produced a participant-observer account of how they operated based on twenty-eight years of forty days per year on average of attending meetings.[18] On the basis of this ethnographic work, Mead and her collaborators produced general insights on human interaction that paralleled small group studies. One collaborator, Paul Byers, a specialist in the photodocumentation of human behavior, produced a careful analysis of interaction among conference participants. He argued that what was relevant about conferences was not their content but the kinds of social interaction that constituted them. Consequently, conference activities were comparable to "the primary form of social interaction among such a so-called simple society as the pygmies of Central Africa."[19]

In this analysis, Byers supplied twenty-five photos to document four minutes of conversation. As was typical of small group studies, the content of conversation was to be ignored. (See figs. 4.1 and 4.2 and accompanying text.) In that way, patterns of group activity in conferences could stand in for generic social phenomena.

FIGURE 4.1. From Mead and Byers, *The Small Conference,* 73. By permission of Walter de Gruyter.

C [Mead]: looks at speaker again; head remains tilted; brows are slightly compressed again

D: has put pipe in mouth; suspends pipe and hand in air; looks probably at B; head somewhat moved toward B

B: Has exhaled smoke but holds cigarette in curled hand at nose/mouth; looks at D (discernible only with magnifier)

A: (speaker) and C now look at each other; A's hand oriented toward C.

The observation that the speaker is looking and gesturing toward or directly at C and that C is looking back with head slightly tilted and brows slightly lowered suggests that the speaker sees some possible disagreement before he hears it. It is significant that B and D are silently in communication with each other and are not looking at A or C.

FIGURE 4.2. From Mead and Byers, *The Small Conference*. By permission of Walter de Gruyter.

    C: speaks, overlapping with A; C and A still looking at each other

    A: holds hands in suspended gesture

    D: has removed pipe from mouth and looks at C

    E: looks toward C

    The spoken interjection by C begins here. But A is signaling his intention to continue by holding his hands in suspended gesture. Note that E looks at whoever is speaking, moving her head back and forth in spectative fashion. Although B never looks directly at the speaker, he manages to signal his dissent, and A becomes increasingly aware of this, as we shall see.

    Among the meetings that formed the basis of Mead's ethnographic study were a series of conferences sponsored by the United Nations. These conferences were themselves devoted to studying conferences (as a social form) and, more importantly, to proposing procedures and arrangements that would make future conferences operate most smoothly and effectively.

    The UN's motivation for studying the culture of conferences came from the conviction that they were the keystone of modern political, cultural, and intellectual life.[20] Brock Chisolm, the director-general of the World Health Organization and director of the UN study group, minced no words in underlining the significance of conferences for the world. As he put it, "many of our most important social functions are carried out in meetings." Meetings had become both "evidence of" and a primary mechanism for dealing with a "new and far greater degree of interdependence." Because of this growing interdependence and the increasing number of meetings, Chisolm noted "many of our most significant and productive relationships with other individuals, and between the groups to which we belong and other groups, are established and developed in meetings." Even more, "meetings," Chisolm asserted, "have begun to replace battlefields as the arenas in which relationships between groups of people are determined."[21] On this view, meetings had become so important that not only had they become a primary forum of social and political life, they were also increasingly the place for resolving conflicts.

Because it took the view that meetings had become the center of society, the United Nations enlisted the aid of social scientists in order to better understand and improve how they functioned. These social scientists included those who specialized in cultural differences such as Margaret Mead and those who were specialized in the study of small group processes. Mead offered her expertise in cultural interpretation while specialists in small group processes brought skills in fine-grained analysis of moment-to-moment conversation.

Such intervention by the social scientists allowed the United Nations to draw on the specialized knowledge that experts possessed. Conversely, time spent with the United Nations project would enable social scientists to develop a new critical dataset that would facilitate building a general model of human interaction.[22]

Practical efforts to improve communication across disciplinary and cultural barriers were quickly translated by social scientists into an opportunity for research on group processes in general. At the UN Conferences on Conferences, social scientists envisioned that they would be "guinea pig[s]" in a social psychological study of conferences in general.[23] Similarly, the Macy Conferences on cybernetics, sponsored by the Josiah Macy, Jr. Foundation and held in New York and Princeton between 1946 and 1953, became an opportunity to exchange ideas with other intellectuals. However, these meetings were also an immediate testing ground for the idea of circular feedback.[24] The first test of cybernetic ideas in the social realm was performed upon the very group of intellectuals who articulated the ideas of circular causation. Consequently, to the luminaries of the social scientific world who participated in the cybernetic conferences, knowledge about themselves was inextricably tied to their subsequent articulations of how humans, culture, and society functioned. Margaret Mead, who had studied the cybernetic principles at work in the early conferences on cybernetics, found this reflexive process so successful that she proceeded to suggest that other conferences use cybernetic principles to organize themselves.[25]

While the members of the cybernetic group made themselves into test

subjects for researching ideas about the role of feedback in society, social scientists also held themselves up as models of human and animal
emotions. This instance of reflexive knowledge appeared in a 1955 reprint of Charles Darwin's *The Expression of Emotions in Man and Animals*. This version of the book appended a conclusion to Darwin's original text composed by modern social scientists.

The last segment of that appendix was a photo essay that provided
modern photographic examples of human emotional expressions (see
figs. 4.3, 4.4, and 4.5). The majority of this essay (and its final section)
consists of a series of photos taken at an academic conference devoted to
human communication in which participants were "the objects of their
own research."[26] Thus conference participants, such as Mead, Marshall
McLuhan, and Reuel Denny (who collaborated with David Riesman on
The Lonely Crowd) were themselves the ultimate illustration of emotions not only in all humans but in animals as well.

FIGURES 4.3–4.4. Photos of Reuel Denny by Simpson Kalisher. From Charles Darwin, *The Expression of the Emotions in Man and Animals, with Photographic and Other Illustrations: With a Preface by Margaret Mead* (New York: Philosophical Library, 1955).

FIGURE 4.5. Photo of Marshall McLuhan by Simpson Kalisher. From Darwin, *The Expression of Emotions in Man and Animals*.

Creativity as Solution for America and the Academy

Although they were central to postwar social science, the Macy Conferences and small group research were only a part of a broader intellectual culture that regularly and habitually used academic culture as a model of other parts of society. In fact, the linkage of intellectuals and the academy to the rest of society occurred outside of theoretical and experimental works in the social sciences. Thus social scientists put into a scientific frame a set of views that circulated widely in casual conversation.

Humanists and physical scientists used academic culture for thinking about national and international issues. At the center of their diagnosis of society's ills and of its cure was a casual treatment of the social world of the academy as microcosm of, and ideal type for, American society.

The nation's problems thus seemed isomorphic with problems within the academy. Consequently, solutions to the challenges faced by the academy could also be solutions to the problems of America and the world. In this intellectual and cultural context, the growth of complexity and specialization in knowledge caused grave concern. It would be read again and again as "fragmentation" of knowledge and would imply fragmentation of the society at large.

Creativity offered the answer to this problem of intellectual fragmentation. Just as social thinkers believed that creativity could generate a coherent national culture, they also contended that it would unify the world of the intellect. Creativity's role in linking the academy and the nation emerged particularly when academics and policy makers addressed one another in formal theoretical terms and even more frequently in relatively informal settings such as at conferences, dinner parties, and in private correspondence. Through these interactions, intellectuals repeatedly emphasized the centrality of creativity for holding together both their own community and also American society at large.

The particular significance of creative thought in the academy for generating a cohesive society appeared in an American Academy of Arts and Sciences (AAAS) conference in late 1963 or early 1964 on the relationship of science and culture that was attended by the leading lights of the intellectual world.[27] At the AAAS conference, concern mounted over the growing specialization of knowledge and over the increased importance of science and technology that had resulted in a society where individuals, lacking sufficient expertise, were divorced both from decisions about the direction of the country and from one another.[28]

But if the fragmentation of knowledge generated by its ever-increasing complexity could cause social disorder, then a potential solution for social disorganization could be integration of the different areas of human knowledge by people who were sufficiently creative. The art historian James Ackerman argued that the specialization of knowledge was not necessarily a problem; the different domains of culture, science, and art,

for instance, cohered at the point of their highest and most creative development. It was only the low-level "technicians" who could not communicate with one another.[29] In fact, to Ackerman it was clear not only
that creativity could facilitate communication between the disciplines
but that creativity emerged from interdisciplinary training.[30] Creativity,
interdisciplinarity, and cultural cohesion, then, each helped to produce
the other.

In the social world inhabited by members of the American Academy
of Arts and Sciences, creativity enabled individuals of different backgrounds to understand one another. Such understanding offered the
hope of a functioning academic community and, therefore, of a unified
national society. Gerald Holton, a historian of science, professor of physics, and the editor of the AAAS conference proceedings, agreed with
Ackerman's analysis. Holton noted that scientists, at the moment of discovery and invention, draw on a repository of ideas or themes located in
a cultural domain shared with other intellectuals. For instance, scientists such as Newton and Bohr had made creative and imaginative use of
such cultural themes in pursuing their physics.[31] Consequently, the cultural disorganization produced by modernity need not exist. Scientists
and humanists who sufficiently exercised their imaginations and creativity could unify modern culture by developing a shared "thematic"
language.

This analogy that linked the academy to the rest of society could be
egalitarian or elitist, inclusive or exclusive, depending on the context.
James Ackerman's argument about the necessity of creativity reserved
true understanding to a select few. Ackerman's view of society was com-
plemented by academics and policy makers who saw the good society
as a managed society. The treatment that Walt Rostow proposed for
dealing with the difficulties of mass society was more control by broad-
minded experts. Richard Bissell, the head of the CIA's Directorate for
Plans, future planner of the Bay of Pigs Operation, and one of Rostow's
undergraduate mentors, agreed and argued that functions of foreign-
policy making needed to be stripped away from bureaucratic functionaries and centralized under a powerful executive. Foreign policy expert
George Kennan concurred with these views as well.[32]

A decade later, as his post as secretary of defense was nearing an end,
Robert McNamara articulated a similar faith. He argued that the most
creative "art" was his own field (management), and that management

was critical to freedom, reason, initiative, and personal responsibility.
McNamara added:

> Vital decision-making—particularly in policy matters—must remain at the
> top. That is partly—though not completely—what the top is for. But ratio-
> nal decision-making depends on having a full range of rational options from
> which to choose. Successful management organizes the enterprise so that pro-
> cess can best take place. It is a mechanism whereby free men can most effec-
> tively exercise their reason, initiative, creativity, and personal responsibility.[33]

This perspective was continuous with the vision of society advanced by
political scientists like Gabriel Almond, who held that "democracies"
functioned best when the mass of the population was disengaged or even
did not vote and in which society was managed by apolitical technocrats
(e.g., people such as McNamara, Rostow, Bissell, Kennan, and Almond
themselves).[34] Like Almond, C. Wright Mills noted the propensity to
equate the "public" with a collection of experts. However, Mills raised
this point in protest not celebration, for he saw this substitution as a part
of the collapse of true communities and democracy and the rise of mass
society.[35]

On the other hand, other intellectuals advocated creativity and noted
parallels between the academy and society using a more egalitarian
mode of analysis. For instance, Harvard's 1945 manifesto on general education compared America to the academy, but it also did so by declaring that every American was an expert. The job of democratic education,
therefore, was to teach all Americans, each an expert, how to communicate with other experts.[36] Likewise, Jacob Bronowski had suggested that
creativity could serve as social glue. Bronowski argued that creative insight is necessary for any person to appreciate a work of art or a work of
science. But unlike Ackerman, Bronowski argued that this form of creativity could be possessed by anyone, not just the elite.[37]

Whether elitist or egalitarian in spirit, the comparison between the
academy and America had both positive and negative dimensions. On
the negative side, problems in the academy could potentially infect
America at large. Intellectuals saw causal links between social disorder,
social complexity, modernity, and disciplinary specialization. On the
more positive side, the academy's virtues offered hope for America as it
could be. In published manifestos and in private correspondence, social

scientists regularly drew analogies between social pluralism in America and their own pluralistic interdisciplinary research collaboration. Such a vision of the positive social consequences of interdisciplinarity often rested on eliding distinctions between the social world inside the academy and the social worlds outside. For instance, academic disciplines were regularly referred to as if they were social interest groups, religions, or cultures of the world.[38]

Calls for finding means of disciplinary cooperation often bled into, or were phrased in, language that echoed the calls for interpersonal, cultural, religious, ethnic, or racial tolerance and pluralism. Just as creativity would save America from being a mass society, so too would it rescue the academy from its own fragmentation. The centripetal forces of the modern world would be overcome by the community of individuals who had the requisite traits of creative open-mindedness that enabled cooperation across social barriers. So too, would the rift in the academy and disciplinary boundaries be healed by creative, open-minded individuals.

Thus, the language used to discuss interdisciplinarity (flexibility, breadth, tolerance, and pluralism) often mirrored the language social scientists used to discuss their hopes for society outside the academy (flexibility, breadth, tolerance, and pluralism). A book that recounted the interdisciplinary efforts at personnel assessment for the Office of Strategic Services during World War II drew an analogy between the ability of social scientists to bridge their differences and Anglo-American political culture.[39] Against the background of calls for the unity in diversity among nations, cultures, or groups within the United States, Robert K. Merton concluded a series of talks for Voice of America by noting the unity in diversity of the social sciences. Similarly, at the University of Chicago social scientists congratulated themselves for their own unity in diversity.[40] Even when looking beyond the academy itself, those who called for toleration discussed the critical importance of members of one profession, such as bankers, getting along with members of another profession, such as scientists.[41]

The University of North Carolina social anthropologist John Gillin compared the disciplines of anthropology, sociology, and psychology to religions. He then suggested that the disciplines had learned to cooperate because they had "found themselves . . . increasingly in each other's company, rather like a collection of Protestants, Jews, and Catholics who keep turning up at the same cocktail parties and who are, at first, put to it to make polite conversation, and then gradually drawn into discussing fundamental issues."[42] As the tone of Gillin's remarks indicates, the religious pluralism of the academy was a relatively novel feature of intellectual life. Exclusion of scholars, especially Jews, on the basis of their religion was a recent and taken for granted aspect in much of the interwar academy.[43] Thus, consensus that the academy should strive to be an inclusive or pluralistic place that functioned outside of a religious, specifically Protestant, framework was a recent development.[44]

Gillin was ultimately ready to extend comparisons between disciplines and social groups to a normative discussion of social structure and politics. In this latter comparison, Gillin used his beliefs about the political and social world to suggest a way of organizing the academic world as if it were a federation like the United States or the United Nations.[45] For social scientists working within the United Nations, interdisciplinary federation was not just a metaphor for the United Nations, it was a positive strategy for building the institution itself, resolving international strife, and producing a community of world citizens.[46]

Gillin was far from alone in referencing the academy as though it were the United Nations. For instance, in his Godkin Lectures at Harvard in 1963, University of California president Clark Kerr made the exact same point. Kerr was quite clear. The university was not an ivory tower. It was a "pluralistic society," a "system of government like a city, or a city-state" or even the United Nations. It was a collection of communities whose interests needed to be coordinated. Once Kerr's lectures were published as *The Uses of the University* in 1963, the casual comparison of the university to a larger social system was available to a wide audience.[47]

Outside the context of departmental organization and interdepartmental relations, the comparison of interdisciplinarity to pluralism played an important role in the social life of the academy. Interdisciplinary activity within a group of academics offered a model not only of the possibility of people with different concerns achieving mutual understanding, but also of the chance of achieving a piece of the genuine, face-to-face community that many saw vanishing with the emergence of modern mass society.[48]

Community Regulation

Despite the pervasiveness of the analogy between the academy and "pluralistic society," it was also not the case that everyone would necessarily be considered equal, or even candidates for inclusion, membership, or, shall we say, "citizenship." The intellectuals' social world was structured in specific ways, and it encouraged certain modes of self-presentation. It selected for certain traits and discouraged others. Those who could best present themselves in the favored ways stood to gain the most.

The autonomous, creative person self-described by social and cognitive psychology found a congenial home in the social world of postwar intellectuals. Traits of breadth, autonomy, creativity, openness, flexibility, complexity, tolerance of ambiguity, and pluralism were associated with an educated mind and were sought out in this world. On the other hand, the opposite traits, narrowness, inflexibility, rigidity, conformity, and authoritarianism, were inversely correlated with education, rationality, and the intellect—and were shunned.[49]

This system of values appeared repeatedly as intellectuals assessed their work and their communities. To the founders and subsequent board of directors of the Center for Advanced Studies in the Behavioral Sciences (CASBS) as well as to Frank Fremont-Smith, the advancement of modern science and the generation of creative ideas required interdisciplinary and relatively informal, unstructured communication.[50] It was for the same reason that Freemont-Smith, as president of the Macy Foundation, sponsored a number of interdisciplinary conference series. The publication of the proceedings of those conferences would, he hoped, demonstrate the importance of informal communication for the production of creative work and provide a model for future conferences.

If conferences could generate creativity, it was also the case that the goal of intercommunication and productive insight would be compromised by people who had the wrong sort of psyche. Such individuals were a threat to the production of community and the generation of creative science because of their "blind spots, prejudices, and over-attachment to or dependence upon an 'authority' or upon too narrowly conceived criteria of credibility."[51] In this analysis, Freemont-Smith echoed the finding of the social psychologists who produced studies of creativity, conformity, and authoritarianism. As with the psychologists at Berkeley's Institute of Personality Assessment and Research (IPAR) who saw prejudice, narrowness, and over-attachment to authority standing in the way of true creativity, Fremont-Smith believed that creative ideas would be fostered by a sense of psychic security among the participants.

The social worlds of intellectuals were regularly critiqued and policed with the tools of social psychology. One case was an interdisciplinary salon on "The Nature of Man" run by Margaret Mead. This salon was supported though Columbia University's interdisciplinary Faculty Seminar program and hosted by Ruth Anshen, an editor of wide-ranging books by leading intellectuals writing about the human condition in the modern world.[52] Evaluating the salon's success or failure and judging members of the discussion group was accomplished by using the very psychological tools that pervaded the literature on the open and closed mind. Thus, in trying to explain the lack of coherence in the salon, Mary Catherine Bateson argued that "communication between the disciplines" required the kind of mind that held society together. That mind possessed the capacity for "tolerance of ambiguity." That is, the smooth functioning of this salon required its participants to possesses precisely the same mental attribute that *The Authoritarian Personality* had identified as a defining feature of the democratic self's thought process.[53]

This evaluative procedure worked to the benefit of some and against others. The creative, broad, flexible, and interdisciplinary person would benefit while those who were narrow, conformist, rigid, intolerant of ambiguity, and disciplinary-focused would be at a disadvantage in these contexts. Both the positive and the negative traits were not distributed in the population or in the academy on equal terms. For instance, the positive traits were linked with cultivation, class, social status, and being male. Fields taken to be interdisciplinary from decision sciences to operations research, small group studies, and cognitive science likewise benefited. On the other hand conformity, rigidity, and narrowness were often attached to those of lower socioeconomic status, to Catholics, to fascists, and to women.[54] Like membership in these social groups, undertaking certain modes of inquiry deemed "narrow" marked a scholar for difficulties in the social world of the elite academy. One group of scholars especially impacted by this rule was behaviorists.

Faculty searches reflected the sensibility that promoted breadth. Even highly disciplinary places like Harvard's Department of Psychology sought to ensure that the senior faculty it hired would be "democratic" leaders, be "broad," be people who could "tolerate," "understand," and occasionally compromise with opposing points of view. This is a remarkable sentiment for a department characterized both internally and externally by its commitment to methodological purity.

Whether individuals' commitments to these values were long held or acquired through institutional defeat, they were sustained enough that the department declined to consider appointing leading behaviorist Kenneth Spence on the grounds of the failings of his personality. After two significant failures to recruit a senior scholar, the subsequent appointment of B. F. Skinner a few years later to the same position indicates that this system of values in this department was short-lived, since the department had initially rejected Skinner because it had concluded that he would "never be a cooperator."[55]

The issue ran deeper than cooperation. According their own students, behaviorists were authoritarians. The main-line behaviorist Kenneth Spence was characterized by his students as authoritarian (88 percent). Only 13 percent said he accepted criticism well and only 6 percent said Spence was open-minded (the trait most commonly treated as the inverse of authoritarian). Much in contrast, Spence's primary opponent, Edward Tolman, a man known for methodological eclecticism, had the opposite results. Only 4 percent of Tolman's students characterized him as authoritarian and 100 percent saw him as open-minded.[56] Even in their own self-conception, behaviorists were antidemocratic. For instance, in the 1930s, behaviorist Clark Hull privately contemplated improving the disorder at Yale's Institute of Human Relations by noting how it needed a "führer" to make the institute more scientific by eliminating "subjectivity, individual idiosyncrasies, and social deviance."[57]

Behaviorists like B. F. Skinner ratified charges of doctrinaire authoritarianism by making a habit of pretending to not understand something a colleague had said if his or her remark included mentalistic language. Skinner's activity was a way of underlining the behaviorist refrain that since the mind and its operations are not visible and measurable, to speak or write about them would literally be to utter *nonsense*—something that should be treated as incomprehensible.[58] Because of the common use of mentalistic language to describe either oneself or others, Skinner's habit would have relegated almost all of the people he spoke with to the category of nonsense speakers. This was surely not a habit that would have endeared him to fellow intellectuals who were committed to both intellectual pluralism and their own intelligence.

William Verplanck, an ally of Skinner's, published a glossary that provided a "translation" of mental terms into a language that behaviorists would recognize as meaningful. This translation, however, simply substituted the physical for the mental. For instance, "frustration" became the operation of preventing the subject from reaching a goal object. Thus, a feeling state became the procedure for restraining laboratory rats under a condition of semi-starvation from consuming food they could see or smell.[59] However, this translation of Verplanck's did more than find a way of treating mentalistic terminology in terms that were considered "scientific." Verplanck also used the glossary to label as unscientific psychologists such as Edward Tolman who failed to carry out the strict behaviorist line.

Due to their unsocial behavior and advocacy of what was taken as antidemocratic epistemology, behaviorists were often not invited to participate in the inner circles of elite academic social life. Compared to their reputation and leading position within psychology, behaviorists were underrepresented in these arenas. They did not receive fellowships at the Institute for Advanced Study (IAS) at Princeton, and those who did spend a year at the Center for Advanced Study in the Behavioral Sciences (CASBS) were either interdisciplinary or were engaged in what was seen as an interdisciplinary effort—such as building mathematical models for the social sciences or unifying the social and biological sciences.[60] On the other hand, individuals engaged in the still emerging field of cognitive science found regular fellowships at the Center for Advanced Study in the Behavioral Sciences and the Institute for Advanced Study.[61] For example, George Miller spent a year at the IAS and later co-authored, with Eugene Galanter and Karl Pribram, one of the seminal texts of cognitive science while at the CASBS.[62] Beyond the level of individual fellowships, these institutions also devoted resources to activities promoted by the institutions themselves for the advancement of cognitive science (but did not do so for behaviorism). The CASBS also supported special projects on cognitive topics in 1963–64 (Education and Cognitive Development), 1964–65 (Psycholinguistics), and 1965–66 (Cognitive Consistency).[63]

A similar pattern held at the American Academy of Arts and Sciences (AAAS). Although he worked within walking distance of the Academy, B. F. Skinner appeared once in its journal, *Daedalus*, in the fifteen years between the mid 1950s and 1970. On the one occasion that Skinner did appear, it was only in the context of being on a panel dominated by people hostile to his views. In contrast, the AAAS had numerous visits and publications from much less famous psychologists.

One reason for this relative disparity—for making special room for an emerging field while slighting behaviorists—may have been the perception of the intellectual community that behaviorists were narrow-minded and unscientific, and intolerant of ambiguity. Social scientists commonly discussed the doing of good science in terms of its breadth and integration. The characteristics of good scientists were their flexibility, ingenuity, creativity, and ability to view problems from multiple perspectives. These cognitive attributes were paired again and again with the social attributes of tolerance and broad-mindedness—precisely the attributes behaviorists were charged with lacking.

The organizers of the CASBS gave typical expression to this way of seeing as they discussed what kinds of people should be invited and cultivated at the center. In a CASBS planning group meeting on December 1, 1952, Henry Murray spoke of experimental psychologists who had narrowed their field of focus and chosen their research areas on the basis of which could be "approached methodologically" rather than on the basis of their importance. Murray noted that "psychologists must be helped to recover from overspecialization—from a 'trained incapacity' to see or work on important problems."[64] In a meeting on December 20, Carl Hovland, an experimental psychologist himself, agreed by noting that experimental psychology "aims at maximum precision, and hence greatly limits and simplifies its problems." For the purposes of CASBS, more damning was Hovland's remark: "the pure experimental psychologist wouldn't be interested in collaboration with other behavioral scientists; he is mainly interested in working with his own field or with physiology and biology."[65]

What was needed was not only breadth or even the interest in working with other behavioral scientists. The CASBS would seek academics who had the proper character, a character that was congruent with the open-minded, democratic, autonomous, flexible, creative self sought by liberal social critics for American character. As the minutes for the December 20 meeting noted,

> [Herbert] Simon suggested that one thing which must be emphasized is the overall "character" which a good scientist should have. Natural scientists have the ability to *handle anxieties* of not knowing answers—they have the ability to *avoid premature closure*, to *tolerate* areas of admitted ignorance. They know how to take it one step at a time and not try to cover up by presenting overall solutions.[66]

Agreeing with Simon was Bernard Berelson, director of the Ford Foundation's behavioral sciences program and perhaps the most important and powerful administrator in the social sciences. Berelson noted that "the ability to *tolerate lack of closure* is related to the confidence in the possibility of getting the answers at all by step-by-step methods. Confidence in method is necessary for an optimal degree of persistence."[67]

The fact that Simon and Berelson framed their comments as they did indicates the pervasive nature of the discourse that equated open-mindedness, creativity, and interdisciplinarity. It also highlighted one of the behaviorists' handicaps. Behaviorists insisted that it was important to be patient in science and that one must not call for practical application in too hasty a manner. Yet, rather than crediting behaviorists with patience, critics portrayed them as either single-mindedly wedded to a unitary approach or with leaping without warrant from narrow, animal-based, laboratory results to explanations of human nature. On this view, behaviorists were not tolerant of the slow pace of their science. They were either irrationally overoptimistic or possibly made so anxious by lack of explanation that they tried to fit the square peg of mechanistic behavioristic methods into the round hole of such complex mental phenomena.[68]

It is important to note that neither Simon nor Berelson were involved in the genres of personality research that produced either the studies of creativity or *The Authoritarian Personality*. Nevertheless, these two echoed many of the same points found in the social psychological literature on authoritarianism and creativity. Specifically, Simon and Berelson pointed to the positive and scientific nature of being tolerant of anxiety and being able to tolerate lack of closure. These two items were the primary means of distinguishing those with open, democratic minds from those with closed, rigid, narrow, and authoritarian minds. And, for the purposes of the CASBS, these traits would serve as markers of who would be welcomed into the elite levels of the academy, and who, like the behaviorists, would be less welcome.

Aside from their intolerance of ambiguity, there was another reason that behaviorists were seen as both fascistic and unscientific. That was because behaviorists were seen as narrow-mindedly conducting research from the perspective of a single discipline. Their methods had served to separate psychology from other disciplines.[69] Behaviorists engaged in what seemed to be disciplinary imperialism, often seeking to take over the domains of other fields.[70] The problems they pursued and the methods they used were chosen not by what was important either from the perspective of the advance of pure knowledge or from the perspective of practical application. Instead, they selected their problems and methods from a restricted domain determined by the discipline, not by nature, truth, or even what was important. Such a narrow-gage approach appeared to many to be a narrow, fascist, prejudiced, or possibly insane mentality. Harvard cognitive psychologist Jerome Bruner, for instance, noted the straitjacketed mentality of behaviorists Edwin Guthrie and B. F. Skinner when remarking on their inability to meaningfully participate in a broad-ranging discussion of the place of learning theory as a part of the constellation of social scientific theory.[71]

In the midcentury academy, this kind of anti-interdisciplinary behavior read as political—in terms that meant more than academic politics. Being open to interdisciplinarity meant being committed to pluralism, at a time when that was *the* marker of a democratic mindset and what made America strong.[72] Along with others in midcentury America, Gerald Holton cited Alexis de Tocqueville to claim pluralism as a central component of American political character. However, Holton disagreed with de Tocqueville by contending that rather than mediocrity, pluralism led to intellectual vitality.[73]

Intellectual Culture: Creativity, Flexibility, and Conviviality

Whether in extended exegesis or offhand remark, when thinking about how to improve national culture, intellectuals repeatedly referenced the forms of life that held academic society together. For instance, perhaps drawing on his experience as director of the Institute for Advanced Study, J. Robert Oppenheimer suggested that social order, meaning, and intellectual unity could be achieved though a series of dinner parties.[74] Similarly, the kind of democratic America the authors of Harvard's *General Education in a Free Society* hoped to bring into being was one held together by communication—of the type that took place at Harvard.

This view of the good society anchored by thoughtful discourse and intelligent communication belonged to a long tradition of political philosophy. It followed, for instance, the work of Condorcet and the more contemporary arguments of John Dewey.[75] Dewey's view of the political arena grounded by active discussion is also central to works that followed the period examined in this chapter, by, for instance, Jürgen Habermas and the works of theorists of deliberative democracy.[76]

Nevertheless, despite overt similarity between the views of midcentury intellectuals and earlier social theorists, Cold War intellectuals often did not draw from their immediate progenitors—at least not in any explicit way. (If anything, these political theorists have been influenced by midcentury social science, not the other way around.)[77]

They looked, in particular, to their own community and the kinds of social engagement that Oppenheimer had called for and presided over at the IAS. Whether thinking about national security, national character, mass society, or the problem of conformity and its cure, autonomy, academics from various fields, along with social critics, policy makers, educators, and foundation officials, often met within a loose network of interdisciplinary modern-day salons. Even when they did not write together, social critics, policy makers, educators, and foundation officials met at conferences devoted to the analysis of these topics. They founded dinner clubs and reading groups, and they attended or founded institutes and took part in themed workshops. "Smokers" at the American Academy of Arts and Sciences and cocktails in individual homes stitched together an interdisciplinary community.[78] These intellectual communities provided a locus and a forum for the exchange of ideas and intellectual practices across disciplinary lines.

The vision of an intimate, interdisciplinary intellectual community as critical to the life of the nation took specific form in think tanks such as the RAND Corporation, CASBS, IAS, discussion groups, and academic societies after World War II.[79] These institutions and others, such as MIT's Center for International Studies, linked scholars to the social and policy world.

This intellectual world was tightly linked to the world of policy via overlaps of individuals and social clubs. Drawing on personal connections formed at finishing schools like Groton, in colleges, especially Yale, Harvard, and Princeton in the late 1930s, and wartime service in the OSS, men such as Richard Bissell, Eugene and Walt Rostow, George Kennan, and William and McGeorge Bundy breached the ivory tower and connected the society of the academy to the world of policy. They linked Cambridge to the Council of Foreign Relations in New York and the so-called "Georgetown Set" in Washington, D.C. The Georgetown Set and its companion "Georgetown Ladies Social Club" composed a world in which lunches, dinner parties, and cocktails formed a community that tied together leaders of the public and covert foreign policy world to leading media figures, such as Philip and Katherine Graham and Joseph and Philip Alsop.[80]

One social group that connected intellectuals to one another and to the government was a dinner club of Harvard and MIT professors that met religiously for fifteen years on the first Friday of each month at the St. Botolph Club in Boston. It included viewpoints from political science, history, physics, industry, engineering, economics, and psychology.[81] The dinner club had arisen out of Project Troy, a top-secret research group organized by the State Department in 1950–51 that was charged with evaluating many aspects of the United States' information policy, from the Voice of America, to methods of psychological warfare, strategies for collecting foreign intelligence, and techniques for countering Soviet signal jamming.

Project Troy was one of a series of interdisciplinary study groups that were organized by civilian and military branches of the government.[82] The idea was for these study groups to solve significant problems by bringing the style of research pioneered during World War II at the Rad Lab and the Manhattan Project. Though based on wartime research, experience in these study groups was one not of regimentation but of community and intellectual fellowship. In fact, while it would be easy to treat the large-scale work during World War II as a depersonalizing experience, physical scientists at Los Alamos and MIT's Rad Lab and social scientists at the Office of War Information looked back on their war experiences as a time of community and fellowship.[83]

Aside from a dinner club, Project Troy led to more formal institutions including MIT's Center for International Studies (CENIS). CENIS was established because the CIA and the Ford Foundation deemed that Project Troy was so successful that it required a more permanent institution that could perform similar functions.[84] Once established, CENIS provided a critical function for Cambridge intellectuals. Since CIA funding was covert, CENIS allowed members such as Walt Rostow and visitors such as McGeorge Bundy to wear the mantle of apolitical objectivity while simultaneously conducting intelligence work. Although CENIS scholars were convinced of their own impartiality, they were nevertheless concerned not to allow the source of their funds to become known to the public lest they no longer appeared objective.[85] Conversely, since CENIS and its members had security clearance (and the architecture to support this), it established as private and closed a space as the St. Botolph club—except in this case, the gatekeeping function was performed by the state security apparatus. As one visiting journalist put it, "much of MIT's center is in a railed-off limbo like those in our embassies and foreign military installations. The visitor must stop there, state his business, and wait till the man he is to see, or that man's secretary, comes to escort him further. Then after his talk . . . he must be escorted out again, departing under the eye of a security guard."[86] Because their discussions were shielded behind the walls of secrecy and security, the people who circulated though CENIS were protected from critical engagement with the wider intellectual community or the public, most especially those critical of U.S. foreign policy and strategic plans.[87]

These study groups, the Project Troy alumni dinner club, and CENIS were connected parts of a web of intimate intellectual exchange that bound the academic world into a community as well as gave their ideas hearing in the highest levels of government. At the monthly Friday dinners of Project Troy alumni, club members could engage in high-level discussions despite their varying disciplinary backgrounds. The point was not so much to learn but to converse properly. "We were never," Jerome Bruner recalls, "in violation of easy conviviality. In fact we drank a great deal. And that surely must have helped."[88] Conversational ease with one another was critical enough that the supper club was "in one respect like the Junta in Max Beerbohm's *Zuleika Dobson*, which met regularly to consider new members, only to blackball any that were proposed."[89]

"What we talked about," Bruner recalls, "was almost without limit. Oddly enough anything was admissible so long as it was not intimate!"[90] Physicist Victor Weisskopf noted that the discussions were "wide-ranging" and "unstructured." They included university affairs, "mere gossip," "the political scene, philosophical topics, and occasionally the latest developments in our fields."[91] Ultimately, it was the form of the discussions that mattered even more than the topics. As Bruner put it, "the effect of these conversations were more stylistic than substantive. . . . The underlying premise [of the dinner club] was that there is no topic so specialized, so arcane, or so rooted in prior knowledge as to be beyond intelligent comprehension and discussion."[92] A fellowship composed of people both intelligent and broad-minded enabled wide-ranging conversation and the bridging of disciplinary boundaries.

Such interchanges were a critical basis of Bruner's social and intellectual existence. As he explained in 1952,

> It's hard to separate my professional life from my social life. We have a good number of dinner parties with close friends. We don't play bridge. We're not a game playing family very much. Pretty much just gabbing, which I dearly

enjoy. Somehow ideas achieve a greater palpability when you can talk with someone else who is interested in them.[93]

Here was a community where intimate intellectual contact could stand against the fragmentation of modern disciplines and modern society.

The dinner club's pattern of building community though intimate, high-level, interdisciplinary discussion in comfortable surroundings was a consistent theme of postwar intellectual life. It was, however, one particular moment of intellectual culture. After the club disbanded in the 1960s, founder Elting Morison and other members in the 1970s and 1980s were still looking back nostalgically on the "wonderful days of the old supper club" and having reunions that "could not compare" with the original.[94]

Atmosphere

In the 1950s, Cambridge intellectual, social, and political life blended at cocktail parties and discussion groups that worked out, among other things, the policies proposed by the presidential campaigns of Stevenson and Kennedy.[95] Intellectual socializing also served as a primary justification for the establishment of interdisciplinary research retreats including the Center for Advanced Studies in the Behavioral Sciences.[96] For the participants of these functions, atmosphere was critically important in establishing the right kind of social environment.

The social environment was also a focus of attention at numerous conferences, including a CENIS conference on the American Style in May 1957. CENIS, its scholars noted, had been quite productive in providing, for the purposes of foreign-policy making, analysis of the characteristics of other societies that might hinder the implementation of American goals. However, this particular conference, on the American Style, was dedicated to looking at the characteristics of Americans and whether those traits inhibited national goals and priorities—goals as defined by CENIS and the attendees of the conference, the leading lights of the academic and policy world.

For this conference, atmosphere and conviviality were critical to community and intellectual exchange. Seeking to record for posterity the feeling at the American Style conference, Elting Morison underlined the "atmosphere within which the deliberation took place":

The days from May 23rd to May 26th, 1957 were, in Dedham, Massachusetts, clear, blue and bright. This kind of weather enabled the members of the
conference, in between sessions, to walk in sunshine across the long lawns,
around the borders of gardens in spring flower, and under the trees of the
considerable arboretum of Endicott House down to an isolated pond. It enabled them to drink sherry before lunch and martinis before dinner together
on a stone terrace in the sunshine. It enabled them to sit at small tables drink-
ing beer or gin fizzes in the twilight talking together as they watched an evening fall on four of their number playing bridge under an umbrella at one
end of the terrace. Everyone talked a good deal about the food at the time
because the food was, in fact, worth talking about. Some adroit administrative intelligence had arranged light but interesting dinners. The weather, the
things to eat and drink, the way people were taken care of, the sense of mo-
mentary protection from the world under discussion produced a set of favor-
ing atmospheres.[97]

According to the participants, sherry, martinis, gin fizzes, and games of
bridge, together with a pleasing natural environment, were important.
They were so because of the effect they had on the participants' communication with one another: "Four days and twenty miles from anywhere,
with sun and conversation, had produced if not a band of brothers or a
company of scholars at least a sense of fellowship among diversely educated men." Fellowship such as they experienced at Endicott House,
the conference attendees argued, would provide a cure for alienation in
modern America and for the fractured nature of its culture, society, and
intellectual life.

The emphasis that these conference attendees put on the relationship
between food, alcohol, comfort, and community was a persistent feature
of postwar intellectual life. The formation and deepening of relationships was often as much of a goal of conferences as the specific and official work to which conferences were overtly dedicated. For instance, at
an interdisciplinary conference of social scientists organized and hosted
by the RAND Corporation, generating "working relations" was considered the primary goal of the meeting, a goal that was even more important than the objective of getting "RAND started on a social science
program that would be useful to national security."[98]

Conference organizers and research groups focused much attention
on face-to-face interaction, on the pressing necessity of forming genuine and interdisciplinary community, on retreat from pressing demands

of the modern world to a secluded location, on proper food and drink, and on the importance of a comfortable atmosphere. In her ethnography of conferences, Margaret Mead exhibited the same concern with their social arrangements (see fig. 4.6). Like the participants of the Endicott House conference, Mead emphasized the importance of providing the right kinds of alcohol with proper regularity. Failing to do so could "wreck a conference."[99] Not only drink but food of the right sort was critical, for "food arrangements could contribute to or detract from the whole conference." Mead's analysis of conference activities showed that some foods were optimal for producing the best sort of conference atmosphere. Specifically, for international conferences, organizers should allow "choice of foods that occur in a simple state, like fruit or nuts, to comfort the stranger who finds local food difficult." Because some conferences are so engaging, "people become so absorbed in what they are talking about that they pay no attention to what they are eating and if very large helpings are served they may end up with an unidentifiable sense of repletion and satiety which they mistakenly attribute to the intellectual fare instead of the meals." In those instances Mead determined that "it is important to have stimulants and snacks available, particularly late at night."[100]

These themes even became an explicit part of the theory and practice of conference planning after the United Nations recognized that with the dramatic increase in the number of conferences and the new ease of long distance travel, organizers would need a set of standards for organizing a successful conference. For the members of the UN group that studied conferences, generating a sense of community among potential conference attendees was a pressing requirement. Depending on the conference topic, community seemed necessary either to advance international relations or to further the integration of knowledge. Either cause, the UN study group agreed, could be served by looking after the comfort of the conference attendees and their ease with one another.[101] Promoting such comfort—whether of the individual or interpersonal variety—would help generate a community of nations or of intellectuals.

The UN study group devoted sustained attention to analyzing which kinds of conference arrangements would best facilitate the primary goal of generating community among attendees. The issues that the study group considered included "cultural" differences among attendees as well as the site of the meeting, comfort of attendees, and extracurricular

FIGURE 4.6. Photo of Margaret Mead by Simpson Kalisher. From Darwin, *The Expression of Emotions in Man and Animals.*

activities. This topic of relaxation received sustained attention as one of the critical components in facilitation of interpersonal interaction. Macy Foundation president Frank Freemont-Smith noted that "it has been known that two people in different cities who had been fighting over some point were invited to a conference and by accident sat next to each other after cocktails and found they were both quite crazy about sailing. Their discussion on sailing led on to an easy discussion in a scientific area."[102] In reply, the psychologist Otto Klineberg agreed, but noted "certain kinds of amenities and social activities are more palatable to some people than to others." He added that "excursions and long walks may be liked by one particular group and sitting round the piano and singing by another."[103] Arranging social activities properly would thus lubricate the intellectual or political work of conferences.

By achieving a proper communal atmosphere, research projects like the Project Troy study group might become dinner clubs that would not only connect creative intellectuals with one another—thereby increasing their creativity—but would also address one of the central problems that intellectuals found in modern society: the distance between the intellectuals and the levers of power. Discussion at the CENIS conference reflected this concern about the relationship between the potentially isolated ivory tower scholar and the government.[104] The American Academy of Arts and Sciences likewise meditated on this issue and scheduled an evening "smoker" in 1953 for a members-only discussion of the issue.[105]

Smokers, dinner clubs, the CENIS conference, and others like it were forums for both identifying and resolving this potential disjunction between "Walden" and the "National Security Council," that is, the break between scholarship and power. These communities connected individuals with one another and to the government, giving them important voices in the direction of the nation, and simultaneously confirming their position as leading academics. These dinner clubs, conferences, and salons not only generated community among intellectuals but also served the function of fostering the social and political standing of individual participants. Further, some of the intellectual socializing that characterized the Cambridge of the 1950s continued in similar form in Washington, D.C., in the 1960s. When he reached the capital, Arthur Schlesinger Jr. established at Robert Kennedy's Hickory Hill mansion a monthly interdisciplinary salon along the lines of those he had participated in during his Cambridge years.[106]

Such social settings offered the chance to examine ideas and to pro-
pose reforms to the academy, America, and even the world. For instance,
the CENIS conference brought leading intellectuals and policy makers
together to Endicott House (a secluded mansion with a pond that was
not far from Walden) to rethink, and perhaps remake, American character and American society. Conference attendees were already or were
soon to serve as members of the National Security Council (McGeorge
Bundy, Walt Rostow, and Carl Kaysen), as secretary of health, education,
and welfare (John Gardner), as government advisors (George Kennan),
or as leading members of the CIA (Richard Bissell, Max Millikan).

Aided by their connections with each other and their access to the
highest levels of government, intellectuals in these communities felt they
could fix major problems of their day. This sense of assurance involved
a combination of confidence in their own abilities and the recognition of
the social power they were allotted. A poem by Kingman Brewster (who
would shortly become president of Yale) gives a flavor of the social and
intellectual world of the Troy alumni dinner club:

> . . . If at time we do sound snotty,
>   Why shouldn't we? We run I Tatti!
> We call Oppie Robert
>   And there's Felix and Dean
> The hand called Learned
>   To us "B" has been,
> It is we who govern the governing boards
>   With a hand that's never seen
> . . . .
> The inside track has allowed us to bet on a
>   Gadget that fools the colorless retina.
> And plastic skins for oil
>   That Nassar couldn't stop
> We shall use to import
>   Our Chateau Neuf de Pape
> What can a status seeker seek?
>   What's left to inherit, for the meek?
> . . . .
> So in a spect that's scarcely circum,
>   At a board that's rarely still

> Let the echoes of St. Botolph's
>> Shake the chandeliers on Beacon Hill![107]

Although this poem's repeated rhymed couplets reads as humor, it was also the case that the poem referred to real facts of the lives of the dinner club members. For instance, they were personally familiar with J. Robert Oppenheimer, Felix Frankfurter, and Dean Acheson.[108] They were also the senior administrators at Harvard (Bundy) and MIT (Julius Stratton). This group had, in their own eyes and in the eyes of patrons and policy makers, the intellectual abilities and the social power to solve significant contemporary problems.

Following their work during World War II, Cambridge academics lived an existence of interconnected social, academic, and political lives. Salons in 1950s Cambridge provided opportunity for academics to become familiar with the soon-to-be President Kennedy. In the case of McGeorge Bundy, such settings were not so much an introduction but a reintroduction; he had attended Dexter Lower School, which had been co-founded by his father, with Kennedy.[109] Academics served as advisors for John Kennedy's election campaign and eventually within his and Lyndon Johnson's administrations. In Cambridge, they might meet at the Harvard Faculty Club, in the Society of Fellows, or at the American Academy of Arts and Sciences. They traveled to New York to advise the Carnegie Corporation or the Ford Foundation. In Washington, D.C., many served in advisory roles for the CIA or on the President's Science Advisory Committee. While there, some retreated to the Cosmos Club for discussion over cocktails before participating in committee meetings of the National Science Foundation or the National Academy of Sciences. In short, this was an intellectual power elite that simultaneously led academic fields, participated in university administration, and served as advisors to the highest levels of government.

Membership in these organizations often overlapped. In playing multiple roles, members of these communities often used their experience in academia to inform their work in public affairs and vice versa. For instance, after moving from being dean of Harvard's Faculty of Arts and Sciences to national security advisor, McGeorge Bundy compared the invasion force at the Bay of Pigs to assistant professors at Harvard who, despite warnings, do not believe they will fail to receive tenure until it actually happens.[110]

Conclusion

Along with Bundy, some of these intellectuals would later provide the title for David Halberstam's Pulitzer Prize–winning book, *The Best and the Brightest*. While this community of intellectuals feared the fragmentation of knowledge and society, they also had a solution: the intimate interdisciplinary social and intellectual interaction that held together their own community. Properly conducted and nurtured, the community of leading intellectuals would heal rifts in the academy and would also provide the model for how modern society could be healed as well. They themselves and their community were what America needed.

The community of elite intellectuals, such as the Troy alumni and the participants of the Dedham conference, placed particular value on a particular form of selfhood. This self was creative, flexible, and broadminded: all traits that were usually taken as synonyms for one another. This form of self was marked as liberal, modern, rational, educated, sophisticated, and cosmopolitan. This form of self gave individuals access to salons, and it enabled members to recognize one another, to converse with one another, and to continue their membership by receiving tenure. Perhaps most significantly, it was that trait that held the community together.

As intellectuals like Gerald Holton and James Ackerman had argued and as psychological research indicated, creativity made community possible by giving people the ability to understand one another. Creativity, the trait central to interdisciplinary community, provided the lubrication that allowed intercommunication and understanding among the brightest individuals even if they possessed different modes of expertise. The creative, autonomous, flexible self would be glue for the community and would, in turn, be fostered by interdisciplinary community. Individuals deemed to possess such traits would therefore be more likely to be included. Those lacking in such traits, whether because of their individual personality or because of their allegiance to certain ideas such as behaviorism, would be less welcome.

Just as this form of self was the answer for the academy, it was the solution for the problems that intellectuals saw in America. That the academy and America seemed to need a common cure was a product of the social scientific tools that intellectuals used to understand both. First, social scientists selected instruments for measuring creativity in which intellectuals served as the standard of measurement. Second, small group studies made conferences seem not unlike any other social situation. Third, intellectuals' solutions for addressing the challenges of modern America and of their own world were congruent because they relied on the same form of selfhood as answer in both instances: themselves. Since they themselves were the very standard of creative, flexible, rational thought, elite intellectuals were also the answer to fixing both the academy and America.

The Human Mind

CHAPTER FIVE

Scientists as the Model of Human Nature

In 1963, Bernard Berelson published a collection of essays by leading scholars in the behavioral sciences. The essays found their origin in a series of radio broadcasts for Voice of America, the radio-based propaganda arm of the United States Information Agency. These programs aimed both to cover the immediate topic at hand and to carry out the general mandate of the Voice of America—showing the virtues of the American way to people around the world.[1] To accomplish this dual aim, the programs' more specific goal was to explain how the various behavioral sciences operated. Their broader goal—unstated but very real—was to demonstrate the connection between American democracy and the objective and scientific study of society.

At the conclusion of his address on psycholinguistics, George Miller remarked, "the scientist is Everyman, looking just as you and I. We go and look for the things we want, and when we find them we find part of ourselves."[2] These comments raise several issues worth close attention. First, Miller collapsed the distinction between the scientist and "Everyman." From this perspective, the psychology of the scientist is the same as the psychology of the human subject. Second, the salient feature of human nature (or of the scientific process) is the process of searching. To Miller, searching provides both knowledge of the world and knowledge of oneself. Clearly, Miller's image of objectivity was not one that required the scientist to stand apart from the object of knowledge.

Miller's conclusion engaged in a double reflexivity, linking the scientist's self to the human subjects studied and, at the same time, connecting self-knowledge to knowledge of the world. While Miller may be unusual in his ability to engage two forms of reflexive argument in the space of two sentences, the mere fact that he engaged in reflexivity at all should not be particularly surprising. Historians have shown us numerous examples of this phenomenon, from Sigmund Freud, to William James, to Gordon Allport.[3] While it might not be surprising to find such reflexivity (some might say "lack of objectivity") in social psychology and psychoanalysis, the same interplay between self-knowledge and scientific psychological knowledge has pervaded even those parts of experimental psychology that have been regarded as the most "objective" and methodologically rigorous. Neo-behaviorists Ernest Tolman, B. F. Skinner, Clark Hull, and Edwin Guthrie regularly worked between their senses of themselves and their scientific investigations.[4] Likewise, the arch-operationist S. S. Stevens took the normative rules of data collection that he prescribed for psychologists and translated them into his studies of audition. By the time he was done, Stevens had produced a theory of hearing in which the brain acts as if it were a scientist following operationist rules of method.

This chapter focuses on a particular moment in the history of human sciences in which this sort of reflexivity played a significant role: the early days of revolution when cognitive science supplanted behaviorism as the hegemonic science of human nature.[5] In the struggle that marked the cognitive revolution, we see little use of Jamesian or Freudian deep and thorough self-examination in efforts to make a science of the human. Rather, behaviorists and their foes regularly traversed the boundary between scientific and folk psychology as a strategy for legitimating their work.

Reflexivity provided the combatants with weapons to attack their foes and also methods and concepts to form their respective sciences of human nature. To enhance their public standing, they sought to make their own thought processes match folk ideas of scientific thinking. They applied the same categories of selfhood found in popular culture and social psychology to themselves. They collapsed the distinction between normative rules for scientific thinking and the actual processes of human thinking. As cognitive scientists like George Miller and Herbert Simon crossed back and forth between scientific descriptions of the human and normative discussions of the best way for scientists to think, they borrowed from the folk and social psychological image of right thinking to inform their own personal and public images.[6] These very same scientific self-images would form the basis for the image of human nature that cognitive science produced.

The (Disciplinary) Politics of Psychological Theory

The widespread practice of treating politics in terms of thinking styles had implications for the disciplinary structure of psychology and for research programs within the field. Because of the low and permeable boundaries between expert and folk psychology, the choice to pursue one kind of psychology rather than another was filled with political meaning. Moreover, again and again psychologists would engage in a casual form of reflexivity, investing not only their models of human nature but also their own thought processes with political meaning.

To many psychologists, pursuit of political change could come about through the development of the right kinds of psychological theories. For instance, in 1950, Theo Lentz articulated a typical political argument for pursuing psychological study of specific forms of human subjectivity.[7] In this paper, Lentz's argument relied on a common juxtaposition of claims. In particular, he linked disciplinary reform (the advancement of social psychology) with political reform (development of world government) with reconceiving human nature (by making world-mindedness a facet of the human mind) with the call for psychologists to be imaginative.

In a 1953 grant proposal, Jerome Bruner articulated a similar argument for cognitive psychology. The "world crisis" required citizens and scientists to understand "reasoning, creative thinking, hypothesis making and testing." Knowledge of such "higher mental processes" was "one of most pressing problems" of the period because these forms of cognition were "necessary for grasping the immense complexities of modern technical and social developments involved in the world crisis." Better knowledge of how higher thought processes work would protect democracy. This knowledge would improve mechanisms for informing both "opinion elites" and the general populace. Bruner went on to describe how other forms of psychology such as behaviorist learning theory and abnormal psychology had not even asked, much less answered, how normal human decision making and creativity operated. And, of course, he offered his own brand of psychology as the best method of research on complex thinking and creativity and thus as the eventual tool for protecting democracy.[8]

Although there were certainly variations in political goals (not everyone called for world government), norms of thought (not everyone called for creativity), models of human nature, and disciplinary goals (not everyone called for cognitive psychology), a wide range of psychologists, social scientists, public figures, and foundation officials made these sorts of links. They tied the promotion of certain "better" forms of human nature to specific scientific models of human nature and to the disciplinary reconstruction of the human sciences.[9]

Although social scientists used their tools of psychological analysis to critique specific social groups or specific modes of thinking, they also regularly used these techniques to talk about their own discipline. In one instance, Milton Rokeach's discussion of psychological ideas indicates the way in which those theories could carry political significance. He noted that the model of human thinking presented in Gestalt theory was that of the democratic citizen, while the model presented in behaviorism and psychoanalysis was appropriate to people who were subjects of totalitarian states.[10]

As is indicated by the language and arguments of Bruner, Lentz, and Rokeach, cultural values were embedded in the effort to develop understanding of the open, autonomous mind. The project of creating a science of the virtuous (i.e., autonomous) mind involved a continual interplay between descriptive and normative accounts of thinking. This meant eliding distinctions between the kinds of thinking that scientists and democratic citizens *should* do and the kind of thinking that people in general *do* engage in. Their arguments involved, at the same time, contentious disciplinary politics.

Psychology, the Science of Behavior or the Science of Mind?

In the 1950s, at the center of psychology's struggle was the question of whether the study of mind could properly be understood as a part of scientific psychology. In a 1958 invited address to the American Psychological Association, the philosopher of science Herbert Feigl noted how psychologists struggled over what their discipline was about:

> "Intuition," "insight," "understanding," and "empathy" have been key words in the strife of psychological movements. These terms are used honorifically by one party, but they are suspect (if not on the *index verborum prohibitorum*) with the other party.[11]

These two parties struggled over whether intuition, insight, and understanding belonged in psychology. Feigl's sense of two parties was by no means an outsider's idiosyncratic reading of the field. George Miller, Eugene Galanter, and Karl Pribram, for instance, divided the field into "optimists" who believed that human behavior was determined by the environment and could be described completely by stimulus-response chains and "pessimists" who believed that other things (such as mental processes) were necessary to explain human nature.[12]

At the end of World War II, and for the following ten years, Miller, Galanter, and Pribram's optimists controlled the discourse on what was and was not scientific psychology. At that time, psychology was anything but the *science* of mind.[13] The "fundamental" and scientific center of the discipline was experimental psychology, which was still dominated by the behaviorist and operationist concerns that made mind an improper subject for scientific study.[14]

The central importance of "learning theory" in the scientific end of the discipline gives another indication of how many in psychology deemed the study of mind to be unscientific.[15] Learning theorists sought to explain how people and other organisms act differently in different circumstances. The main branch of learning theorists followed in the footsteps of J. B. Watson and Edward Thorndike. These psychologists included E. R. Guthrie, Clark Hull, Kenneth Spence, B. F. Skinner, and their followers. They, their students, and their theories dominated experimental psychology, at least in numbers.[16] Although it is important to recognize the differences among these psychologists, they shared a fundamental perspective on the nature of human (and animal) subjectivity. For them, most, or all, of what one needed to know about psychology could and should be explained on the basis of the environment's effects on the subject. From this perspective, organisms "learned" to solve problems not from "understanding" but through random trial and error and the association of particular behavioral responses with a reward or other stimulus. Although there were certainly other approaches to learning represented by Gestalt theorists or Edward Chance Tolman or David Krech (né Krechevesky) that emphasized the importance of insight, hypothesis, and cognition in learning, these were distinct minority positions.[17]

It is certainly the case that large sections of psychology—particularly clinical and educational psychology (the fastest growing components)—were concerned with mind.[18] But social and clinical psychology had, at best, marginal status as scientific endeavors. Those who studied mind may have been psychologists, but their status as scientists was questionable in their own community. The more psychologists were concerned with mind, the less they qualified as scientists within the discipline.[19] As the psychologist E. Parker Johnson put it in 1956:

> Practically everyone who is *not* a psychologist knows that [psychology] is the science or study of the mind, and anyone with a dictionary may easily confirm this. . . .
>
> But, oddly enough, many modern psychologists refuse to accept this definition. Why? . . . The word mind . . . [is] by its very definition beyond the ken of science which, by its very definition, is built on the observation of observable events. . . .
>
> Many protest, indeed, that scientifically speaking there is no such thing as the mind to be studied![20]

Two critical assumptions were necessary for this argument, the definition of mind and the definition of science. With these two definitions taken as given, there was little compatibility between mind and science. But those two definitions were under attack even as this paper was published.

At the end of World War II a broad range of academics called for developing a science that could account for the autonomous, the creative, and the rational aspects of human nature. This effort involved a struggle with proponents of behaviorism and (somewhat less) of operationism and positivism—those who saw humans as (mere) products of their environment and their basic drives (such as hunger). Drawing on the politics of thinking, a primary strategy for advocates of the science of mind was to attack the thought processes of behavioristic psychologists. These attacks made use of the normative categories of thinking drawn from social and folk psychology.

Behaviorists were charged with exhibiting the antisocial interpersonal attributes of the authoritarian personality. However, their opponents within experimental psychology also ascribed the authoritarian's *cognitive* deficits to behaviorists. Specifically, the features of the closed-minded, conformist person appeared as characteristics of the behavioristic psychologists. In these critiques, the center of scientific psychology (which was primarily behavioristic) appeared as uncreative, narrow-minded, rigid,[21] and dogmatic;[22] in short, it appeared to be governed by an ideology that confused methodological rigor with true (i.e., creative) science.[23]

The primary reason offered for using epithets such as "narrow" and "ideological" to describe operationist, behaviorist, and positivist psychology was because of its reported aversion to and slighting of the study of the mind. Some psychologists, for instance, argued that "*narrow* operationism" had limited the "freedom" of psychology to pursue its "ultimate purpose, the scientific understanding of man's cognitive behavior."[24]

Behaviorists and psychologists who took a strict operationist view of scientific method demanded a relatively direct or mechanical correspondence between the physical world and its representation in scientific theory. To B. F. Skinner, theory was not necessary in science, only direct recording of the world.[25] Other psychologists noted that the physical world was complex and could not necessarily be apprehended directly. Consequently, the best way to approach it scientifically was *not* to depend on simplistic and positivistic ideals. On this view, because the world is complex and does not present itself to investigators, it would be best to understand that the best kind of scientists are not passive recording devices. Instead, the best scientists were seen as tolerant of nature's ambiguity and were able to use their creative and intuitive mental faculties to represent it. Here we see not only a supplement to a form of empiricist "mechanical objectivity," but the stronger view that such empiricism was not scientific at all.[26] Further, according to Else Frenkel-Brunswik, one of the contributors to *The Authoritarian Personality*, a good way to distinguish a fascist from a democratic scientist would be to see if they lauded simple empiricism (these were the fascists) or if they believed that science involved an active mind and the tolerance of ambiguity (these were the believers in democracy and the true scientists).[27]

Just like the ethnocentric, closed-minded people described in *The Authoritarian Personality* or Gordon Allport's *The Nature of Prejudice*, behavioristic psychologists were pictured as conformists, intolerant of difference. Of operationists, one psychologist noted, "their discussions and criticisms have produced a social climate in which the psychological theorist may hesitate to present theories which contain non-operational definitions."[28] Another commented, "it has been noted that psychologists seem over conscientious and even compulsive in their efforts to be simon-pure [*sic*] and scientific almost to the point of fetish. Colleagues suspected of indiscretions are ostracized and avoided."[29]

Critics of behavioristic psychology regularly suggested that it was a religious phenomenon. One of the standard procedures of behavioristic psychology—the assumption that rats could stand in for humans or other organisms as experimental subjects[30]—came under attack as religious dogma.[31] Donald Hebb participated in this critique, noting that psychologists deviating from the formula of stimulus and response theory could place themselves in a "larger demonology." "As for 'insight,' 'purpose,' 'attention,'" Hebb continued, "any one of these may still be an invocation of the devil, to the occasional psychologist."[32]

This characterization of behaviorism was by no means a simple matter of internal disciplinary debate. The picture of behaviorism as rigid, religious ideology spilled outside of the discipline of psychology into academic journals outside the field. In a famous review of B. F. Skinner's behaviorist study of language, the linguist Noam Chomsky contended that Skinner's work had only the mere trappings of "objectivity." As Chomsky put it, Skinner had only engaged in "play-acting at science" and had done nothing more than use seemingly scientific terms as "paraphrases" for the popular words used to describe language. Skinner's work, on this account, was a "dogmatic" exercise in "general mystification." Chomsky's review also added that behaviorists' attachment to their theories had blinded them to evident facts and even caused them to forget established science that conflicted with their beliefs.[33]

In an article in *American Quarterly*, Lucille Birnbaum noted behaviorism's "one overarching dogma," the view that "all human functions, including thinking, can be observed and described in terms of stimulus and response."[34] She linked this dogma to "rigid" childrearing practices. Birnbaum also noted that critics had argued that such childrearing practices produced people with "rigid and compulsive personality types [that were] directly attributable to early strict conditioning."[35]

A similar critique spilled into popular nonfiction. To the popular novelist and nonfiction writer Arthur Koestler, Gestalt psychology represented a "Renaissance" that was aborted by neo-behaviorism's "Counter-Reformation."[36] For Koestler, who was well versed in the history of science,[37] this tripartite chronology could very well have been a retelling of the chronology of the scientific revolution. Thus Watsonian behaviorism was the medieval dark ages, Gestalt psychology was the Renaissance and the age of Galileo, and the neo-behaviorism of B. F. Skinner, Kenneth Spence, and Clark Hull was the church of the Counter-Reformation that silenced Galileo. Koestler's metaphor, then, not only associated behaviorism with the Renaissance church but also linked it to a commonly held negative image—that of a church that upheld a dogmatic, reactionary, and antiscientific ideology.[38]

When the National Science Foundation organized a large-scale, three volume survey of psychology as science, its editor, Sigmund Koch, devoted his epilogue to an extended critique of the reigning methodology of behavioristic psychology. In Koch's terminology, both behaviorism and the use that psychologists made of philosophy of science (operationism and logical positivism) could be subsumed under a single heading, the "Age of Theory."

Koch commented again and again on the "ideology"[39] or "code"[40] of the "Age of Theory," its "reigning *stereotypes*,"[41] *"lack of realism"* (in contrast with "increase in realism" as the age waned),[42] and its "narrow" approach.[43] Koch noted the Age of Theory's "hypothetico-deductive prescription,"[44] its "doctrine,"[45] its *"programmatic"* thinking style and attachment to a "facile" mythology of perfection,[46] as well as its "autism" and "autisms."[47] In the parlance of the psychological theory of the time, autism meant lack of connection with reality and often implied lack of creativity.[48]

Noting that psychology sought security and respectability in following "fashionable theory of proper science," Koch argued that "the dependence of the Age of Theory on *prescription from extrinsic sources* is but the most recent chapter in a consistent story of such *extrinsic determination of ends and means.*"[49] Psychology's ideological subscription to an external and simplistic vision of science had narrowed its range and restricted its creativity.[50] Behavioristic psychology had relinquished its own autonomy to a philosophy of science that even philosophers themselves no longer believed.[51] Koch's assessment of the Age of Theory operated by a similar system as the one that the social psychologists used in their analysis of the prejudiced or authoritarian mind. Both critiques pointed out a lack of autonomy, a tendency to follow rules imposed by others, a lack of realism, and narrow, stereotyped thinking.[52]

Philosophy of science was at the center of this debate within psychology. Whether following a formal school or a more informal understanding of science, psychologists were highly attuned to philosophical issues.[53] Casual references to modern philosophers and philosophical issues that peppered articles in psychology journals indicate the familiarity psychologists had with philosophy of science.[54] In fact, psychology paid more attention to philosophy than did the other sciences. As Herbert Feigl put it in an invited lecture to the American Psychological Association, "The majority of physicists want to unmuddle themselves without the aid of philosophical clarifiers. But I have found psychologists and social scientists much more hospitable."[55]

There was symmetry between behaviorists and anti-behaviorists. While behaviorists such as Kenneth Spence commonly welcomed Feigl's positivist brand of philosophy of science, anti-behaviorists subscribed to an anti-positivist vision of science.[56] The normative account of scientific practice the anti-behaviorists adopted was one that emphasized the insightful aspects of science. Rather than functioning in the role of collating data, to anti-behaviorists the scientist's mind had to be active and creative.

Men who articulated the anti-positivist vision of science were vested with enormous cultural authority to speak for the nature of the scientific endeavor. These men included leaders of the scientific establishment such as James Bryant Conant, J. Robert Oppenheimer, Warren Weaver, and Jerrold Zacharias.[57] They were joined by best-selling science writers such as Jacob Bronowski and Arthur Koestler.[58] Sociologists such as Bernard Barber and Talcott Parsons articulated similar anti-positivist visions of science.[59] Likewise, historians of science, including Thomas Kuhn (who was a protégé of Conant), attacked the positivist vision of science.[60] All of these figures argued that science was a process that involved creativity, insight, ideas, and invention as much as the collection of data.

Jacob Bronowski's vision of science indicates just how closely the critiques of behaviorism and positivism were bound. For him, positivist and operationist philosophers failed to grasp the creative nature of scientific work. Attacking the behaviorist flavor of operationist philosophy and the rigidity of logical positivism, Bronowski argued that they flew "in the face" of both historical and contemporary evidence. Scientists, but not operationists and positivists, he said, know that science requires creativity. From this point about the nature of science, Bronowski appended an argument about the nature of human thought:

> The world which the human mind knows and explores does not survive if it is emptied of thought. And thought does not survive without symbolic concepts. The symbol and the metaphor are as necessary to science as to poetry.[61]

In Bronowski's eyes then, all thinking and science rely on the creative use of symbolic concepts. He took his vision of cognition and proper science to be opposed to that advocated by positivist philosophers of science.

Oppenheimer concurred with Bronowski. "Truth," he proclaimed in a 1955 invited address to the American Psychological Association, "is not the whole thing; certitude is not the whole of science. Science is an immensely creative and enriching experience; and it is full of novelty and exploration."[62] Oppenheimer cautioned psychologists against borrowing from obsolete classical physics the mistaken view that the physical world is determinate.[63] Even more, he warned psychologists against quantification for its own sake, noting that such fascination with numbers had been typical of and appropriate for Babylonian prophecy and magic. More modern sciences, Oppenheimer suggested, should be pluralistic enough to value descriptive naturalistic approaches. Tellingly, Oppenheimer explained to the psychologists in his audience that the cognitive-developmentalist Jean Piaget deserved respect despite his lack of statistically robust results.[64]

Oppenheimer's arguments would have been useful to the anti-behavioristic psychologists. His argument against rigor for its own sake, suggestion that zeal for quantification was superstitious, and call for methodological pluralism gave ammunition to anti-behavioristic psychologists.[65] Jerome Bruner, for instance, echoed many of Oppenheimer's points in a hostile review of a book by Kenneth Spence.[66]

Critics of behaviorism offered a positive counterpart to behaviorism's narrowness. This positive version of psychology would value and reward creative insight among its practitioners rather than seeing merit in rigorous methodology alone. In this version of psychology, psychologists would be autonomous of narrow positivist philosophy of science, independent of external influence, open-minded, flexible, realistic, interdisciplinary, and creative.[67]

Scientific Thinking as the Content of Cognitive Psychology

In this section, I turn to a direct challenge to the *content* of behaviorism, a challenge right on behaviorism's home turf: the scientific study of normal (and universal) human nature. At stake was whether human nature could or could not be completely accounted for by S-R connections. In the 1950s, an array of scientists joined forces from several fields to critique behaviorism by arguing for the existence of thought or behavior that was autonomous from stimulus. Central to this endeavor of creating a cognitive rather than behavioral psychology was proving that human behavior was creative and not simply the product of experience.[68]

To those who insisted that thinking could not be explained solely by conditioning, this project extended beyond the claim that cognition was an irreducible aspect of normal human nature. Instead, there were very specific modes of thought that they ascribed to the normal human: human cognition was supposed to operate much like the thinking of a particular sort of person—the good scientist. But as discussed above, there was not unanimity in America about how to conduct proper science. When cognitivists compared human thinking to the scientific process, they were quite selective in their choice of the model of science. They adopted a vision of science that emphasized the creative and insightful nature of the scientific process.

From the earliest days of cognitive science, studies of human mental processes treated thinking, perception, and language themselves as relying on scientific methods such as hypothesis formation and theory construction. When the pioneering generation of cognitive scientists said that everyday higher mental processes were like scientific thinking, they focused on creativity in both forms of thought.

The 1956 *A Study of Thinking* by Bruner, Goodnow, and Austin (BGA) explained everyday cognitive processes by comparing them to scientific thinking. BGA wrote "the development of formal categories is, of course, tantamount to science-making":[69]

> Let us take as an example of concept attainment the work of a physicist who wishes to distinguish between substances that undergo fission under certain forms of neutron bombardment from substances that do not. . . . This kind of problem is hardly unique. The child seeks to distinguish cats and dogs by means other than the parent's say-so, the Army psychiatrist seeks out traits that will predict ultimate adjustment to and performance in the Army.[70]

While these psychologists saw science in everyday cognition, their example was more focused than the simple linking of types of problem solving. Not just any form of science would be the model of human thinking. They selected only certain aspects of the scientific process to compare to thinking: inference, invention, problem solving, making hypotheses, and model construction.[71]

BGA saw their account of science as opposing the dogma of "naïve realism" in which science is a "voyage of discovery" that seeks to "discover the islands of truth." In contrast with this vision they argued that "science and common-sense inquiry alike do not discover the ways in which events are grouped in the world; they *invent* ways of grouping."[72]

This emphasis on the inventive nature of science drew support from both the biological and the physical sciences. BGA cited Ernest Mayr's point that "species are not 'discovered' but 'invented.'"[73] The physical sciences taught "the revolution of modern physics is as much as anything a revolution against naturalistic realism in the name of a new nominalism."[74] From the perspective of this nominalism, they asked: in what sense do the categories "such as tomatoes, lions, snobs, atoms, and mammalia exist?" The answer was that "they exist as inventions, not as discoveries."[75] The category *atoms* is invented by scientists and the category *tomatoes* by everyone—but neither is discovered.

The nominalistic lessons of modern physics had two sorts of implications. On the one hand, these lessons were relevant to the argument about which sorts of scientific thinking were appropriate as metaphors for everyday cognition. In this regard, the claim was that nominalistic philosophy of science was better than realist philosophy of science as a model for human nature. On the other hand, the reference to nominalism could also serve as a critique of behaviorism. This was because stimuli were equivalent to people not when they were objectively, measurably equivalent to the experimenter, but when people constructed psychological categories that grouped the stimuli together.[76] This point could make meaningless the behaviorist effort to relate observable and measurable stimuli and responses.[77] As Bruner put it in a paper for the Institute of Unified Science in 1951,

> Let us begin by stating a heuristic theory of perception. We shall assume that the organism is always set or tuned or expectant; he is, in short, ready for certain classes of stimulus events to occur. The tuning of the organism, and we shall discuss its determinants presently, we shall call an hypothesis. It is a predisposition to organize and classify the perceptual field in a certain way at a certain moment. Stimulus information enters the prepared organism. We use the term stimulus information rather than stimuli for what we wish to denote here is not the energy characteristics of stimulation, but the cue characteristics provided by stimulation—its signaling value. . . . The data of the scientist are not the raw cues of stimulation, but the perceptions of the scientist which occur when those cues confirm perceptual hypotheses which he has acquired. In this important sense, then, the scientist's data are not found, but created.[78]

On this account, data, stimuli, and responses do not exist independently of expectations. Both the psychologist and the people he or she studies do not experience the world in "raw" form. Human perception is so thoroughly laden by prior hypotheses and theory that it is impossible for any (human) scientist to make purely objective pointer-readings.[79]

While the implications of nominalism for behaviorism were not explicitly drawn in *A Study of Thinking*, Bruner did make this last point explicit in his review of a book by the behaviorist Kenneth Spence only a year later.[80] In this sense, nominalism was not merely the proper model of human thinking, it was also a better model for psychological research. The argument, then, was that it *is* human nature to think nominalistically, and that it is *good* for psychologists (and other scientists) to think nominalistically. Thus, there was an equivalence of good scientific thinking and normal human thinking. An implication here is that the only people who held to the dogma of realism were naïve philosophers and behaviorists. And realism, since it was dogma, was abnormal or ill—or in Koch's terms, autistic.

While *A Study of Thinking* focused on categorizing the different ways that people go about understanding the world, Eugene Galanter and Murray Gerstenhaber's 1956 article "On Thought" drew on the conceptual model-building aspects of science. In order to support the claim that there could be a scientific study of thinking, this article argued that thinking and understanding were much like building an internal model of the world. These models would be either like maps or like the three-dimensional scale models scientists and inventors constructed to understand and represent large-scale physical phenomena. They asserted that thinking occurs when "an aeronautical engineer . . . manipulates a model of a new airplane in a wind tunnel." Galanter and Gerstenhaber extended the technical nature of the model analogy of thinking by suggesting "the environment will be a 'machine,' or 'mechanism.' . . . The process by which the behavior of the mechanism is predicted is called 'thinking.'"[81] Thus the answer to the behaviorist charge that thinking could not be studied by science was the assertion scientists think when they use models of the physical world and that human thinking is akin to scientific thinking.[82]

Whether they came to the study of mind from psychology or other disciplines, for the early cognitive scientists creative thinking was praiseworthy but not exceptional. Instead, theory construction and creative problem solving was the cognitive scientists' model of everyday thinking and problem solving. Learning was not so much a process of acquiring facts about the world as of developing a skill or acquiring proficiency with a conceptual tool that could then be deployed creatively. It was, thus, akin to the process of doing creative science envisioned by patrons like Charles Dollard.[83]

According to MIT linguist Noam Chomsky, the acquisition and use of language was an active and creative process. In his eyes, a child learning a language was not acquiring specific words so much as operating like a scientist by actively developing a theory of how to speak properly. The "theories" that children created were nothing other than the grammar of the language in question. Chomsky also held that adults require similar theories in order to produce and comprehend sentences.[84]

This view of language users and learners as scientists was not neutral with respect to either linguistic theory or philosophy of science. According to Chomsky, the linguist or psychologist looking to account for the "actual behavior of speaker, listener, and learner" would fail if he or she followed the purely empiricist rules of scientific method as described and advocated, for instance, by B. F. Skinner.[85] Indeed, the "mechanisms" or "theories" of grammar that Chomsky saw each individual as possessing could never be observed directly but only be inferred from their behavior.

While Chomsky's view of language set constraints on the proper method for scientists seeking to account for language, it also framed humans as following certain forms of method as well. Specifically, native "hypothesis-forming" abilities enable children to rapidly learn grammar, a process which Chomsky described as constructing an "abstract deductive theory" or "an extremely complex mechanism" for producing or recognizing proper sentences.[86] Chomsky supplemented his claims for the active and thoughtful nature of learning by citing the neurologist Roger Sperry's argument that even simple conditioning requires insight.[87]

The connection between human cognition and the form of cognition that cognitive scientists engage in appeared most starkly in Chomsky's hierarchical categorization of linguistics theories in "Three Models for a Description of Language" (1956). The first step of Chomsky's hierarchy was the set of theories based on "finite state" languages or grammars: that is, languages which could be produced by a finite-state machine. A finite-state language was one that could be produced by a machine (or organism) with a finite number of internal states and which produced a word when it made a transition from one state to another. If there was a machine in which certain transitions were possible but others were not, then that machine would produce words only in certain orders, but not in others. Chomsky identified this machine as a "finite-state grammar."

In the 1950s, finite-state languages appealed to human scientists of the behavioristic stripe for two reasons. First, finite-state languages had the advantage that they did not require the linguist to postulate that humans think or that they understand what they say. All that was needed was that the speaker be a finite-state machine. Second, the finite-state machine's transition rules were structurally much like the stimulus-response chains at the center of the behaviorist program. Behaviorists often explained complex physical behavior as a stimulus that, much like a conditioned reflex, produced a specific muscular response or twitch. Such a response could act as a stimulus for a second response that, in turn, might function as stimulus for a third response.[88] Finite-state languages consequently provided the opportunity to transfer this model of behavior directly to language production. Just as a given response would serve as a stimulus for the next response, so would a word in a finite-state language serve as a stimulus for the subsequent word.

Having explained the nature of finite-state languages, Chomsky proceeded to demonstrate how they and behaviorism were inadequate. He noted that English allows one sentence to be embedded within another sentence. He then proved that finite-state languages cannot account for this essential feature of human language.[89]

A second demonstration of the poverty of behaviorism appeared on Chomsky's presentation of his two other models of language: phrase-structure and transformational grammar. Chomsky underlined that one significant reason why these models, and therefore theories of language, were better than others was that they could identify the fact that sentences like "The shooting of the hunters" and "They are flying planes" were semantically ambiguous. This is to say that both humans and cognitive scientists like Noam Chomsky were sufficiently open-minded and creative to recognize that it is not so clear if the hunters are doing the shooting or if they are being shot. On the other hand, the incorrect theories of language were intolerant of ambiguity in exactly the same way that authoritarians were. Because of their intellectual handicaps, they artificially turned ambiguous situations into ones of certainty.

Chomsky thus implicitly set up a pair of opposed sets, each with two terms. On the one hand, there were the behaviorists who, much like the authoritarians described by Else Frenkel-Brunswik, were unable to cope with ambiguity. On the other hand, there were normal people and cognitive scientists who, because of their creativity and open, democratic minds, could tolerate and even celebrate ambiguity.

Chomsky's views of language, learning, thinking, and the ways to study them both typified early cognitive science and catalyzed much later work in the field. During the decade between the mid 1950s and the mid 1960s, Chomsky collaborated with Harvard psychologist George Miller in developing a cognitively oriented field of psycholinguistics,[90] and Miller would himself promote linguistic theory that was tolerant of ambiguity as a central plank of his theories of human cognition and action. One such example of such celebration of linguistic theories and therefore of linguists who could tolerate ambiguity appears in one of the seminal texts of cognitive science, *Plans and the Structure of Behavior*, coauthored by Miller, Eugene Galanter, and Karl Pribram.[91]

Computers, Cognitive Science, and Creativity

One critical way that cognitive scientists made their work compelling was the adoption of the computer as a tool for demonstrating the possibility of a rigorous science of mind. In so doing, practitioners in cognitive science often discussed the human mind as if it were a complex machine or a computer capable of reasoning, hypothesis formation, and creative insight. These cognitive processes were taken as innate human abilities that were necessary for both learning and perception.

To many cognitive scientists, the computer provided a useful metaphor to combat S-R behaviorism. George Miller, for instance, recalls that the metaphor allowed cognitive psychologists the opportunity to have a mechanism to support their views.[92] If the computer could demonstrate higher thinking, then surely it would not be pure speculation to attribute those thought processes to people. There could, therefore, be a defensible science of thinking. The computer metaphor was also used to make an anti-behaviorist point and emphasize the way in which human nature is creative and (partly) autonomous of the environment. Donald Hebb, for instance, argued that the "computer analogy" developed by Miller, Galanter, Pribram, and Donald Broadbent "can readily include an *autonomous* central process as a factor in behavior."[93]

Although Hebb made a common, cognitively oriented interpretation of the computer analogy, it is worth noting its paradoxical framing.[94] Specifically, why did he believe that human autonomy could be illustrated by a machine? As Miller, Galanter, and Pribram argued, the mechanism that psychologists choose as a model of humanness does not necessarily force a particular vision of that nature.[95] It may have been the case that computers did things that could not be predicted, but committed behaviorists like B. F. Skinner did not conclude that human cognition is autonomous from that result. Indeed, on looking at the information processing metaphor central to cognitive psychology, Skinner made the same contention that he had of other mentalistic modes of psychology: that it could "be reformulated as changes in the control exerted by stimuli."[96] Prior failures by psychologists to perfectly predict human behavior had not convinced behaviorists that organisms are autonomous, and the computer, by itself, changed little.

What did matter more was the kind of mind that cognitive science modeled and the cultural and political role that mind played. Early cognitive scientists and artificial intelligence researchers selected quite specific features of human nature to model with computer programs. For the cognitive scientists Newell, Simon, and Shaw, and Miller, Galanter, and Pribram, there was a clear route to convincing their audiences that mind did in fact exist and that it was possible to study it scientifically. This was to build models of the forms of human thought that Americans saw as requiring higher reasoning. The cognitive scientists thus selected quite specific and widely recognized problems to model. For instance, Herbert Simon and Allen Newell built a program that could solve a logic problem that had recently been featured on a television quiz show.[97]

The computer models cognitive scientists built used heuristic (rather than strictly logical or deterministic) methods to play chess, re-derive the proofs in Russell and Whitehead's *Principia Mathematica*, or produce novel proofs of the theorems in Euclid's *Elements*.[98] The picture of human thinking that cognitive scientists inscribed in their computer models depended on the accounts of science provided by Henri Poincaré, Michael Polyani, and George Polya,[99] and shared much with the anti-positivist one developed by men such as Conant, Oppenheimer, Kuhn, Bronowski, and Zacharias.

Notably, Herbert Simon opined in 1958 that Bruner, Goodnow, and Austin's use of "strategies" to describe thinking was "the nearest thing in the psychological literature to the use of programs to describe behavior."[100] In the time since these early works, cognitive science has continued to use computer models to ascribe such scientific methods as hypothesis making, theory construction, and inference to everyday thought.[101]

However, despite the frequency of comparison between the human and the computer, it would be a mistake to conclude that cognitive science involved showing that the human mind operated like one. Several of the most important works in the history of cognitive science indicated just how the human mind is unlike a mechanical device. John von Neumann, one of the architects of computer science, devoted attention to the differences between the analog processes of the human brain and the digital processes of computers.[102] Noam Chomsky devoted the first part of his typology of linguistic theories to showing that the information theory and mechanical, finite-state languages promoted by engineers and computer scientists could not account for human languages including English.[103] Chomsky also contended that the machines were not useful even as aids for human linguists in framing or judging linguistic theories.[104] It was no accident that computer scientists and engineers took Chomsky's work as an attack.[105]

Like Chomsky, George Miller was both aware of mechanical analogies and distanced humans from machines. In the most highly cited article in the publication history of *Psychological Review*, "The Magical Number Seven Plus or Minus Two," he surveyed the literature that had studied human psychology from the perspective of information theory.[106] Miller had been an early entrant in this area by helping introduce his colleagues to communications engineer Claude Shannon's paper, "A Mathematical Theory of Communication."[107] When psychologists, Miller included, initially looked at information theory, they approached their problems by considering humans as information channels in Shannon's sense. So they asked about human channel capacity, the rate at which humans might take input from the environment, the rate at which they might output information, the rate at which they might transfer information from one part of memory to others—or the rate at which they would pass information from the sensory apparatus to memory.[108] Miller would even consider memory to be a communication channel from the past to the present.[109] In summarizing this research, Miller noted that the span of absolute judgment is about seven. And the span of immediate memory is about seven as well. Considered purely as machines, then, people are pretty weak.

Miller then followed with the truly important and innovative part of his paper. He noted that although immediate memory is limited in the number of items it can hold, the amount of information contained in each item is flexible. This is to say that individuals might hold seven binary digits, seven letters, or seven words in immediate memory. As Miller put it, immediate memory is like a coin purse that can hold seven coins no matter if they are pennies or dollars.[110] Miller would end up arguing that it is virtually impossible not only to determine the size of a person's long-term memory; even more, it is hard to estimate the amount of information that people think with—since how that information is encoded is the determining factor.[111] In this regard, humans are not at all like machines because humans change their qualities depending simply on how they think.[112]

Where linguistics and psychology showed how people are not like machines, the history of another component of cognitive science also shows that it would be a mistake to conclude that its account of the human mind arose from its study of the computer. Indeed, several early works in artificial intelligence were not initially computer programs at all but rather procedures carried out not by machines but by people. For instance, when Herbert Simon announced his "invention" of a "thinking machine," the "machine" he referred to was not an inanimate object. In fact, what Simon had done was, first, take down a transcript of a person completing a proof and, second, devise an algorithm that reproduced that transcript. That algorithm was a set of step-by-step rules through which novel proofs of mathematical theorems could be invented. Simon then assigned the algorithm to his wife, children, and graduate students. Each member of the group would perform one step of the proof and then pass the results to the next person. When this group of people had produced a proof, Simon declared that he had invented a thinking machine. His "machine," then, was not an inanimate object made of vacuum tubes, memory relays, and wire. It was, as he put it, a "computer constructed of human components." The only thing artificial about it was that it was a group of people replicating the presumed thought processes of an individual person. It was only subsequently that Simon programmed an actual machine to reproduce the activities of humans.[113]

If anything, Simon learned about the computer after, not before, learning about people. In this regard, Simon's experience was far from unique. Robert Abelson's model of ideological thought processes drew from a real-world example: Barry Goldwater's filtering of all information into a single master script on the evils of communism. As with Simon, Abelson's research started with a model of human activity and made its way first to paper before ever being tested or "run" in silicon. Marvin Minsky, Seymour Papert, and David Gerlenter likewise all conducted research in computational psychology using pencil and paper to design models of the psyche that depended as much or more on prior views about the mind as on the structure of the computer.[114]

The lack of essential connection between computers and the specific models of human nature produced by cognitive science is apparent when we consider that computers are flexible instruments. Because of their flexibility computers played little role in providing a definitive answer to the debate that raged over human nature: does creative, autonomous thought exist? Instead of weighing on one side or the other, the computer served as a tool for both.

Just as Herbert Simon used the computer to simulate the rational, autonomous, cognitive, creative version of human nature that Jerome Bruner and George Polya told him existed, so too did behaviorists use the computer to simulate the reactive, reflex-driven version of human nature. Consider the models of human thinking proposed by Clark Hull, E. G. Boring, Saul Gorn, and Howard Hoffman.[115] In every one of these cases, psychologists developed models that strengthened the behaviorist vision of human nature. Clark Hull, for instance, compared learning to making a series of connections linking stimulus and response on a telephone switchboard and thereby emphasized its noncognitive, noninsightful aspects.[116] Gorn designed a computer program and Hoffman built an electrical device (an "analogue lab") to simulate S-R learning. In both cases, the models indicated that the behaviorist vision of human nature, rather than being at stake, was assumed to be true.

E. G. Boring's robot model of human nature was likewise dependent on the stimulus-response model. Having failed Norbert Wiener's challenge to give a single example of any human mental function that computers could not perform, Boring proceeded to outline all of the characteristics that the computer should have if it were to be a good model of human nature.[117] The remarkable feature of Boring's mechanical model is the specific characteristics of humanness that the computer was supposed to mimic. To Boring, a computer or robot would be a good model of human nature if it exhibited the properties of stimulus-response learning developed by behaviorists.[118] Unlike Newell, Simon, and Shaw, Boring did not argue that a computer could be considered a good model of human nature if it could produce novel solutions to mathematics problems or play chess.[119] In other words, unlike cognitive scientists, Boring used the computer model of mind to argue *against* mentalism.

Miller, Galanter, and Pribram pointed out that the behaviorist, anti-cognitive implications that Clark Hull drew from his analogy between a telephone switchboard and organisms were not necessitated by the model itself. They noted that until shortly before they wrote, switchboards needed human operators to work.[120] Thus, they showed that the switchboard metaphor of thinking and learning could be used to imply that human thought processes were necessary to properly link a stimulus to a response. Of course, Hull did not take this option. His model emphasized the nonthinking aspects both of telephone switching and of learning. Although Miller, Galanter, and Pribram noted that the telephone switchboard did not force behaviorist conclusions, they did not also point out a related conclusion: much as Boring had demonstrated, the computer model does not require a cognitive vision of the human mind.

Set the task of designing something like a Turing test, behaviorists and cognitivists framed very different sorts of questions to put to the computer. Boring looked to see if the computer could simulate what he took to be human nature by following stimulus-response rules of learning. By contrast, the cognitivists looked to see if the computer could mimic their own preexisting vision of human nature. They asked if the computer could play chess or solve problems using heuristic methods. The way that these two groups of human scientists understood themselves and their own thinking was at the center of this difference.

That Boring, Hull, Gorn, and Hoffman looked at machines and saw a way to support behaviorists' claims about organisms says more about these psychologists than it does about the organisms or the machine models they worked with. Likewise, that Hebb, Miller, Galanter, Pribram, Newell, Simon, and Shaw looked at computers and saw a model of autonomous human cognitive processes says more about these scientists than about either computers or people. Whether behaviorist or cognitivist, the meaning that scientists read into mechanical metaphors depended on highly value-laden visions of human nature. It turned on what they already knew about thinking, human nature, themselves, and the scientific process.

Conclusion

This chapter has examined the elision of a normative vision of "right thinking" with a descriptive account of "thinking." Cognitive scientists inscribed a highly political and value-laden notion of proper thinking into their descriptive accounts of human thinking. In their struggle with behaviorism, contests over the structure of the world (i.e., human nature and the nature of mind) were often contests over proper scientific methodology at the same time. Conversely, statements about whether rats or computers exhibited insight were also claims about the kind of thinking in which scientists themselves ought to engage. Psychologists who ascribed insight to rats, computers, or people were likely to see insight as an important component of the scientific process. On the other hand, behaviorists who denied rodent insight and focused on learning through trial and accumulated experience were more likely to frame the empirical experience as the foundation of proper scientific method—of which they were practitioners.[121]

Making a science of the mind involved developing a cognitive psychology—an image of normal human nature that was universal, independent (at least in part) from the environment and instinct, creative, autonomous, and tolerant of ambiguity. Despite the centrality of the computer as a tool for the development of the field, it would be a mistake to conclude that knowledge about the computer determined what cognitive scientists said about humanity. Instead, as in the case of George Miller and Noam Chomsky, they used information technology to show what was unique about humanity. For his part Herbert Simon used the computer to model what he already knew about problem solving by individuals and groups.

The work of cognitive science involved, for the most part, dropping the normative descriptions of better and worse kinds of people and the personality types described in political discourse and social psychology. Rather than contrasting the open and closed mind, the democratic and authoritarian, the creative and conformist personality, cognitive scientists looked at computers and modeled the open-minded, creative, flexible, and heuristic thinking processes that they deemed to be characteristic of human nature.[122]

Although lacking language that identified better and worse forms of thinking, this science of autonomy substituted normalizing for normative terminology. Instead of identifying better and worse forms of humanness, it identified as universally human the specific forms of human nature that political discourse, social psychology, and anti-positivist philosophy had already marked as good. Thus, the better forms of human thinking constructed by social psychology (democratic, broad, open, flexible, creative) became the only forms of human thinking. The

person portrayed by this psychology was fundamentally the same regardless of social situation, personality, or culture. By implication, any
modes of thought that did not fit this model were consigned to a pathology worse than illness—to non- or sub-humanness.

The permeable boundary between expert and non-expert knowledge
of human nature afforded cognitive scientists a collection of tools that,
used reflexively, could further their research programs. The politics and
social psychology of right thinking in Cold War American culture gave
anti-behaviorists techniques to turn on their own discipline and with
which to mark themselves as open-minded and behaviorists as authoritarian ideologues. Anti-positivist philosophy of science offered cognitive scientists not only a defense of their own kind of thinking but also a
model for how humans in general think. As George Miller put it in the
Voice of America address discussed at the beginning of this chapter,

> A scientist . . . searches through ideas as well as through objects in order to
> find what he seeks. And he does not look indiscriminately—always he carries
> an image of what he seeks. . . . He is looking for something that matches up
> to his image of what the world must be, something that meets a test he him-
> self imposes, something that has meaning only in terms of the standards he
> lives by.

Concluding his address Miller remarked, "the scientist is Everyman,
looking just as you and I. We go and look for the things we want, and
when we find them we find part of ourselves."[123] Miller and his fellow
cognitive scientists did just that—they looked for human nature by holding an image of what they were looking for in their minds. The image
they held was none other than their own self-image.

CHAPTER SIX

Instituting Cognitive Science

Cognitive scientists drew two benefits from equating human thinking and the thought processes of scientists. On the one hand, scientific cognition functioned as a model of all human thought in the same way that fruit flies, tobacco mosaic viruses, nematode worms, yeast, and laboratory mice have served as models of the genetics, biochemistry, and physiology in all organisms.[1] In this way, the personal, informal, and folk knowledge that cognitive scientists held about how they and their peers thought could become a resource for context-free scientific knowledge about human thought processes.

On the other hand, this connection between scientific cognition and human cognition also functioned as a resource for the practical pursuit of further knowledge in cognitive science. What cognitive scientists knew about how humans thought could move beyond basic science and be applied in a number of contexts. Each applied context offered an opportunity to use cognitive scientific knowledge to improve how individuals thought and learned. This chapter examines what happened when cognitive scientists applied George Miller's argument that "the scientist is Everyman" to themselves at Harvard's Center for Cognitive Studies (CCS).

When the CCS was founded by Jerome Bruner and George Miller in 1960, they applied their expertise in how humans think and learn to designing a research environment that would maximize their own chances for acquiring knowledge about the world—specifically, about the nature of human thinking. The Center provided the possibility to develop cognitive science as a discipline as well as to apply cognitive theories to cognitive scientists themselves.

The Center, recognized by cognitive scientists as the first institutional home for their field, trained or opened its doors to most of the first and second generation of cognitive scientists.[2] Not only did almost all of the early leading cognitive scientists either visit, work at, or receive training at the Center, but the work visitors conducted during their time at the Center proved critical to the development of cognitive science. For instance, Ulric Neisser wrote the first textbook in cognitive psychology during his stay, and Noam Chomsky developed his work on the deep structure of grammar and on the history of linguistics in *Aspects of a Theory of Syntax* and *Cartesian Linguistics* respectively while at the Center.[3]

The cognitive psychology that Bruner and Miller practiced saw human learning as, fundamentally, the acquisition of new structures of thought and of new tools with which to think.[4] In this account of mind, what was important was not simply facts that people learned or scientists discovered. Rather more significant were the procedures, forms of mental representation, and heuristic methods that enabled individuals to have original forms of ideas, novel hypotheses, and techniques for investigating the world.

Miller and Bruner applied this vision of the human mind in shaping their Center in the hope that the institution they founded would facilitate the construction of new theories and new scientific tools while establishing the disciplined study of human cognition on a stable foundation. At stake in the Center for Cognitive Studies, therefore, was not only a program for studying human cognition, but the application of cognitive theories to the design of a creative research culture.

Drawing on their expert knowledge, personal experience, and the values of the intellectual culture they inhabited, Bruner and Miller conceived of the Center as an interdisciplinary institution in order to make it home to an innovative and creative research environment. The Center included people from the fields of psychology, linguistics, philosophy, biology, mathematics, anthropology, pediatrics, history, psychiatry, and psychoanalysis among its researchers, visitors, and directors.[5]

Shortly after founding the Center, Miller and Bruner wrote the dean of the faculty about its guiding principles: "It was an essential feature of the Center, as originally conceived, that it would be broadly interdepartmental. The slogan, only half in jest, was that the cognitive processes are far too complex and important to be left to psychologists."[6] In this memo, these two psychologists boldly declared the inability of their own discipline to understand the very subject matter of the institution that they directed. By adopting this position, Bruner and Miller situated themselves in an intellectual culture that placed a high value on interdisciplinary social scientific research. By telling their dean that the Center would be "interdepartmental," Bruner and Miller underlined the point in the next sentence: the Center's business was important. To social scientists, their patrons, and university administrators, important issues required interdisciplinary solutions and, conversely, interdisciplinary work indicated the importance of particular research topics.

The Center's "interdepartmental" structure served more than rhetorical purposes. It was also a highly esteemed strategy for acquiring knowledge—in this case, knowledge about thinking. But despite its design, the Center retained its initial research culture through only about half its decade-long life, despite continued interest by Miller and Bruner in establishing and maintaining the Center's interdisciplinarity and despite the continued membership of scholars from numerous disciplinary backgrounds.

Once scholars at the Center had successfully established reliable systems of research, the character of their social and intellectual interaction changed from what I shall call *interdisciplinary* to what I shall call *multidisciplinary*. Reflecting this transition, the Center's initial *interdisciplinary* atmosphere incorporated the view that "a Center is a place where people know each other's business."[7] On the other hand, a few years later, after the Center's atmosphere had changed into a *multidisciplinary* research culture, its members would comment on how they no longer knew one another's work and that there was less cross-field communication.

The Center's shift from an interdisciplinary to a multidisciplinary research culture resulted from a transformation in the ways its members exchanged their intellectual tools. Its initial interdisciplinary intellectual economy was characterized by researchers acquiring and learning to use unfamiliar tools. By contrast, the subsequent multidisciplinary economy involved researchers working in parallel, using their own tools and rarely peering into someone else's toolkit. For the sake of expository clarity, in this chapter I will use "cross-disciplinary" as a neutral category to refer to either interdisciplinary or multidisciplinary situations.[8]

Understanding the different research cultures at the Center for Cognitive Studies is a matter not so much of merely identifying or cataloging all the different departments from which its members originated, but more one of investigating the nature and kinds of tool exchanges that constituted its intellectual culture. These exchanges were an important part of the center's intellectual work, and the means through which the people at the Center negotiated their social and intellectual lives with one another. It was the particular character of these exchanges and therefore the Center's research culture that made the Center at once distinctive and a place that the cognitive scientists who passed through it would recall and seek to replicate.

Working across Departmental and Disciplinary Lines: Tooling-Up the Social Sciences

In founding the Center for Cognitive Studies, Bruner and Miller aimed to address what they took to be an imbalance in their field: the neglect of mind produced by behaviorism's hegemonic control of psychology. Bruner and Miller constructed their Center as they did in part because they believed that both the success of behaviorism and its exclusion of mind from experimental psychology resulted from use of tools imported from physiology—tools originally designed for studying the body and not the mind.[9] For Bruner and Miller, understanding psychology's history this way implied that their efforts to make a science of mind would depend on finding or inventing new tools to support their work.[10]

Seeing research problems as an issue to be solved by locating or inventing the appropriate tools was itself a (meta) research strategy that had a history in the personal lives of Bruner and Miller, in the popular philosophy of science of the midcentury social sciences, and in the content of their psychologies of thinking. Like many of their generation, the directors of the Center for Cognitive Studies took the experience of World War II as ratifying the productivity of solving specific, practical problems by assembling a set of research techniques from a range of disciplines.[11] For his part, Miller was engaged in work in Harvard's Psychoacoustic Laboratory during the war. The work there ranged from engineering earplugs to designing codes for best comprehension under conditions of high noise interference. Bruner's own experience had been shaped in the context of morale surveys and public opinion surveys conducted for the Office of War Information (OWI).

In the years following the war, both Bruner and Miller had a sabbatical at the Institute for Advanced Study (IAS). Bruner was appointed to Harvard's Social Relations Department while Miller worked in MIT's Lincoln Laboratory[12] and spent a year at the Center for Advanced Study for the Behavioral Sciences.[13] Each of these locations was organized along cross-disciplinary lines, and Miller saw his time in those places, particularly his sabbatical years, as presenting opportunities not only for research and writing but also for learning new techniques. For instance, Miller's year at the IAS was motivated by his desire to learn mathematics and the techniques of game theory from John von Neumann.[14]

Although Bruner's and Miller's involvement in cross-disciplinary work reflected significant aspects of their personal research styles,[15] their choices to operate in such a manner were in no way idiosyncratic. Their pattern of cross-disciplinary collaboration was part of the background environment fostered by the structure of government funding, wartime demands, Cold War politics, and an intellectual culture that attached high value to interdisciplinary research.

This system provided significant financial, institutional, and methodological resources for Bruner and Miller and helped to produce psychological work that took advantage of tools developed in other disciplines. For instance, Miller brought information theory and linguistics to psychology;[16] Bruner used his experience in opinion research for research on perception[17] and applied John von Neumann's game theory in his work on concept acquisition.[18] It also conditioned Miller's rejection of behaviorism and his move to adopting a cognitive psychological view of human nature.[19]

Founding the Center for Cognitive Studies

In the winter of 1959–60, Bruner joined forces with George Miller and turned these research and fundraising strategies toward founding a new research center. The Center for Cognitive Studies they built was specifically aimed to address the limitations of Bruner's and Miller's respective Departments of Social Relations and of Psychology as well as to serve as a "unifying force, a neutral place where students from both Social Relations and Psychology could meet and educate each other."[20] Neither of their departments was concerned with how normal individuals think. The Department of Psychology, dominated by the behaviorism of B. F. Skinner and the psychophysics of S. S. Stevens, left little room for investigation of mental processes. In Social Relations, Bruner had colleagues interested in what happened among groups of people or, in the case of Henry Murray, in the nature of the abnormal mind.

Their personal preferences for varied approaches to psychological problems encouraged them to cast a wide net in the search for effective research methodologies. Accordingly, they designed their Center with the belief that "the study of cognitive processes . . . [was] clearly not an enterprise that [could] be entrusted to any single discipline, that [could] be developed from any limited or specialized point of view."[21]

Mutually unhappy with the state of affairs at Harvard, Miller and Bruner looked to build a research center that, in significant respects, mirrored an industrial research and development group Bruner had previously studied in order to understand the psychological processes of creativity. For the group that Bruner studied, innovation was facilitated by using unconventional methods, energetic interpersonal interchange, and operating "out back," beyond normal institutional structures.[22] Likewise, the Center for Cognitive Studies (CCS) that Bruner sought to build with Miller would operate both methodologically and physically outside the Departments of Social Relations and of Psychology. Indeed, the CCS would be located on the periphery of the Harvard campus at 61 Kirkland St., well away from university offices, laboratories, classrooms, and dormitories (see fig. 6.1).

In establishing their Center, Bruner and Miller began by approaching John Gardner of the Carnegie Corporation and McGeorge Bundy, Harvard's dean of the faculty. From Bundy they secured approval and space on Kirkland Street for the Center. Gardner, who had been instrumental in supporting research on creativity and in founding interdisciplinary social science research units all over the country, helped them to a grant of unrestricted funds of $250,000 over five years. Bruner and Miller proposed to use Carnegie funds much as the Laboratory of Social Relations had, as a resource that would support "exploratory ventures" and would allow the "freedom to gamble."[23] They would shortly bring the yearly funding of the Center to over $500,000 with grants from NIH, NSF, the United States Office of Education, and the Defense Research Projects Agency. Most of the money went for equipment, students, and postdoctoral fellows. Harvard contributed space (initially $78,000), administrative support, portions of Bruner's and Miller's salaries, as well as portions of the salaries of many of the visiting senior fellows who would frequently hold teaching appointments in the Department of Psychology or in the Department of Social Relations.[24] Eventually, Miller and Bruner were so successful in attracting funding that they had to devote administrative at-

FIGURE 6.1. 61 Kirkland Street.

tention to making sure that they spent their funding fast enough as well as to deciding which of several grants to bill various expenses to.[25]

As a nondepartmental unit, the Center aimed to maintain a research culture distinct from that of the Departments of Social Relations and of Psychology. Its wealth allowed rapid investment in personnel and in flexible, multipurpose tools including a PDP-4 computer and a fully operational laboratory inside a twenty-seven-foot, twelve-ton International Harvester truck that allowed experimentation in the "field." With the flexibility of being a nondepartmental unit came the consequent lack of ability to make faculty hires, to admit students, or to confer degrees. Those powers always remained with the Departments of Psychology and of Social Relations. Thus, while the Center had members, those members always had student or faculty affiliation within departments at Harvard or at other universities. The only exceptions to this rule were postdoctoral researchers whose salaries depended on Center funds or on limited-term fellowships that they brought to Cambridge.

For Bruner, Miller, and the other members of the Center in the early 1960s, the effort to find useful methods involved setting up a particular research culture grounded by the exchange of research tools. Without standardized techniques with which to study cognition, and working within the pattern of eclectic methodological search as the standard method of research problem solving, the Center operated with the explicit intention of bringing scholars together. Their proximity would enable psychologists, linguists, and philosophers, among others, to learn from one another by trading tools. Such tool exchange fostered an intellectual economy and social culture characterized by negotiation and argument. As the people at the Center exchanged tools with one another, they both redefined their fields of study and cemented their identities as innovative interdisciplinary scholars.

Interdisciplinary activity at the Center for Cognitive Studies was not, for the most part, conducted by people from departments with different names (e.g., psychology and anthropology) collaborating on single projects. (There were, however, exceptions; for instance the work of Miller and Chomsky.)[26] Rather, interdisciplinary research at the Center for Cognitive Studies mostly occurred as scientists borrowed techniques and learned methodological approaches from their colleagues to apply to their own research problems.

This approach found repeated expression in public and private forums in the Center's early years. For instance, Jerome Bruner saw the goal of the Center as promoting interaction of close, but autonomous, research programs:

> What we need to do is get together and hammer out not a joint idea, but some pattern in which our various ideas are represented and are rubbed against each other. The important thing about a federation or research cluster, or whatever you want to call it, is that each person do exactly the research that he or she wants to do, but that all the research done be made as related as possible to other work being done.[27]

The relations Bruner advocated were meant not only for the purposes of social and intellectual conviviality but also to help generate and foster eclectic methodological sharing and borrowing. In discussing his own field of psycholinguistics, George Miller made a similar point about the importance of methodological exchange. Pointing out how interdisciplinary research operates "shoulder to shoulder," Miller, like Bruner, envisioned interdisciplinary research projects organized in a way that investigators would be both continually aware of one another's work and seeking to acquire methodological tools from one another.[28]

The Center's annual reports to its patrons noted how such an interdisciplinary research culture based on tool exchange played out on a daily basis.[29] The significance of the Center for its many researchers was the opportunity to invent new research practices and to exchange research tools with colleagues.[30] Indeed, scholars at the Center for Cognitive Studies made a regular practice of peering into and borrowing from their colleagues' toolkits. The work of Jerome Bruner's group on developmental psychology and on perception made use of information-processing models as well as grammatical theory drawn from those in Miller's group. For instance, Rose Olver, a student of Bruner's, investigated the "grammars" that children at different ages use in grouping objects (e.g., by perceptual features vs. by function).[31] Similarly, George Miller's research on the psychology of language drew on tools developed in Bruner's group and on those invented by Noam Chomsky. When investigating the way people learn grammars, for instance, Miller relied on Bruner's model of concept acquisition in *A Study of Thinking*.[32]

In conducting very different research programs, Miller and Bruner each used a critical piece of Noam Chomsky's linguistic program, the "embedded sentence." These are grammatical structures in which one or more whole sentences are contained within another sentence. For instance "she thanked the producer" can be inserted in "the producer discovered the novel," as follows: "The producer, whom she thanked, discovered the novel."[33] This particular structure was a critical part of the argument Chomsky had developed in favor of his own brand of linguistics. Whether arguing for or against particular theories of language, Chomsky made a practice of finding or creating particular sentences that could be accounted for in one kind of theory but not in another.[34] In this instance, Chomsky's target was linguistic theory based on "finite-state grammars." Since finite-state grammars do not produce embedded sentences, but embedded sentences are a part of English, embedded sentences were part of the proof that finite-state grammars are insufficient to account for the nature of the English language.[35]

While embedded sentences were a means to decide between two sorts of linguistic theory, these sentences were also a foray into the battle between cognitive scientists and behaviorists. George Miller noted that behavioristic psychological theories, which explained learning through the connections between stimulus and response (S-R), fit extremely well with linguistic theories using finite-state grammars. As he explained to his students, S-R theories would account for a child learning grammar in such a way that the first word he or she says in a sentence is a stimulus and the second word is a response. Similarly, the second word is a stimulus for the third word—and so on. Miller concluded that the phenomenon of embedded sentences undermined both finite-state linguistic theory and behavioristic psychology.[36]

When Miller used embedded sentences to classify and compare human thinking to computer information processing, he was making use of a critical piece of Chomsky's instrumentation.[37] In addition, those sentences, Miller knew, were themselves an argument for the existence of human mental processes and thus were convenient tools for studying those processes in detail. Miller argued that since recursion is necessary for processing embedded sentences, and that since people experienced difficulty in comprehending and correctly remembering these sentences, then, for the most part, human mental processes were nonrecursive.[38]

In his own research program of investigating the psychology of vision, Jerome Bruner took similar advantage of this sentence type. Using a special camera designed by Norman Mackworth for recording precisely where people look (see fig. 6.2), Bruner adapted embedded sentences to develop his longstanding effort to show that vision is controlled by mental processes.[39] This research program was an argument against a more empiricist and behaviorist position on perceptual processes that sought to explain them completely by the structure of environmental factors. For Bruner, Chomsky's embedded sentences and Mackworth's eye camera became tools to demonstrate that patterns of direction of gaze became erratic when reading syntactically complicated material. And since syntax was supposed to be mental rules, this finding could support the longer-term goal of Bruner's research program: showing that mental processes control vision.

Negotiating and Arguing: Creating the Center's Atmosphere

The interdisciplinary exchange of tools that characterized the Center's early years came with a definite social structure and mode of interper-

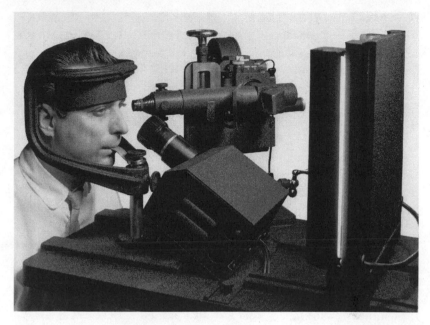

FIGURE 6.2. Stand-mounted eye marker camera. By permission of Harvard University.

sonal interaction. The variety of disciplinary backgrounds represented, and the intense, ecumenical features of the Center's daily life, generated a multifaceted, if chaotic, intellectual environment. Remarking on the "kaleidoscopic quality" of the Center's activities, an annual report echoed Bruner's earlier application for funding from the Laboratory of Social Relations:

> Research at the Center is opportunistic; as good ideas occur, we try to take advantage of them. Freedom and flexibility to explore in whatever appears to be the most promising direction is, we believe, an indispensable factor in good research. It does result, however, in an over-all program that looks somewhat loose and untidy when one tries to summarize it for an outsider.[40]

This eclecticism of research interests, when combined with efforts to exchange methods, generated a research culture that Bärbel Inhelder characterized as pluralist.[41] An annual report reflected: "It may well be that the intensity of discussion and activity reflects the wide spectrum of interests we comprise."[42]

Arguments formed a central and constitutive feature of the Center's milieu—a characteristic that enhanced the Center reputation. Recently, Donald Norman recalls, "the first couple of people I met I immediately started violent arguments with. . . . We just fought and fought and fought in the most positive of ways. And the end result were a series of quite important papers."[43] Bruner has noted that people requested to attend the Center as visitors specifically because of its well-known hallway arguments.[44]

The combination of numerous approaches and intensity of discussion came with a fluid pattern of social interaction. After his visit to the Center, Masano Toda wrote:

> Something terribly active was booming at the Center and I have never seen such a noisy place. Besides the weekly colloquiums, the seminars held three times week, the Thursday staff luncheons, which were already lots of activities, informal discussions and debates were going on everywhere—a group of people were talking on cognition in the second floor corridor, one person sitting on the railing precariously keeping his balance, another sitting firmly on the floor, while another group of people was running subjects in the basement.[45]

Movement from the hallway and lab into the lunchroom brought a pause in computer testing or quizzing experimental subjects, but argument did not rest. For instance, recalling Noam Chomsky's visits to the Center, Donald Norman recounts:

> Chomsky would have lunch all the time at the Center for Cognitive Studies. . . . There would always be fights over Chomsky's utterances. Chomsky would say something and then [Jerry] Fodor would say "What Noam meant was . . ." And Jerry Katz would say, "No, what Noam meant was . . ." And everyone would take their turn saying what Noam must have meant. And Noam would never say a word after that.[46]

Learning from one another involved much arguing. Exchanging and refashioning research ideas, methods, and tools involved discussing, fighting about, and negotiating how cognitive scientists or people in general might use ideas, words, and concepts.

Underlying this culture of eclecticism and argument was a sense that it was both productive and an essential part of the Center's research program. They believed so on the basis of their experiences in the daily life of the academy and also through research into the psychology of creativity. Before the Center had been founded, Bruner engaged in research on the psychology of the creative process and, like others engaged in similar research, had determined that creativity in groups is facilitated by free-flowing discussion and argument.[47]

Miller and Bruner repeatedly pointed out that variety and flexibility in communication were essential features of their, or any, research center. They wrote in a grant application:

> A Center is a place where people know each other's business. It is not obviously by what means this is assured. It takes the form of conversation, at lunch, in the corridors, in weekly colloquia, in the two or three informal seminars that are always in being at the Center, organized either around a visitor's interests, around a hot topic, or around some coalition. It is not by chance that some forty of the publications that have come from the Center are collaborations among its members. The Xerox and the mimeograph are probably critical instruments in the exchange that leads to such collaboration.[48]

If they were unsure of the means by which they could assure the success of communication at their center, Bruner and Miller proceeded by facilitating the forms of communication, seminars, and colloquia that they mention in their grant proposal.

The success the Center had in generating a productive atmosphere and in drawing support from private and government patrons was itself exciting. Miller gave voice to both the productivity of the Center's research and its ability to elicit patronage:

> I have a sort of Paul Bunyan-like vision of Bruner and Miller sitting high on a hill, surveying thousands of people busily asking one another questions, typing out the answers, and mailing out manuscripts while a mobile laboratory brings in truckloads full of data that it dumps into the computer while acres of beautiful girls sit loyally at their typewriters alternately hitting the keys and drinking coffee, and faithful slaves bring periodic installments of newly-minted money from Washington. It is all a little mad and nothing at all like an ivory tower.[49]

Similar comments on the Center's culture of "flexibility," "freedom," and "intensity," its "energetic," "heated" atmosphere, its productive arguments, and the "kaleidoscopic," "loose," "untidy," "opportunistic," "flexible," and "wide spectrum" approach to problems appeared repeatedly in grant proposals, annual reports, and personal communication.

The Center's spirit of methodological and disciplinary inclusiveness did not mean complete openness, however. The recruiting process involved inviting "to the Center only those who eschew intellectual blinders." Voicing interest in eclecticism, they commented that "one mark of our success must be the way few of the research programs we have described fall neatly into any one of our categories."[50] This comment, and others made by members of the Center about their efforts to avoid "blinders" and methodological narrowness, could easily have been understood as a means of marking the difference between their work and that of behaviorism. For behaviorism had been a way that experimental psychology had marked its scientific status and autonomy as a discipline.[51] By the 1950s, however, behaviorism had come to be labeled by its opponents as a narrow, rigid, dogmatic, and authoritarian system. Thus, the Center trumpeted both its openness and its eclecticism—each of which served as a rhetorical device for distinguishing it from behaviorism.

If the Center's culture was internally open, it was not necessarily open to outsiders. The in-group feeling and pattern of behavior that was facilitated by closeness and frequency of contact, arguing in the halls and over lunch, and regular informal seminars, had the effect of increasing solidarity within the Center while also excluding other people. For instance, when outside speakers visited to give a paper, interaction among audience participants was at least as important as the speakers' presentation. After the presentation was complete, the audience members posed questions, which were directed as much to the other audience members as to the speaker. Typifying this mode of interaction, rising from his seat in the front of the auditorium, a senior member of the group would usually pose the first question—facing the audience and with his back to the speaker. And significantly, the speaker's response was considered relatively unimportant.[52] It was the other members of the Center who were significant; these were people who would be active as sounding boards for developing research ideas.

There was no substitute for being a fully participating member of this intellectual community. For outsiders, missing out on discussions at the Center meant more than lost opportunity of early appraisal of research results. It also could mean inability to make use of the research tools used at the Center, since some of their methods were never fully and completely discussed in print. The Center's own members were well aware of the difficulty of knowledge transfer through print. For instance, Leo Postman, an early collaborator of Bruner's, complained to Bruner that after moving from Harvard he was unable to get *his own* instruments to work because of insufficient local technical support in his new location. In addition, Norman Mackworth was invited to become a member of the Center because his published work on eye cameras was not rich and detailed enough for the others at the Center to make use of it without his direct involvement. Actually, that extensive help from Mackworth was necessary for others to make use of his tools is indicated by the fact that his working at the Center on a once-a-week consulting basis was eventually converted to full membership.[53]

Even the Center's research results were fully available only to insiders. Published work by the Center's scholars often cited internal unpublished studies, research reports written for patrons, the Center's own annual reports, papers given at the Center, mimeographs distributed at the Center, and/or personal communications.[54] Randy Allen Harris has noted that a similar phenomenon occurred in Chomsky's group at MIT:

> Developments spread rapidly. Everyone spoke in the hallways, attended the same colloquia, and saw each other's papers long before they reached publication. They also saw many papers that never reached publication at all, the notorious *samizdat* literature that still characterizes work at MIT: arguments and analyses circulated in mimeograph (now electronic) underground, never making their way to the formal light of day but showing up in the notes of important works that did. This situation, quite naturally, infuriated (and infuriates) anybody trying to follow the theory but failing to hook into the right distributional network.

Harris noted in addition that "an even more frustrating manifestation of this in-group attitude was the large number of notes referring to remarks made 'in a lecture at MIT,' or, even, to observations made in 'personal conversation.'"[55]

Thus, a critical feature of the Center for Cognitive Studies was its culture of intimate and energetic interpersonal interaction. By all accounts, whether retrospective or from the period of the Center's existence, this self-catalyzing atmosphere was one of the Center's most important features.[56] But, the economy of arguing and of rapid tool exchange characterized the Center's research culture for only about half of its life.

The Transition from Interdisciplinarity to Multidisciplinarity

At about the midpoint of its life, the research culture of the Center for Cognitive Studies shifted significantly. This metamorphosis involved both modes of research culture and patterns of interpersonal interaction. After the first few years of rapid expansion, the initial economy of tool exchange that had been the foundation of the Center's interdisciplinary culture ended. As the workers at the Center began to regularize their research techniques and reduce the types of questions they asked each other, the Center's initial economy was supplanted with a new pattern of exchange: *multidisciplinarity* defined by a new intellectual economy and an associated new set of social relations.

This shift of culture first appeared with a brief comment in the third annual report: "We are aware that our various discussion groups have tended to have less overlap of membership."[57] Within a few years, this early notice of decreased intellectual cross-fertilization expanded into a more lengthy analysis of difficulties with interdisciplinarity. The Center's members stopped remarking, as they once had, on the frequency of their collaborative work. Bruner and Miller told the NIH in 1965:

> During the past five years we have necessarily condensed and crystallized our interests. . . . So inevitably, the Center has moved increasingly more to concentrated research on fewer topics. . . . The danger of being interdisciplinary is that one ends up a prey to that easy eclecticism by which one swallows all and digests nothing.[58]

This crystallization of research interests and increasingly focused research programs would bring less communication among the different researchers. As a consequence, research at the Center moved from an interdisciplinary collaborative exploration to a multidisciplinary environment where work was conducted more in parallel than jointly.

As they found reliable research techniques, the practical need for an eclectic spirit at the Center declined. Once the Center's researchers had settled on coherent research tracks, they were no longer as interested in the interdisciplinary exchange that had initially been the foundation of their own research programs. As this mutual withdrawal and concentration occurred, the Center's members could no longer support the wide variety of activities they once had pursued. No longer would the weekly seminar on research methods in perception spend half a year discussing the book *Art and Illusion* by the art historian Ernst Gombrich. Significantly, the Center's younger members began to chafe against the very techniques Bruner and Miller had initially designed to foster the Center's early eclectic, exploratory, and interdisciplinary spirit. They now found the weekly seminars a time sink and superfluous to their own research activities.[59]

Without an instrumental need for interdisciplinary discussion, the drawbacks of the Center's initial "untidy" and "loose" intellectual economy seemed to outweigh the benefits. Early on the Center had sought philosophers to join their ranks—eventually including Nelson Goodman, Israel Scheffler, and Marx Wartofsky. But once its research atmosphere had shifted modes, the Center's leadership became less concerned with "find[ing] a philosopher with the patience to keep us aware of the epistemological issues we might bark our shins on."[60] As the research programs at the Center gained momentum, the people at the Center focused less on acquiring tools from one another and more on using the tools each already had.

This transformation from an interdisciplinary to a multidisciplinary environment depended on a variety of factors, including changing research priorities, changes in the Center's funding patterns, and changes in the Center's location relative to the other behavioral sciences. But perhaps the most tangible basis for the shift in the Center's culture was its physical relocation in January 1965 from its original location at 61 Kirkland Street to the 11th and 12th floors of the newly built William James Hall (see fig. 6.3).

The Center's home at 61 Kirkland Street had played an important role in generating its collegial atmosphere. In several significant ways the physical characteristics of the original building helped generate a cohesive social and work culture among the Center's members. From the very start, even before the Center's membership had reached full size, 61 Kirkland Street was small enough to generate a cozy and intimate feeling that produced "high morale."[61]

While the Center's morale depended on an intimate setting, its communal sense was similarly derived from it geographic distance from both the Social Relations and the Psychology Departments. When the Center moved into William James Hall with the other behavioral sciences, it became clear that its communal mood had depended in part on separation from Social Relations and Psychology and its location on the periphery

FIGURE 6.3. William James Hall

of campus. An annual report noted "how dependent affective processes are on environmental support." It continued, "we encounter different people in this building from those we saw before, and our visiting back and forth with them has diminished the sense of social identity—the 'in group' feeling—that we had when we worked in greater isolation."[62]

While integration with Psychology and Social Relations had a serious effect on the community ethos of the Center, the physical structure of William James itself also played a significant role in the declining sense of group cohesiveness. Don Norman recalls:

> In the original building of the Center everybody was on different floors but it was such a small building in a house that the stairway was not a barrier. Since there were only a few offices on each floor, you always had to be going up and down the stairs. In William James Hall, however, there were huge floors. And in fact you would seldom move from your end of the floor to the opposite end, let alone up or down the stairs.[63]

Thus, the relocation to William James not only ended the Center's isolation, but it also, by the very layout of the new offices, drove the Center's members further from one another. In fact, the division between the two floors was severe enough that recently several members of the Center failed to remember that the Center had more than one floor.[64] Clearly, they did not spend much time moving from floor to floor.

There were several specific architectural features other than elevators that made in-group communication difficult. The most significant was the layout of offices and labs on each floor. Each floor of William James Hall is a long rectangle with windows on the long sides, north and south. Offices were arranged on both north and south walls. Labs and seminar rooms ran up the center of the floor, between the offices on the north and south walls. Consequently there were two long hallways sandwiching the labs—one lay between the north offices and the labs and its counterpart lay between the labs and the south offices (see fig. 6.4).

This floor plan produced three significant patterns of movement and interaction. First, no one sat in an office across a hall from anyone else— across the hall was always a lab. Second, because the only route from the north to the south hall was a cross-wise hall in the center of the building, there was almost no movement between the two major halls. Third, because the elevator bank was in the center of the building (located on the short hall connecting the long north and south halls), researchers never needed to travel more than one quarter the length of the total hallway space on their ways in and out of the building. So, even researchers who worked on the same floor would very likely not run into one another unless they had specifically made plans to do so.

This floor plan significantly reduced the amount of informal interac-

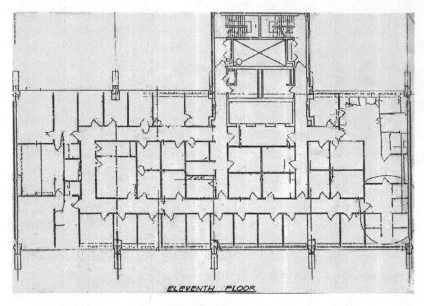

FIGURE 6.4. Floor plan of the Center for Cognitive Studies in William James Hall. By permission of Harvard University. Enhanced by the author for clarity.

tion among the Center's researchers. There was also an organizational feature that contributed to the architecturally driven decline of discussions between people who were not working together. Some time after the move to William James, office occupancy was rearranged so that all of the people working in George Miller's group were on one floor and all of Bruner's people were on the other. For George Miller, in retrospect, the move to William James Hall and the changes it brought "destroyed" the Center's previous collegial atmosphere.[65]

Changes in policy and attitude from the Harvard administration also damaged the Center's interdisciplinarity. Within Harvard, McGeorge Bundy, dean of the faculty when the Center was established in 1960, had been supportive of extradepartmental research groups and in reducing the power of departments—a policy that worked in the Center's favor. On the other hand, Bundy's successor, Franklin Ford, found his predecessor's approach too be too chaotic and disorganized. Bruner noted in 1971, "in all these years, for example, there was never anything except the most grudging willingness to accept the Center for Cognitive Studies and (after Bundy) the Deans always politely inquired how much longer the untidiness would continue."[66] Just as the deans after Bundy were hostile to the Center's independent existence, so too was Harvard's president, Nathan Pusey. He opposed proposals by Bruner and Miller to establish an endowment for the Center and to add two faculty members in the regular departments who would be regular contributors to the Center.[67]

Change in patronage policy similarly restricted the Center's members' early opportunity to focus attention on building and exchanging tools. Regretting this change, Bruner explained to one of his patrons "a changed policy at the Center." The Center had originally been organized in terms of "coalition and collaboration" such that "a given experiment grew out of discussion in seminars or in the informal give and take of the corridor of a laboratory." As a consequence it was difficult for the Center directors to detail which experiment was due to which grant. However, Bruner noted that a shift in patronage policy had meant that funding agencies had become more interested in funding "particular" studies rather than "generic program[s]." So significant was this new mode of patronage that it undermined the Center's pattern of interdisciplinary give-and-take and even made it difficult to recruit postdocs to join the newly narrowed sponsored research.

In light of these changes, Bruner proposed reshaping the Center so as to "limit research appointments to a smaller number of post-doctoral fellows who come to us with their own grants. In this way, I shall be able to narrow the range of studies being carried out to those that I originate myself and to accept two or possibly three post-doctoral fellows."[68] These changes in funding, personnel, and administration meant a decreased scope of research at the Center and less day-to-day flexibility. With researchers committed ahead of time by their grants to projects originated by Bruner, there would be less opportunity for the interdisciplinary tool exchange of the Center's early years.

Also affecting the Center's atmosphere were local political events, which further damaged its early "in-group" feeling. Drawing on their experiences in Bread and Roses, a community-based, feminist consciousness-raising group, the women in the Center, from administrative staff to graduate students and postdocs, formed a support group in order to address the sexism they saw in the Center.[69] This group was possibly difficult for Bruner to cope with since he had been a longtime supporter of and advocate for women in psychology. For instance, one of his students, Rose Olver, reports that it was only through Bruner's efforts at arguing the equality and similarity of women and men that she was able to secure an appointment in 1961 as the first woman on the faculty at Amherst College.[70] But, in the context of the late 1960s, the arguments that had supported Olver—that female psychologists were essentially no different than male psychologists—had a different valence. By that point, failure to recognize difference could be construed as sexist.

Of course, political mobilization at Harvard was not restricted to the Center. Other events at Harvard would also strain the Center's communal atmosphere. On Wednesday, April 9, 1969, Harvard and Radcliffe students occupied the administration building, University Hall, and ejected the administrators in protest against Harvard's involvement in the Vietnam War and its policy of giving academic credit to ROTC candidates. The Harvard administration responded by asking the Cambridge police to end the protest. The next morning, the Cambridge police forcibly removed and arrested the students who had occupied the building, clubbing many in the process.[71] These events were followed by a student strike and several months of strife between the administration and the student body over a variety of issues including the disciplinary measures that would be taken against the students who had occupied University Hall, the reaction of the administration, the violence of the Cambridge police, and the original ROTC issue. At the end of the process, bad feelings against Nathan Pusey and his handling of the situation remained. While these events led eventually to Pusey's resignation, their wake also strained relations between the Center's younger members and its leadership. On this note, Bruner remarked to a colleague that these events had led to a loss of the Center's collegial quality, since the "younger postdocs are so full of hostility to anybody in a senior role."[72]

Because of changes in the research space, Harvard departmental politics, protests against the Vietnam War, rising feminist consciousness, and changes in funding structures, the group of scholars at the Center for Cognitive Studies spent less time speaking and working with one another than they had previously. Yet this transformation in the character of the research atmosphere depended as much on developments within the research itself. In fact, as a direct consequence of its research success, the Center lost its in-group spirit, its sense of being arrayed against the world, and of being engaged in a battle of the faithful against oppressive behaviorist dogmatism.

If the younger postdocs were "hostile to anyone in a senior role," it was because they represented the "Establishment." The Center had actually become both intellectually and institutionally part of the mainstream. Just as the Center physically moved from the outskirts of Harvard at 61 Kirkland Street to being an integrated part of Harvard's behavioral sciences in William James Hall, so too did its research agenda move from the edge to the center of what was seen as acceptable science.

There were several markers of the new success and mainstream status of research on cognition. Ulric Neisser, one of the Center's visiting fellows, published *Cognitive Psychology* in 1967. As the first textbook on the topic it indicated the newfound confidence held by those working in cognitive science. Neisser opened the book with the comment that "a generation ago, a book like this one would have needed at least a chapter of self-defense against the behaviorist position. Today, happily the climate of opinion has changed, and little or no defense is necessary."[73] As institutional markers of this success, Bruner and Miller chaired Harvard's Department of Psychology and were elected presidents of the American Psychological Association. In addition, their students secured significant jobs, and increasing journal space was devoted to cognitive questions.[74]

Although the research programs at the Center for Cognitive Studies were successful and played a significant role in making space for discussion of mind within experimental psychology, and although the Center's researchers would go on to leading roles in cognitive science—founding journals and founding research institutions often on the model of the Center—the long-term effect of the Center's work and culture would play out outside the Center.[75] As an extradepartmental research unit, the Center depended on soft money and the continual grant-raising efforts of its leaders. Eventually both Bruner and Miller tired of Harvard and their administrative responsibilities (as directors of the Center and also as successive chairmen of the Psychology Department). After George Miller departed for Rockefeller University in the fall of 1967 and Jerome Bruner for Oxford in 1971, the Center, lacking sufficient support from the Harvard administration for continued existence, closed its doors in 1972.

Research on cognition had become easier in part because of the Center's success. But with this success came less concern with the interdisciplinary effort involved in learning to use unfamiliar tools. Looking back on these events, Bruner reflected:

In retrospect, I think that as time went on, fatigue pushed us toward more intrinsic problems. . . . The task of intrinsic revision was sufficiently demanding to shoulder aside much of the more extrinsic, interdisciplinary preoccupation that had initially inspired us. That extrinsic work took on the character more of an avocation than of vocation.[76]

Once they were faced with several successful research programs, the Center's directorship could not maintain both group cohesiveness and multiple modes of work. They remarked at the time:

We are suffering the wages of success. . . . We are no longer the closely knit "primary group" that we once were. We no longer know each other's business as well as we once did. . . . There has been more tendency toward autonomy among developmentalists, among psycholinguists, and so on.[77]

This was a remarkable change in a place where the directors had argued that an essential feature of a research center was precisely the ability of its members to know and participate in one another's work.

Conclusion

Once the Center helped turn cognition into a respectable research topic, its members lost much of the social and intellectual glue that had come from their earlier sense of being embattled. Without this sense, without the intimacy afforded by isolation in a small house at the edge of campus, and without funding and administrative support, the Center's members became more multidisciplinary than interdisciplinary in orientation, focused less on one another's toolkits, spent less time exchanging ideas and arguing with each other, and concentrated more on pursuing their own research.

Ultimately, the productivity of its researchers eventually played a role in undermining one of the Center's primary purposes—to be a seedbed for the creative fashioning of tools to study the mind. Because of the ethos that equated interdisciplinary tool exchange with creativity, an ethos that the Center built into its initial structure, the Center's eventual multidisciplinary research culture represented a failure that was produced in part by the Center's own productivity.

In the arc of its life, the Center for Cognitive Studies embodied several significant transformations in the nature of the academy. Most broadly, the Center for Cognitive Studies was part of an explosion of interdisciplinary units in universities in the second half of the twentieth century. For these organizations, as for the Center for Cognitive Studies itself, extradepartmental status was linked both to filling the interstices between traditional disciplinary scholarship as well as to advancing scholarship deemed important, creative, and innovative.

At the level of its own field of study, the Center at Harvard was instrumental in establishing cognitive science as a permanently multidisciplinary field. Programs such as that at the University of California, San Diego were built by Center alumni on the model of the Center for Cognitive Studies. Later, George Miller would participate in a study by the Sloan Foundation that both identified interdisciplinarity as a central characteristic of cognitive science and also helped frame the foundation's measuring stick for determining the distribution of the funds it would supply to help establish programs in cognitive science around the nation. Perhaps because of these trends, cognitive science maintains itself as fundamentally and inherently multidisciplinary. Unlike areas such as molecular biology that began as interdisciplinary collaborations and then matured into full-fledged disciplines, cognitive science remains proudly multidisciplinary. In so doing it retains the marks of the intellectual values that equate creative research, open-minded inquiry, and interdisciplinarity.

CHAPTER SEVEN

Cognitive Theory and the Making of Liberal Americans

"What is human about human beings? How did they get that way? How can they be made more so?" These were the motivating questions for "Man: A Course of Study" (MACOS), an elementary social studies curriculum organized in 1964–65 by Jerome Bruner with funding from the National Science Foundation (NSF) and the Ford Foundation. In running this curriculum project, Bruner would assemble organizational practices, pedagogical expertise, and psychological knowledge that pervaded Cambridge intellectual culture and that he had personally developed over the previous twenty years.

MACOS was an applied science project that offered an opportunity to bring cognitive scientific knowledge to the wider world. Cognitive psychology provided MACOS a pedagogical program that guided the sequence of its course materials, its mode of presentation, and even the course's desired outcome: that is, the kinds of learning that the fifth graders enrolled in the course would hopefully achieve. In shaping learning, MACOS undertook a more ambitious plan: making its own students more fully human.

As with the Center for Cognitive Studies and cognitive science more generally, the participants of the MACOS project were an interdisciplinary group. MACOS also adopted one of the central intellectual components of cognitive science: the equivalence of people and scientists. It did so by treating the classroom as a place to do social science and elementary school children as though they were little scientists. Children were to be taught to become social scientists and to think better by giving them as nearly as possible the raw materials of social science, including unnarrated ethnographic films of Netsilik Inuit and of free-ranging baboons. Due to the commitment to seeing all humans as scientists, the answers to the course's first two questions (What is human about human beings? How did they get that way?) would be given by the most recent findings in the social sciences.

By treating students as little social scientists, MACOS also provided one implicit answer to the third question: how to make people more human. The answer was to make people more scientific. MACOS made its students more human by inculcating a scientific attitude in them.

The course's third question—how can people be made more human?—indicates both its normative character and the politics associated with the sciences of human nature. MACOS's project of making people more human—or at least asking how they could be made so—emphasized how queries about the nature of human nature were moral and political even when framed in ways that were, nominally at least, strictly empirical. The course would seek to project and implement a future for people and their societies that would make all people more human. This project necessarily involved identifying some characteristics as essentially human and others as not and then seeking to emphasize and reinforce the former.

Previous chapters have examined how both academic society and the centrist political culture of the early Cold War period valued and promoted specific mental attributes as, simultaneously, rational, democratic, scientific, and creative. With the help of cognitive science, these attributes came to serve as more than features of the elite intellectual or of the American character as desired in the salons of Cambridge, New York, and Washington, D.C. Cognitive science had made these attributes characteristic of all humans.

MACOS helped the codification and propagation of this cognitive worldview well beyond the circle of people who read or were affected by research and popularizations of cognitive science. At its peak in 1972, MACOS was taught to 400,000 students in 1,700 schools.[1] It was further used as a training device in teachers colleges. Thus, even teachers who were not going to be teaching MACOS would imbibe its view of learning, the mind, and human nature.[2]

The ability of MACOS to draw both federal and private patronage and its rapid uptake in schools around the country need to be understood against the background of pedagogical thought and curricular programs in the 1950s. Viewed in this light, MACOS stands not so much as an idiosyncratic development but rather as an outgrowth and consolidation of existing intellectual, pedagogical, institutional, and political movements. Bruner's work on this social studies curriculum was part of a wider effort undertaken by research scientists to design courses in their own fields for precollege education.[3]

Education for National Defense: The National Science Foundation

To understand the aims, goals, and pedagogical method of MACOS, we must begin with the politics of education in the early Cold War. In the early 1950s, precollege education came under assault on a number of fronts as many Americans felt that schools were failing in their job of properly educating children. Much of the animus was directed against the curricular programs and pedagogical designs of the progressive education movement and its inspiration, John Dewey.

Dewey had seen educational reform as a route to making American society more fully democratic. He envisioned an educational program that looked to develop autonomy on the part of pupils by equipping them with a set of mental skills. The classrooms themselves were intended to be laboratories of democracy. They would be communities that would train children to be democratic citizens by teaching them what to Dewey was the scientific attitude—active learning, creativity, tolerance of different ideas, and free inquiry. Classes would be structured by beginning with the actual experience of the children. Through practice in such activities as woodworking, sewing, and cooking, children would learn the academic and disciplinary fields such as reading, writing, and mathematics that were the backbone of traditional education.[4]

Pedagogues who supported "progressive" education carried Dewey's banner, if not his ideas. In the 1930s and 1940s, progressive education focused on courses that involved practical "life-skills" and "adjustment" to the modern world. These curricula, however, paid less attention to two critical aspects of Dewey's educational views. First, teaching practical topics would not, as in Dewey's program, serve as techniques that would lead to competence in academic skills. Instead, topics such as how to be a good consumer, getting along with other people, and the protocols of dating were ends unto themselves. Second, progressive education often substituted the children's adjustment and conformity to society for Dewey's idea that schooling would be a means to enact social reform.[5]

Life adjustment and progressive education helped stimulate a conservative and reactionary political movement at both the grassroots and elite levels. Because these programs did not focus on traditional subject matter, critiqued traditional teaching methods as authoritarian, focused on discovery and active learning, emphasized child-centered orientation, they all seemed to conservative critics to be a corruption of the proper topics, methods, and ends of education. Conservative pedagogy emphasized students be taught respect for authority rather than self-reliance and focused on basic information rather than methods of discovery.

In the early 1950s, conservatives attacked progressive education as undermining American values. Allen Zoll, an anti-Semite and former member of a fascist group, argued that schools were failing because they increased delinquency and taught relativism, not moral absolutes. At the time, Zoll was president of the National Council for American Education, a group which had ten thousand members and connections to over four hundred local organizations. Zoll was joined by others, such as author Albert Lynd, in the argument that God needed to be at the center of American public education. Conservatives at the local level as well as national groups including the Daughters of the American Revolution, Veterans of Foreign Wars, and the American Legion looked to ban progressive educational curricula and books from school libraries on the grounds that they were indistinguishable from communism. According to grassroots conservatives, including in school curricula such topics as the United Nations, UNESCO's work on international understanding, the New Deal, Eleanor Roosevelt, or General George Marshall was a dimension of communist plots designed to undermine America. Organizing against such presumed communist plots, on the one hand, formed a seedbed for the emerging conservative movement and, on the other, drew support from well-connected conservatives such as William Buckley Sr. and William Buckley Jr. who published a journal, *The Educational Reviewer*, devoted to naming textbooks and topics for schools to blacklist.[6]

Even the very methods that some schools had developed to generate consensus within the community about educational aims disturbed conservatives. Specifically, conservatives identified the use of "group dynamics" to achieve consensus as un-American.[7] Their estimation that such efforts were political activities was, indeed, correct. Social psychologist Kurt Lewin and his followers had developed group dynamics during World War II as a method of increasing intergroup understanding and reducing anti-Semitism so as to improve morale on the home front. It was subsequently adopted in communities around the country as a technique for reducing interracial and religious strife.[8] Experience with research on group dynamics also underpinned the liberal social vision of social scientists like Margaret Mead and others who populated the salons discussed in chapter 5. They believed that a major cause of conflict within and between nations was lack of understanding. It was to encourage understanding that Mead and other contributors to UNESCO activities so vigorously supported small group meetings in localities as well as international conferences.[9] The Los Angeles school system likewise supported group dynamics to reduce international tensions.[10]

At the local level, the combination of pressure from parents and support from national right-wing organizations ended in banning UNESCO publications in Los Angeles and, in a widely discussed case, the firing of Willard Goslin, Pasadena's progressive superintendent, who had proposed integrating the school system and raising taxes to pay for the increasing size of the student body in the early wave of the baby boom.[11] The local Chamber of Commerce participated in driving Goslin out because its members believed that school integration would depress home values.[12]

At the national level, the House Un-American Activities Committee published a pamphlet warning of communist designs on the schools. HUAC's description of communist pedagogy was essentially the same as the conservative account of progressive education—that progressive education was both authoritarian and a disorderly chaos.[13]

Other Americans joined conservatives critiquing the education establishment without, however, agreeing with the diagnosis or cure that conservatives had chosen. For instance, while federal support of education scored behind only communism and progressive pedagogy in the demonology of conservative education critics, Congress was concerned enough about the state of science education in the early 1950s that it repeatedly appropriated more money for the National Science Foundation's precollege education division than the NSF had requested. Typifying this pattern, in 1956, the NSF requested $3 million for training high school teachers. The House Committee on Appropriations responded by allocating $9.5 million with the stipulation that the money *must* be spent on high school teacher training. This increase, coupled with the restrictions on funding, placed the NSF in the strange position of requesting that the Senate *reduce* the amount the House had allocated.[14]

Also concerned about the state of American schools were educators and intellectuals like Robert Maynard Hutchins, James Bryant Conant, and Arthur Bestor, a historian at the University of Illinois. Bestor was perhaps the most famous critic of life adjustment curricula. In *Educational Wastelands* (1953), he argued against the practical and anti-academic focus that had come to characterize high school education.[15] Bestor, Conant, and Hutchins all opposed the conception that the value of education lay in its ability to deliver skills useful for life. Instead, they both pushed a vision of education which functioned to generate insightful, liberated minds.

By the middle of the 1950s, there was movement from within the scientific community and within the upper echelons of the NSF to address the perceived problems in precollege education. What was significant was the perspective that the best way to teach students science was to teach them to think like scientists. To the authors of these curricula, this point had several implications. It was, first of all, part of an effort to highlight the humanistic and creative nature of the scientific process. As with the general education programs, these curricula emphasized not the learning of content but rather improving thinking skills and disciplining the mind. According to this view, doing science was not rote learning or memorizing facts as much as developing an aptitude for systematic inquiry, judgment, and the evaluation of evidence. Real science involved creative judgment. On the other hand, the opposite of proper thinking was rote memorization or unreservedly swallowing as truth statements from authority figures.[16]

What distinguished these programs from the NSF's initial entries into education was that they did not necessarily aim to produce more scientists for the nation. Instead of a pure instrumental end, the virtues that the NSF curriculum projects sought to foster were social and political. They were aimed to make the nation more liberal and democratic. The first of these curricula, the Physical Sciences Study Committee (PSSC), was organized by Jerrold Zacharias, an MIT physicist who believed that teaching Americans to be more rational and to think more like scientists would lead them to think properly and resist being "molded" by Joseph McCarthy.[17] This was an issue close to his own heart. In part because of his testimony that J. Robert Oppenheimer be allowed to keep his security clearance, Zacharias had been enough of a target of McCarthyism that he developed a consulting career outside of his work at MIT as an insurance plan in case he was fired from his university post.[18]

Zacharias was not the only scientist who took proper education as a matter of fighting McCarthyism. For instance, in a review of a book on Willard Goslin's firing as the Pasadena superintendent, James Conant characterized Goslin's opponents as "reactionary" "forces hostile to public education."[19] Noting their "orgy of misrepresentation," Conant compared Goslin's critics to the illegal lynch mob in *The Ox-Bow Incident* (1943). If that was not inflammatory enough, Conant argued that Goslin's critics were like the organized "totalitarian" mob that in 1791 had been "stirred up," in the name of "Church and King," to burn the house and scientific laboratory of Joseph Priestley, the prominent English scientist, philosopher, and Unitarian theologian who had defended the French Revolution. In this regard, Conant noted that the arson and the treatment of Goslin was simply a "blind fury provoked by a few people bent primarily on doing mischief in the name of conservatism and patriotism." After comparing conservatives to royalists, Conant went on to show that both the Communist Party and Goslin's attackers saw John Dewey as intolerable. As Conant put it, the Communist Party and conservatives shared "an identical technique—the arousing of emotions by words which are repeatedly so twisted as to have only evil connotations in the minds of certain types of readers." For Conant, conservative attacks on the schools were not only irrational, they were activities engaged in by a dishonest minority seeking to enhance its own power. Conant's views were backed up by studies that showed the groundless nature of specific charges made against Goslin.[20]

While Conant saw attacks on Goslin as ideologically driven falsehoods, leaders of the science curriculum movement sought to inoculate Americans against the pathological separation from reality that Conant diagnosed. The vaccine would be inculcation of scientific thinking in American citizens. Properly inoculated by science in elementary and secondary school, young Americans would grow up to build a truly liberal democratic culture based on open-mindedness and respect for both facts and reality.

Given science education's preeminent position in this vision, it was critical to adopt the proper view of what science actually was, how it operated, and how science should best be taught. Most directly, this perspective meant eschewing teaching methods that were then common. First, Zacharias felt that involvement of research scientists in curriculum design would turn pedagogy away from memorization of received facts or a rigid notion of scientific method and refocus classroom activity toward teaching students by giving them a closer feel of what it was like to actually do science as a researcher when the correct answer was not already known. In so doing he took a page from the 1945 programmatic statement, *General Education in a Free Society* (the "Red Book") discussed in chapter 1.[21] Second, PSSC avoided what Zacharias and other university scientists understood as a false and simplistic view of scientific method common in schools, which they blamed, unfairly, on John Dewey.[22]

In effect, this philosophy of science meant designing educational programs that were, implicitly at least, anti-behavioristic. Zacharias's curricula emphasized discovery and inquiry-based learning and was associated with cognitive and gestalt psychology. Behavioristic learning theory, on the other hand, emphasized learning by trial and error. For instance, when the behaviorist B. F. Skinner designed machines that might replace teachers, the object of his curricula was not to generate insight or understanding in his students, nor did he organize the course to use students' insight as a means to the end. Instead, the goal was simply to have students produce the proper answer when given a particular prompt. The leaders of the scientific community largely recoiled from rote instruction and, even more, found memorization of given facts and formulas contrary to the core of what they saw as the most important feature of the scientific mindset: discovery. Discovery-based learning, they argued, gave students "a far greater grasp of the basic principles and procedures, the habits of mind and methods of approach, of a particular field than has been true in the more conventional courses of instruction in science and mathematics." Discovery-based learning was so important to this curriculum movement that when its leaders summarized the previous six years of their work in 1963, they contended that the emphasis on discovery was even more important than was the fact that they and other university faculty had been involved in the movement.[23]

After the Soviet launch of Sputnik in 1957, this view of science pedagogy drew a massive increase in federal patronage. Americans explained Sputnik less by reference to Soviet accomplishment, or by blaming the scientific, military, industrial, and government establishment for inadequate planning and production, but instead by focusing on the inadequacy of American education. This view, although it caused a defensive reaction among some educators, sharpened the perceived need for federal funding of education. The perceived crisis of education overcame the concerns of Republicans like Eisenhower, who had previously opposed federal aid to schools, except for building projects, on the grounds that the national government would inevitably dominate this area intended for local and state control, and it led to the passage of the National Defense Education Act of 1958 (NDEA).[24] The passage of this legislation marked an expansion of public support for national involvement in education. With this new space to operate, the NSF expanded its activities in curriculum design. Over the next few years, it funded the Chemical Bond Approach (CBA), the Biological Science Curriculum Study (BSCS), and the Chemical Education Material Study (CHEMS).[25]

In 1959, Jerrold Zacharias and Francis Friedman (also in physics at MIT) orchestrated the entry of the National Academy of Sciences (NAS) into the curriculum reform effort. With support from the NSF, the RAND Corporation, and the United States Office of Education, the NAS organized a ten-day conference at Woods Hole "to discuss how education in science might be improved in . . . primary and secondary schools."[26] Of the thirty-four attendees, seventeen were leading researchers in physics, mathematics, and biology. Only two were high school teachers, and there was only one representative of teacher-training schools. There were also seven psychologists in attendance, one of whom, Jerome Bruner, served as director.

That conference included a disproportionate number of nonbehaviorists as well as an underrepresentation of behaviorists in the field of psychology. The nonbehaviorists were Bruner, George Miller, and Richard Alpert from Harvard, Donald Taylor from Yale, Robert Gagne of Princeton, Lee Cronbach from the University of Illinois, and Bärbel Inhelder from the Institute Rousseau. Kenneth Spence was the only behaviorist of any stature in the group.

This effort led in 1960 to Bruner's *The Process of Education*, which argued that any subject could be taught in a serious way to children at any age. This statement articulated a pedagogical philosophy that had been brewing throughout the 1950s. It built, for instance, on sentiments that underlay the development of the New Math and the School Mathematics Study Group (SMSG), the NSF curriculum that most widely distributed the New Math approach. This approach to teaching mathematics sought to bring to elementary and middle school students the fundamental principles of mathematics, including set theory, abstract algebra, and topology.[27]

The New Math, SMSG, PSSC, and the other NSF curricula embodied and propagated throughout the nation the principle that everyone could and should think like a scientist—that is, a scientist as understood by the leaders of this movement: an antipositivist, a creative, open-minded, humanistic individual.[28]

The NSF spent over $200 million to support the development and propagation of fifty-three curriculum projects between 1956 and 1975 following these principles. Aside from these projects, the NSF also reached into schools via its summer institutes for teachers, spending over $350 million on teacher training.[29] These institutes served the function of disseminating the new curricula and pedagogical methods developed by the NSF. By 1975, 80 percent of NSF teacher-training activities were directed toward materials and methods developed under NSF funding.[30]

As a group, these curricula distributed the cognitive and liberal view of human nature throughout the nation. Further, Bruner's book *The Process of Education* (1979) sold over 400,000 copies between 1960 and 1979.[31] Many of these copies were distributed to teachers in training. In fact, a survey held in the early 1970s found that *The Process of Education* was the second most widely assigned book for teacher training.[32] By 1980, another survey found that *The Process of Education* was fifth in the list of texts rated to have had "major influence" on the American curriculum since the turn of the twentieth century.[33] Both Bruner's work and NSF curricula influenced many teachers. In turn, those teachers brought the cognitive viewpoint to generations of elementary and secondary school pupils.

A Pedagogical Psychology for the Science Curriculum Projects

By 1960, Bruner had been thinking seriously about the role of mental processes in education for several years. Since his tenure at the Institute for Advanced Study, Bruner had been exploring concept attainment. This work, so clearly related to educational questions, had culminated in *A Study of Thinking* (1956).[34] In the second half of the 1950s, he and the others at the Harvard Cognition Project moved subsequently to exploration of the role of blocks (impediments) in the learning processes of elementary students. This work, conducted primarily through observing school settings rather than through the laboratory exercises Bruner was more expert in, had turned out to be a dead end.[35]

Perhaps most significantly for the natural scientists he worked with, Bruner held views of learning and mental activity which supported their image of what counted as true science. For instance, Bruner found in his work on thinking that people learn most efficiently when working to code and categorize the information they encounter. He argued that his subjects operated most effectively not when engaged in rote memorization but instead when actively and thoughtfully operating on their experiences. Perhaps even more important to Bruner was noting that when people see their task as more than rote memorization, they are transformed by the efforts they make to "recode" the information they encounter.

The idea of "recoding" as a psychological process came from George Miller's work on information theory as a tool for psychological research. Miller's finding was enthusiastically adopted by Bruner, who began to investigate how humans change themselves when they recode.[36] He found that people learn strategies for discovering the "defining attributes" of categories. Critically, Bruner argued, once they achieve success in one case, "new instances can be recognized with no further learning and memory of the instances already encountered need no longer depend on sheer retention."[37] This point was significant because it highlighted the effect that learning had on people. It literally made them better thinkers.

However, for this most critical transformation to occur, the learning had to take place in the right way. The student could not be in a situation in which memorization of facts was the understood goal. Instead, the goal had to be the extraction of underlying principles through which information already known could be recoded and future information predicted.

Unlike the behaviorist theories of learning, Bruner's cognitive model paid particular attention not to the content of material learned but to the learner's mindset and to the underlying structure of the subject matter. As Bruner argued, "when one goes beyond the information given, one does so by virtue of being able to place the present given in a more generic coding system and that one essentially reads off from the coding system additional information either on the basis of contingent probabilities or learned principles of relating material."[38] Thus, arguments for education by learning subjects in depth, and for learning through acquisition of modes of thinking rather than by learning enormous amounts of material, found expression and support in Jerome Bruner's branch of experimental psychology.[39]

As well as depending on contemporary cognitive psychology, the new science curricula also relied on pedagogical methods that differed from two other approaches: behaviorism and the pedagogy of John Dewey and his followers. Bruner and others in his circle like Zacharias consistently highlighted the difference between their own pedagogical visions and those both of behaviorism and of the progressive educators inspired by Dewey. Specifically, Bruner and Zacharias inverted the progressive educators' efforts to teach science and other fields by exposure to practical activities or to the technologies of everyday life. Zacharias and the PSSC team thus struck discussion of everyday technologies from their curriculum to focus directly on theory first.[40] For his part, Bruner advocated teaching children about the world through direct exposure to the highest levels of research in various fields. Thus, for instance, the team that designed MACOS looked to teach grammar to children by introducing them to the concepts of phrase structure and transformational grammars.[41] These topics were, at that very moment, under active research at the Center for Cognitive Studies by George Miller, postdocs, and Noam Chomsky, who was then on leave from MIT.

For the NSF science curricula the strategy was the opposite of the progressive, Deweyian model of teaching children about physics by undertaking projects on a farm.[42] This is why children in the New Math curricula would, from an early age, learn concepts from the cutting edge of mathematics including set theory and abstract algebra. As Jerome Bruner put it in a lecture explaining the obsolete nature of Dewey's pedagogical and social theories, it was counterproductive to teach by starting from the experiences and interests of the children as Dewey had suggested. Dewey, Bruner argued, misunderstood "what knowledge is and how it may be mastered." Defending the New Math, Bruner argued: "if set theory—now often the introductory section in newer curriculums in mathematics—had to be justified in terms of its relation to immediate experience and social life, it would not be worth teaching."[43] As we will see below, Bruner maintained the same anti-Deweyian pedagogical approach in MACOS, where the primary way that students would come to learn and think was by starting with the unfamiliar.

Each of the features central to Bruner's experimental cognitive work entered into his summary of the Woods Hole conference. In his experimental work and in *The Process of Education*, Bruner placed enormous emphasis on the importance of teaching structure rather than content or specific skills. Teaching through structure would not only help facilitate

understanding and retention, it would also help students be more proficient in applying their knowledge to new situations.

But, for Bruner, structure functioned in several other ways as well.
Emphasis on structure, Bruner argued, was a good way of developing a
student's interest in the subject material, which was the ideal motivating
technique because inherent interest was better "stimulus to learning . . .
than such external goals as grades or later competitive advantage."[44]

Emphasizing the structure of knowledge also led Bruner to several
other curricular prescriptions. First, access to the deeper structure of a
field's knowledge would facilitate such intuitive operations. Second, in
what has been the most noted claim in *The Process of Education*, Bruner
postulated "any subject can be taught effectively in some intellectually
honest form to any child at any stage of development."[45] There was an
instrumental reason for making this claim—and hoping that it was true.
Bruner suggested that children would be most proficient at learning and
understanding if taught the deep structure and formal properties of a
field—even if taught in an intuitive, nonformal method. For instance,
Bruner pointed out that children have access to "certain topological notions, such as connection, separation, being interior to . . ." before they
have understanding or the ability to deal with either Euclidean or projective geometry.[46] Second, Bruner argued that not only could children
understand anything at any age, but that the more intellectually rich the
material they understood at an earlier age, the faster they would learn
at later stages. Not only could young children be taught complex ideas
early, but doing so would improve their understanding and "mastery" of
subsequent material.[47]

Where Bruner's experimental work had supported the claim and
goals of such educational theorists as James Bryant Conant, Jerrold
Zacharias, and Arthur Bestor, Bruner's specific pedagogical statements
in *The Process of Education*, which arose in part out of his experimental
work, gave further voice to their views. For instance, a significant consequence of Bruner's emphasis on learning through structure was his repeated emphasis on the need to involve experts from the field in question in the education of elementary and secondary school students.[48] As
Zacharias had argued, the best way to teach was to expose students early
on to deeper ideas and ways of thinking in each discipline. And who,
Zacharias argued, was better than subject area experts, such as research
scientists, to point out the essential structures of a discipline's knowledge base?

Building a Social Studies Curriculum Using Cognitive Science

In June 1962, under the sponsorship of the President's Science Advisory Committee's education panel (which Zacharias chaired and of which Bruner was a member), Educational Services Incorporated, and the American Council of Learned Societies, Jerrold Zacharias organized a meeting at Endicott House in order to consider how to bring the science curriculum reform efforts to social studies and the humanities. Before the conference started, reading materials were circulated to all attendees. These materials included proposed curricula and Jerome Bruner's *Process of Education*, which attendees referred to as "the Gospel of Saint Jerome."[49]

This conference was motivated by an earlier ESI-sponsored event that was devoted to improving education and thus economic, social, and political development in Africa. At the conference on African education, it was clear that David Page's methods for teaching the New Math and techniques of the natural science curricula offered good direction for African education. On the other hand, American academics had not developed any comparable curricula or pedagogical methods for the social sciences and humanities. It was this lack that the 1962 Endicott House conference would seek to remedy.[50]

The meeting began on a divisive note in its very first paper. The sociologist Robert Feldmeser criticized social studies classes for their orientation toward "indoctrination" and efforts to "defend or extol the virtues of our society."[51] He noted that this kind of curriculum was antiscientific because its content was neither true nor false. He added that this "antiscientific" approach to social studies was a consequence of insufficient attention to the behavioral sciences and to too much dependence on the field of history.[52] "We shall make no progress," he stated, "in transforming the social studies into social science until we slaughter the sacred cow of history."[53] The historians reacted defensively. Zacharias, dismayed by the difficulty he faced getting accord among social scientists (and having faced less political problems among natural/physical scientists), called on the MIT historian Elting Morison to help broker an accord. Morrison asked Bruner for help and Bruner suggested a mode of work which he, Zacharias, and Morison had found productive—interdisciplinary studies.[54] Indeed, the mode of interdisciplinary work Bruner proposed was at that moment working in spectacular fashion at his own Center for Cog-

nitive Studies. It had also been both instrumentally successful and personally gratifying to Bruner, Zacharias, and Morison to work together
on Project Troy and belong to the dinner club that grew out of it.

Working out an interdisciplinary curriculum meant more than solving turf battles among the assembled teachers and scholars at Endicott House. Conference attendees recognized that the interdisciplinary
"value system" adopted at the conference itself was one they sought to
inculcate in the children who would ultimately take the curriculum they
produced. That set of values included self-awareness, tolerance, the "development of the capacity of suspended judgment; the ability to see how
the world looks from points of view other than the one we acquire by
birth," awareness of the fragility of civilization, and the need for social
control mechanisms to protect it.[55]

Courses would be structured by focusing on a particular historical epoch in depth, or "post-holing." Students would be encouraged to
view their material from a variety of disciplinary perspectives in order
to think about literature, religion, art, social structure, and family life.
They would learn essential "underlying concepts or principles of a field
or discipline." Textbooks of predigested facts would be left by the wayside in order that students not be forced to learn "great gobs of knowledge by rote."[56]

The social studies curriculum would also adopt pedagogical methods
from the new and successful courses in natural sciences such as PSSC.
Even when digging a post-hole into the past, students would not learn
in a "linear" fashion and would learn inductively from the "real," "raw"
materials of social science. They would "sort out" the "random" materials and put them in "meaningful order." For instance, in a proposed
course on the Atlantic world for eighth graders, textbooks and even lectures would be replaced by "navigational instruments, diaries, charts,
maps," and "contemporary documents."[57] Students would learn from
these "honest historical materials" about life in the seventeenth and eigh-
teenth centuries by, for instance, using only the instruments and knowledge of the time to "sail an imaginary ship from England to America."[58]

Students would be encouraged to understand that there were no
"right" answers, but instead to think like professional academics and see
their material as the data from which to draw one or more likely conclusions.[59] This approach served several simultaneous functions. First, it
allowed the smoothing of disciplinary conflict between the conference
attendees since no one disciplinary perspective would be favored or ban-

ished from their curriculum. Second, it fulfilled pedagogical goals that Bruner and Zacharias had long championed. Specifically, the post-holing approach fit well with Bruner's emphasis on the role of structure in learning and thought. Thus, rather than teaching history or social studies as a series of events, the post-holing approach would focus on training the mind by emphasizing the deeper structure underlying the available data. Third, the sequence of each proposed social science course fit with recent experimental research by Bruner and colleagues at the Center for Cognitive Studies on the nature and sequence of children's cognitive development.[60]

In addition, children would be taught to see the structure through grappling with their subject matter in the same way that academics are supposed to. The project's interdisciplinary structure meant that they would come to see how there were many (disciplinary) ways to view information and many plausible interpretations. They would learn to think with the particular disciplinary toolkits and thus the "disciplined capacity to see" or "habits of mind" of active researchers in the social sciences or humanities.[61]

This was just the style of teaching that Zacharias had promoted since he initiated PSSC; he had sought to teach physics by encouraging students to think like physicists, to be active learners, to learn to make conclusions from sometimes insufficient information. The group of scholars and educators assembled at Endicott House also hewed to one of the anti-Deweyian pedagogical principles used in PSSC: it specifically rejected beginning teaching from children's day-to-day experience and using the "concentric circles" approach to social studies. Instead of starting with the familiar home environment and moving in stages outward toward the local community, nation, and the world, social studies would be taught first by using contrast and the unfamiliar.[62] As Bruner put it, there was no reason to assume that the child mentally "lives" at home. He or she is just as likely to be engaged in fantasy and imagination as engaged with immediate surroundings. Bruner added a metaphor that would become a refrain: the fish is the last to discover water. He used this point to argue that students could potentially be most interested in and best taught through unfamiliar material.[63]

Following the conference at Endicott House, a committee of a dozen senior scholars, educators, and staff at Educational Services Incorporated looked into translating the Endicott House ideas into specific humanities and social studies curricula. In order to do so, members of this committee decided that that they first had to articulate the fundamental nature and dynamics of human society and history. But in order to accomplish that task, they needed a common "language" with which to speak to one another.

In order to design this interdisciplinary language, committee members met weekly to discuss and debate which principles of human nature and society they took as a given:

> The discussions that ensued were lively, sometimes sharp, but rarely acrimonious, and always entertaining. Indeed, the weekly sessions were in danger of taking on many of the aspects of a club for intellectual gentlemen, except that there were no overstuffed chairs and no one had gout.[64]

Ultimately, the ESI's Social Studies Program would incorporate most of the features discussed as critical at Endicott House and at the Woods Hole conference in 1959. Social studies curricula should not only teach facts but engage the students actively in thinking about the important moral and social issues of the time.[65] This awareness and engagement would be conducted via encouraging students to think for themselves.

The committee also decided that the discovery-based pedagogical methods would be at least as important as any particular topic that ESI ended up generating.[66] However, it was not only the students who were to learn to think for themselves. So too were other teachers and curriculum developers. The curricula planned by ESI were not intended to be the definitive answer to social studies, but rather were to be "model systems" which would serve as inspiration for other pedagogues.[67] In that way, the social studies curriculum would transform the minds of students, teachers, and curriculum planners.

The social studies curriculum project was divided into elementary, junior high, and senior high divisions. The elementary section would focus on the evolutionary development of humans and their culture until about 2500 BC. The junior high course would examine the development of political culture under the heading of "from subject to citizen." Meanwhile, the high school curriculum centered on the interaction of science, technology, and society in the United Kingdom during the industrial revolution.

A separate division, which came to be called the Instructional Research Group (IRG), was established for evaluating the three curricula. Following the pattern of other ESI curricula, the social studies curricula would be treated as drafts and evaluated for their success. In the case of social studies curricula, success would be measured by the appropriateness of specific course units and pedagogical methods for the particular stage of students' psychological development. Thus, the IRG was run by psychologists and headed by Bruner.[68]

Full-scale work on the social studies curricula began in 1963 with grants to ESI from the NSF ($513,360) and the Ford Foundation ($795,000).[69] With this money, the Harvard anthropologist Douglas Oliver, who was director of ESI's Elementary Social Studies Project, engaged Asen Balikci to help him produce a series of films on the Netsilik Inuit. Oliver hoped that Balikci, who had just completed a dissertation in anthropology on the Netsilik, could help him depict the Netsilik as they had lived when encountered by Knud Rasmussen in 1923. Oliver saw these films as a beginning stage of an "evolutionary" elementary school sequence which would open with presentation of hunter-gatherer cultures and move successively through increasingly "complex" societies, culminating with ancient Greece in the sixth grade.[70]

At this point, the Harvard anthropologist Irven DeVore entered the fray. He argued against Oliver's historical evolutionary approach to human culture. DeVore suggested as an alternative that the curriculum focus on the common features of human societies. In DeVore's scheme, the point of contrast would be not between humans but between humans and primates and baboons in particular, which he had been studying since 1959.[71]

After the death of his wife in the spring of 1964, Douglas Oliver resigned from his position as director of ESI's Elementary Social Studies Program. With Oliver's departure, Jerome Bruner filled his position. On assuming leadership of the elementary school curriculum, Bruner dropped the focus on the evolution of human culture that Oliver had emphasized. Instead of focusing on progress of cultures, the course would emphasize the common aspects of human nature. He also directed the course to use DeVore's suggestion of highlighting the distinctive aspects of human nature by contrasting it with animals' nature.[72]

In 1964–65, Bruner took sabbatical leave from Harvard and focused his attention on the ESI curriculum. Bruner would manage the project much as he and George Miller had been managing the Center for Cognitive Studies. Much of the Center's early success rested on its fluid, vigorous interdisciplinary activities, its content discussions, arguments, and seminars involving participants from a wide variety of fields. Bruner would seek to replicate this atmosphere in his curricular work. He organized a group which not only could design a multifaceted curriculum, but one that was itself representative of a wide variety of backgrounds.

By all accounts, Bruner achieved his vision of interdisciplinary collaboration on the MACOS design team. Much like the interaction Bruner had fostered at the Center for Cognitive Studies, the MACOS design team worked in an environment that was enlightening and invigorating. Howard Gardner would recall:

> The intellectual atmosphere was tremendously bracing for all of us. Bruner attracted and, by his own magnetism, maintained the enthusiasm of a highly talented gaggle of individuals from disparate spheres of life: administrators, game designers, filmmakers, psychologists, primatologists, linguists, anthropologists, and educators of diverse stripes rubbed elbows across desks during the day and got together frequently, though less formally, at nocturnal parties, at which many names famous in Cambridge and occasional visiting artists and politicians also appeared.[73]

Bruner's emphasis on exploration would be embodied not only in the curriculum but in the curricular design process as well. By organizing an interdisciplinary perspective, by bringing together a group of designers from a multiplicity of backgrounds and encouraging freewheeling discussion, Bruner hoped to spark innovation. As it turned out, this strategy was quite successful.

New ideas and approaches extended not only to the psychological idea which underpinned instruction but also to presentation and pedagogical methods. For instance, Edward Field, a poet, translated Rasmussen's accounts into blank verse; Amy Greenfield, a professional dancer, choreographed a Netsilik hunt which involved having students stalk "seals" by crawling across a gymnasium floor seal-like on their stomachs (to decoy the prey); and Carter Wilson wrote fictional stories based on Rasmussen's accounts.[74]

Timothy Asch, who had experience in anthropology and filmmaking, worked with movies to help generate feelings of immediacy within the student:

> By rear-projecting scenes from the Kalahari on a huge translucent screen and accompanying these images with a tape recording of desert sounds, Asch took children on a trip through the Kalahari seasons of the year and exposed

them to the cycle of a Bushman day. Seated on the carpeted floor of their darkened classroom, the students . . . could imagine themselves side by side with the people they were studying, hearing the soft Bushman language, absorbing the sounds of the Kalahari environment, and listening to a Bushman musician playing mournful songs on his hunting bow.[75]

Another method was the "think-in" (a name suggested by Howard Gardner), a formal mode of small group discussion that would produce improvements in students' cognitive abilities and social skills. The think-in involved a small group of students working to collaboratively answer a question related to course material. For instance: "How would you learn the !Kung language?" Students and think-in leader were supposed to make sure that discussion remained on topic. Their discussion was recorded. The last stage of the think-in involved individual students listening to the tape and summarizing the discussion. This activity was, in significant ways, much like the habits of a product design group that Bruner had previously used as the very model of creative collaborative thinking and problem solving.[76]

Think-ins had a number of benefits. The activity did more than train students to be better conversationalists. It also gave them practice in figuring out strategies for answering questions and practice in distinguishing relevant from irrelevant information. Think-ins also helped socialize shy students into participating in group activities.

For the curriculum development team, the think-ins and their transcripts served as means of getting feedback on how well the rest of the course was working.[77] Once MACOS was fully developed, some of these experimental techniques continued. Teachers were trained to conduct formal interviews of their students that were supposed to serve simultaneously as a means of evaluating the success of the course and of teaching students how to clarify their ideas.[78]

By the end of the summer of 1965, Bruner and his colleagues had completed most of the curriculum for MACOS. Holding to the principles articulated in *The Process of Education* and fitting with general epistemological views of cognitive science, the goal of the course would be to do more than teach information. It would also seek to develop a specific sensibility in the students. As Bruner put it:

It is only in a trivial sense that one gives a course to "get something across," merely to impart information. There are better means to that end than teaching. Unless the learner also masters himself, disciplines his taste, deepens his view of the world, the "something" that is got across is hardly worth the effort of transmission.[79]

That sensibility, worldview, and disciplining of taste that MACOS sought to inculcate was a joint product of its subject matter and pedagogical methods.

The course centered on three questions "What is human about human beings? How did they get that way? How can they be made more so?" The course material did not give explicit answers to these questions. However, because of the course's commitment to discovery-based learning, these questions were the guiding framework for the course and were asked of students as well.

The course design did point students toward certain kinds of answers. A fundamental principle was human equality. This perspective was grounded by cultural relativism and by the common biological characteristics that humans shared. Comparison and contrast with other species such as baboons, herring gulls, and salmon would be used to highlight what was unique about people and what all people share. Exposure to a foreign culture, that of the Netsilik, would be a way of making children critically aware of their own culture. This was an application of the principle that Bruner had articulated at the Endicott conference: that the fish is the last to discover water. Just as fish need to leave water to see it, children need to leave their own specific culture to comprehend culture as a generic human creation.

MACOS provided several clear and unmistakable answers to what was human about human beings. One was the unique ability to use symbols.[80] Since a primary goal of the course was fighting racism in America by teaching children tolerance, a second answer to the question of human nature was simply that humans have culture and that each culture is equally deserving of respect.[81] For instance, children were encouraged to see the value in the Netsilik, in their cosmology and mythology, and to see these as functionally equivalent to cosmology and religion in modern America. A third goal of the course was to show the coherence and connectedness of human cultures and ways of life. The course's interdisciplinary structure helped promote that end.

The course held that because the characteristic feature of humans is the possession of culture, the differences between people could not be understood according to the principles of natural selection or biology. Instead, to answer the question of how people came to be human, MACOS took the view that "the five great humanizing forces are . . . tool-making, language, social organization, the management of man's prolonged childhood, and man's urge to explain."[82] Tools and culture and not biology were now responsible for human nature and evolution.[83]

MACOS adopted a mentalistic approach to explaining these humanizing forces. Thus, tools were more than implements for helping solve a particular task. MACOS underscored how the specific form of tools was less important than the function that they performed for the toolmakers and users. In turn, tools were to be seen as expressions of culture and could be entirely immaterial—e.g., language. On this view, individual tools were expressions of a specific "cultural program."[84] Similarly, "man's urge to explain" was treated as a need to provide symbolic expression in myth, religion, or science.

MACOS focused on the common features of human societies by looking at the interrelationships of language, technology, mythology, social organization, and childrearing. Bruner argued: "One good example of Eskimo narrative or Eskimo poetry, if skillfully handled in class, can show the child that the problems of an Eskimo are like our problems: to cope with his environment, to cope with his fellow men, and to cope with himself."[85] Thus the connections among parts of a single culture would point to connections between that culture and other cultures.

Taking this approach served the cognitive goals in Bruner's pedagogy. Encouraging students to find the common aspects of different human societies would move them to discovering for themselves the deep structures of human culture. Bruner argued that by watching unnarrated films of Netsilik hunting, family interaction, travel, and toolmaking, by being encouraged to think actively about the material at hand, students would find the relationships between family, myth, childrearing, and technology. They would then be able to "go beyond the information given" about the Netsilik and apply this knowledge to their own society and to the human condition.

To the question of what could be done to make people more human, MACOS offered an immediate answer: they could take MACOS. As Peter Dow, one of the curriculum designers put it, humans had become "in part" their own makers and therefore those who took MACOS would learn how "individual lives, as well as society itself, may be reshaped in order to maximize the attainment of a fuller humanity." According to Dow, "one way man might learn to become more human is to begin by trying to know himself, both by examining his biological kinship with the other animals with which he occupies the world, and by studying the unique patterns of behavior that he shares with men in other cultures."[86]

One tactic MACOS adopted to the end of making people more human was in its very operation and pedagogical methods. Bruner and MACOS posited the view that the distinguishing feature of humans is that we learn.[87] In the context of a curriculum development project, this commitment provided an implicit answer to the question of how people could be made more human. The answer was simple: they could learn more, better, or more effectively.

To improve social studies teaching and therefore the project of making students more human, MACOS took on the philosophy and methods that had been pioneered in PSSC and the New Math and that had been further honed in the Woods Hole discussions, in *The Process of Education*, and in the Endicott House conference. The natural scientists, social scientists, and teachers involved in these efforts had focused on discovery, intuitive learning, and teaching students as though they were scientists. These methods would enable students to develop and improve their uniquely human abilities of creative symbolic thought. Exposure to a variety of "real" materials would not only elicit natural human interest in learning but would also enable what Bruner had characterized as the fundamental human impulse to discover the deeper structures that underlay and connected human culture and knowledge. Bruner distinguished different modes of learning. Some were more likely to allow students to "go beyond the information given" by engaging in that particularly human capability of hypothesis-making and symbolic manipulation.[88]

MACOS retained all of these pedagogical innovations and added twists that were designed for the social sciences. Where the natural sciences programs sought to introduce students to the creative aspects of scientific inquiry, intuition, and inference, MACOS included its own modules designed to cultivate the mindset of the social scientist. Students were given booklets that were imitations of Irven Devore's field notes on baboon behavior and asked to draw inferences from the basic observations. Another assignment asked students to be ethnographers on the kindergarten playground and take careful notes on the occasions and causes of fights, to tabulate the results, and draw conclusions about the origins of conflict.[89] Further, MACOS added its own particular emphasis on instilling a tolerance for uncertainty and ambiguity in its teachers and students. Such a capability would increase humanity by improving cognitive abilities.[90]

If there was a feature that most distinguished MACOS from the other curricula designed by ESI, it was the effort to keep the course culture self-similar at all levels. Thus, the experimental and freewheeling conversations and investigations that characterized the curriculum design group would subsequently be replicated among the teachers who would offer the class around the country. Consequently, when teachers learned about MACOS's unusual content and pedagogical techniques, they were taught to interview their students and, just as the IRG had done, to observe other classrooms, conduct pedagogical experiments, and confer with other teachers about how best to promote desired intellectual skills in their students. Just like MACOS students, the teachers were not to be trained through rote learning. They would learn to teach MACOS thorough the experimental and discovery-based method. Holding to the principle that people learned best when they had ownership of knowledge rather than simply seeking to absorb wholesale facts handed down from an authority, the people who designed MACOS expected that teachers would ultimately "cast aside" MACOS for courses they designed themselves.[91]

While MACOS was implemented with the end of enabling teachers to be pedagogical theorists and curriculum designers, it similarly looked to enhance the cognitive abilities of the children who took the class. And, just as the MACOS designers were not to be the ultimate authority on pedagogy, so too were teachers encouraged to structure their classes so that students would see them as resources, not authority figures.[92] In this way, it sought to develop skills of conversation and reasoned inquiry among the students.

Conclusion

As a piece of applied social science, MACOS was extraordinarily successful. At the broadest level, MACOS, like the NSF-supported curricula in the natural sciences and mathematics, was a project in the public understanding of science. It brought recent and leading social science to a broadly distributed group of schoolchildren and their teachers. Those who came in contact with MACOS had the opportunity to learn how to use conceptual tools prevalent in the social sciences such as the analysis of structure and function. Along with the other science curriculum projects, MACOS sought to transform the minds of its students and therefore the nation. Social and natural science when taught in the proper way would enable pupils to become responsible citizens. For curriculum designers from the National Academy of Sciences to the American Council of Learned Societies, understanding that learning was a discovery-oriented process rather a matter of memorization would enable Americans to separate truth from falsehood and to critically evaluate evidence, and this would in turn inoculate Americans against the kind of wrong-headedness, reactionary politics, and dishonesty that typified McCarthyism.

The connection between politics and science went deeper for MACOS than it had for other science curricula. Because of its adoption of a pedagogical theory structured by cognitive science, MACOS was structured in a way that fostered a liberal sensibility. It did so by seeking to apply cognitive science to improve learning. Instead of emphasizing memorization of facts and truths derived from authority, MACOS taught its students mental self-reliance. In so doing MACOS aimed to foster tolerance, encourage reason, and promote an attitude of self-confidence and the questioning of received authority. It was these open-minded attributes and not the ability to memorize that cognitive science had marked as distinctively human. Thus, by cultivating the skills of inquiry and discussion and the trait of flexible thinking, MACOS at once made Americans more liberal and more fully human—at least as defined by cognitive science.

The Divided Mind

CHAPTER EIGHT

A Fractured Politics of Human Nature

This book has examined the culture of open-mindedness of the early Cold War period. This culture treated the open-minded American, the open-minded intellectual, and human nature as interchangeable. It was because of open-mindedness's multiple roles that it not only marked the boundaries of acceptable centrist political views but also offered the possibility of applying knowledge about one domain of human affairs to another. For this reason intellectuals from Margaret Mead to Daniel Bell to Clark Kerr could treat the academy as if it were America, intellectuals could apply what they knew about academic conferences to social groups all over the world, academics and administrators could treat interdisciplinary thinking as if it were social tolerance, psychologists could equate creative scientific thinking with normal human cognition, and those designing curricula, whether for general education or MACOS, could seek to support American democracy by cultivating open-mindedness and making students more human.

For a time, this was a robust and self-sustaining cultural, intellectual, and political system that underpinned a sense that America had reached cultural and political consensus. Yet in the decade between 1965 and 1975 open-mindedness moved from serving as an element of cultural cohesion to one that divided Americans one from another. This chapter examines how the cultural web sustained by open-mindedness unraveled at two ends of the political spectrum. Disputes over how to assess intellectual merit, over empirical claims about humans and society, over the relationship of the university in society, and over the very nature of human nature tore apart the cultural web that held Cold War centrist pluralism together.

The first half of this chapter focuses on the late 1960s and examines how the political culture of the Cold War period came apart when the Left enthusiastically adopted the open mind and pried it away from centrist political positions and from the social scientists, intellectuals, and policy makers who espoused them. This transition in the politics of open-mindedness threatened the claims to expertise of centrist policy makers and intellectuals whose open-minded objectivity was now in question. The shift in the political meaning of open-mindedness also ended the once easy equation of the academy and America and divided self-professed liberal intellectuals from one another. Some would continue to promote the virtues of open-mindedness and move to the left. Others would reject the new politics associated with the open mind and, no longer able to control the debate, move out of liberalism into neoconservatism.

The second half of this chapter turns to the politics of the open mind in the early 1970s. While the Left adopted open-mindedness as a political weapon in the 1960s, in the 1970s the New Right used attacks on the vision of open-minded reason that had linked the academy, centrist culture, and human nature as an organizational and movement-building strategy. I focus on the reaction to the place where the culture of open-mindedness was at its most concentrated: MAN: A Course of Study (MACOS). The course had distilled a vision of open-mindedness into a program for improving the world by making its students "more human." Conservatives responded that MACOS's cognitive vision of humans as, by nature, open-minded embedded an anti-American, liberal, and secular humanist agenda. They used opposition to MACOS as a strategy for building their own political muscles and succeeded in reducing the course adoption rate by seventy percent, shutting down federal support for training teachers on the course's pedagogical techniques and ending the National Science Foundation's support of precollege science education more generally.[1]

The Success of the Open Mind and Its Leftward Movement

The frames and methods of social analysis developed during the early postwar years flourished in the 1960s. Concerns about the fracturing of society and democracy, alienation, conformity, and the loss of community and individual autonomy that had animated social criticism of the 1950s continued unabated into the succeeding decade. However, the tenor of that social thought shifted by engaging a more directly critical stance or by bringing to prominence criticism that had been less visible in the 1950s.

Typical of this change in tone was Kenneth Keniston's 1960 analysis of social alienation, which adopted many of the analytical tools and sources of 1950s social science. Like his predecessors who had found America in Harvard common rooms, Keniston evaluated the national society by interviewing Harvard undergraduates.[2] However, Keniston emphasized that he wrote not to praise but to criticize the primary features of American society and their effect on the individual man.[3] As Keniston saw it, the complexity of jobs and the economy, and a "fractured," decaying, disintegrating culture and society led to selves and society that were not "integrated." This was Keniston's diagnosis for the causes of ennui and an alienation that he insisted was a new feature of American life.[4] From this perspective, what was novel about Keniston's approach was neither its data set nor its analytic frame, nor even the belief that the complexity of modern life led to fracture. It was, quite simply, the tone of his critique that set Keniston apart from his predecessors.

Joining him were Herbert Marcuse and C. Wright Mills, who continued the dominant patterns of liberal centrist 1950s social analysis by marking free, autonomous thought as a—if not the—most important goal of social life.[5] But instead of identifying America's special claim to autonomous thinking, they underlined its conformity. Already in 1955 Mills had argued that the autonomous mind was critical for democracy. On his analysis, intellectuals' lack of autonomy meant that "knowledge does not now have democratic relevance in America."[6] As with centrist critics of authoritarianism, Marcuse critiqued behaviorism as inimical to freedom and, more specifically, to autonomous thought. However, unlike men such as Jacob Bronowski and Arthur Koestler, Marcuse used his attack on behaviorism as a critique of modern America itself. For Koestler, criticism of behaviorism functioned as a discussion of an abnormal, but not pervasive, malady of social life. On the other hand, for Marcuse, behaviorism was a mechanism through which the totalitarian aspects of modern society, America included, had been achieved.[7]

Members of the growing student movement likewise adopted the themes and analytic frames of centrist social thought. The New Left critiqued rigidity, conformity, and dogmatism while calling for flexibility, individualism, and freedom.[8] Much like the 1950s experts on creativity, Students for a Democratic Society (SDS) contended that specialization was a danger to democratic culture. It saw in the creative, autonomous character not isolating individualism but the ability to unite a fragmented self and to form true community that would be the foundation of a healthy national political culture.[9] In its manifesto, the Port Huron Statement, SDS noted again and again the complexity of modern life and of modern knowledge and reasoned that this complexity led to apathy as well as the loss of democracy.[10] Berkeley's Free Speech Movement (FSM) echoed this point by arguing that America was a "fragmented society."[11] And, much like Marcuse, members of the student movement contended that positivist philosophy and behaviorism were common in the university, identifying this philosophical presence in the academy with problems faced by America at large.[12] SDS and centrist and liberal intellectuals and policy makers thus shared much. Their disagreement appeared most vividly not so much in their expressed goals for what America could be, nor of the foundation of genuine community in creativity, but in their account of the distance between America as it actually was and that ultimate goal.

The Virtues of Social Criticism

During the 1950s, social scientists marginalized significant social criticism both from the Right and the Left by using the tools of social psychology to mark criticism as irrational. Personal commitment to effecting political change and action, a desire to make the world a better place, or certainty on political matters—all of these attitudes had been signals to centrist intellectuals of the authoritarian, reactionary, and closed minds possessed by both Communists and members of the radical Right. Social psychological instruments buttressed this perspective with demonstrations of the irrationality of those who diverted from centrist politics.

In the 1960s, the very dispassion that centrist scholars had treated as equivalent to reason came under fire. Only two years after Milton Rokeach published his book *The Open and Closed Mind*, the authors of the Port Huron Statement criticized the kind of inaction that Rokeach's values demanded. When they called for social and political change, Port Huron's authors criticized the Americans who "regard[ed] the tempo-

rary equilibriums of our society and world as eternally functional parts."
They railed against "the message of our society that there is no viable alternative to the present" and criticized "most Americans," for whom "all
crusades are suspect, threatening."[13] Those positions reversed the kind
of moral calculus that had animated Rokeach and other centrist social
scientists who marked desire to make a difference in the world as an indication of pathology.

This was not a new point but a popularization of an argument that
C. Wright Mills had made the previous decade. Mills had argued that, as
they were commonly construed in the 1950s, the traits valued by centrists
were dangerous to democracy and themselves partly to blame for mass
society. According to Mills, what was branded as "open-mindedness,"
"reasonableness," and "tolerance" was actually nothing of the sort. It
was actually little more than "professional disinterest," "vagueness of
policy," and "lack of involvement in public affairs."[14]

Martin Luther King later bolstered this kind of attack on dispassion
and accommodation of the status quo. His "Letter from the Birming-
ham Jail" of 1963 as well as a 1967 invited lecture for the American Psychological Association underlined the necessity of not accommodat-
ing the existing political and social system. In each case, King made the
point that the greatest impediment to the Civil Rights movement was
the liberal desire to pursue change only through the channels of patient
discussion and debate. King criticized psychology's focus on social "adjustment" and urged that some features of the world, such as inequality,
should not be adjusted to. He proceeded to call for the establishment of
"The International Association for the Advancement of Creative Maladjustment." "Through such creative maladjustment," King argued, "we
may be able to emerge from the bleak and desolate midnight of man's in-
humanity to man, into the bright and glittering daybreak of freedom and
justice."[15]

According to the perspective of the student Left, one of the problems
with centrism was precisely its strategy of labeling as "undemocratic,"
"totalitarian," or "authoritarian" people who simply "say what [they] believe is 'right.'" The Left derided an orientation that demanded that "any
statement no matter how outrageously principled, just and true it might
be . . . has to be hedged." According to this critique, hedging would be
necessary even for statements like "I think black people are the same as
everyone else." (Indeed, so central was hedging to the early Cold War
period that it had been codified by Rokeach as the most important cri-

terion of open-mindedness.) The student Left contended that centrists' commitment to even-handedness was so great that they could deny such simple propositions as "it is possible to decide what is right" in the service of turning a blind eye to America's problems.[16]

Students were trying to expose liberal centrism as an ideology or mode of power (rather than a description of reality or rationality). But the threat to the liberal policy framework came as much from the intellectual ranks as from leftist radicals. In fact, some of the strongest arguments against holding to a policy of dispassionate debate came from intellectuals. They engaged in a thorough critique of the foreign policy establishment's modes of analysis and issued a direct challenge to the claim of intellectuals and government experts that American foreign policy was based on rational, objective, open-minded analysis. The cognitive scientist Noam Chomsky noted that measured discussion over the most effective tactics of the war in Vietnam was, at base, not significantly different than German intellectuals of the 1930s dispassionately analyzing whether it was "technically feasible to dispose of millions of bodies" or if it was "true that the Jews are a cancer eating away at the vitality of the German people."[17] According to this argument war advocates or even those who managed to calmly discuss the Vietnam War were like German people of 1930s, complicit in the activities of the subjects of the Nazi state unless they objected, resisted, or protested.

The force of Chomsky's point came from a call for the application of the set of universalist values that centrist intellectuals and policy makers had professed. Advancing one of the central principles of open-mindedness, logical consistency, Chomsky demanded that both America and the other nations of the world be analyzed with a uniform set of tools. Though he himself called for consistent application of moral and political principles, Chomsky noted that intellectuals failed this test of consistency. He added that the foreign policy establishment shunned those who maintained a commitment to universal analytic principles—and indeed, that the action of applying presumably universal analytical tools to America would mark someone as emotional and not objective. Chomsky contended that the best way to be treated as a serious, rational, and objective foreign policy expert by the establishment was to withhold criticism of America by *not* to being consistent in applying the principles of moral and political analysis.

This line of attack was especially damning for liberal intellectuals and policy makers. Although those like Walt Rostow and Ithiel de Sola

Pool had held up universalism as a cherished value, Chomsky's under-lining of their inconsistent treatment of American actions undermined their claims of objectivity or reasoned analysis of foreign policy.

According to modernization theorists like Rostow and Pool, a pri-mary reason that America stood as the preeminent example of moder-nity was its "universalism." The so-called modern (American) person navigated the world by using universalistic, "rational," or "cognitive" standards rather than particularistic standards based on the relationship of the evaluator to the individuals or objects evaluated. Thus, to judge American actions according to a different set of standards than those that applied to the rest of the world was to adopt a premodern mindset. Doing so demonstrated one's lack of a key trait setting off the American mind as preeminently modern and rational.[18] Consequently, Chomsky's charge that individuals like Rostow were not applying consistent modes of analysis to United States would, by the standards of Rostow's own work, mark the leading proponent of modernization as a premodern, un-American, irrational mind.

Chomsky's criticism of foreign policy experts was a part of a larger cri-tique of centrists' mental processes. Just as McGeorge Bundy and Sidney Hook had criticized leftists and Communists in the 1950s for their dis-honesty and ideology, Chomsky and other leftists charged liberals with a blinkered, distorted view of the world and with blatant hypocrisy in their support of the Johnson administration and its prosecution of the war in Vietnam. They found in centrist ideology precisely the same features as those that policy makers and liberal scholars had once diagnosed among those on the far Right or Left. Leftists offered catalogues of examples of centrist intellectuals' dishonesty—such as McGeorge Bundy's risible assertion that the Johnson administration deserved support from those who wanted a restrained foreign policy because the bombing of North Vietnam had "been the most accurate and the most restrained in mod-ern warfare."[19] Of course, by this time, leftists were far from alone in seeing members of the administration and its defenders as liars. Even mainstream sources like the *New York Times* reported on the dishonesty of the administration and of scholars like Schlesinger who advised it.[20]

A rising wave of social scientists, members of the New Left, and even members of Congress challenged the self-presentation of centrist social-scientific scholarship as objective, apolitical, and nonideological.[21] The issue was widespread enough that it became a matter for congres-sional investigation after a government-sponsored research program on counterinsurgency came to light.[22] James Thomson, a former East Asian specialist at the State Department, noted that the architects and theorists of the United States counterinsurgency strategy in Vietnam were "technocracy's own Maoists," *a new breed of American ideologues who see Vietnam as the ultimate test of their own doctrine.*"[23] In accounts such as these, centrist social science simultaneously stood in the way of truth, presented a false veneer of objectivity, adopted only the semblance of real science, and carried out antidemocratic projects in the process.[24]

Chomsky added an additional critique that not only questioned the expertise of centrist behavioral scientists but also diagnosed their mentality using another criterion that social psychologists had adopted to identify closed-minded reactionaries. As he put it, the centrist members of social and behavioral sciences were engaged in a "desperate attempt to imitate the surface features of sciences that really have significant intellectual content." Their failure to develop knowledge with "significant intellectual content" led centrist social scientists to be both particularly prone to errors of fact and to slavishly produce reports that served the government's programmatic interests. Thus, the behavioral scientists were rather like the authoritarians and narrow-minded conformists described by social psychologists: their ideology substituted for direct access to the real world.[25]

This kind of critique was characteristic of the intellectual life of the New Left and emergent radical thought.[26] It appeared in their leaflets, in reading materials associated with teach-ins, and in essays published by left-wing social critics and intellectuals. It served as introduction to and as continuing themes in public lectures. In this literature and at these events, left-wing social analysis emphasized that it "emphasize[d] relevant questions" and brought to bear "empirical evidence," while centrist social science was depicted as both "facile" and presumably lacking empirical grounding.[27]

According to the Radical Education project of SDS, centrism had turned social scientists into cheerleaders for America and into technicians rather than scientists.[28] Such critics charged that centrist apologies for America depended on mental contortions similar to the adoption of epicycles in support of geocentric cosmology.[29] On this view, centrist social theorists were not only immoral; they were unscientific or, at least, devoted to epistemological schemes obsolete for 350 years; they were bureaucrats, not free thinkers, even when they were employed in the academy. They had abandoned "the true function of social scientists, and indeed all intellectuals," which was "telling the truth, evaluating it, and acting like free men."[30]

Such attacks on policy makers' claims to reason and rationality became staples of leftist critiques.[31] The supporters of American politics, especially of the war in Vietnam, received precisely the same criticism that centrists had leveled against McCarthyite and communist authoritarians in the 1950s. According to the new view, the preeminent examples of intolerance of ambiguity, rigidity, and conformist thinking were the architects and planners of the Bay of Pigs and the Vietnam War, including McGeorge Bundy and Walt Rostow.[32] The authoritarian, "paranoid" style of American politics described by Richard Hofstadter was no longer best found in individuals who supported populist movements or the John Birch Society. The paranoids were now members of the foreign policy establishment who jumped from small pieces of evidence to "hysterical" and fevered imaginings about an international communist conspiracy.[33]

Taken together, these arguments effectively turned tools of centrist social criticism against the people who had fashioned them. While centrists had maintained their objectivity, democratic character, and clearheaded nature, the Left contended that centrist social scientists had all of the moral and cognitive failings that had once been reserved for communists and right-wing authoritarians: antidemocratic character, lack of autonomy, rigid thinking, the substitution of ideology for freedom of thought, and a crippled access to reality. By this point, the social and behavioral sciences had traveled a long way from where they could self-confidently assert that their research was part and parcel of democratic society.[34]

While left-wing intellectuals and radicals critiqued centrist policy makers and social scientists for their ideology, new developments in social psychology indicated that centrist Americans at all levels were closed-minded. By the 1960s, the virtues of the open mind that social scientists had once found only among individuals who were content with the status quo now seemed best located among some of those most critical of American society and especially among members of the left wing. Conversely, the mental handicaps that had been found mostly in those outside of the mainstream were now found in centrists; however, the Left now seemed free from those same maladies.

Social scientists, indeed, began to find that the most politically active and committed individuals—from Black Nationalists to Vietnam War protestors—were better off mentally than centrists. For instance, literature on "cognitive dissonance" moved from examining members of marginal religious cults, members of the John Birch Society, or individuals induced under experimental conditions into holding inconsistent views to studying supporters of Lyndon Johnson. As a consequence, even support of a sitting president could acquire the taint of pathology.[35]

In such analyses, opposition to the Vietnam War and U.S. foreign policy was inversely correlated to both authoritarianism and dogmatism. Those who were most democratically minded and least dogmatic and authoritarian were also those who believed that the war was "unnecessary," "unjustified," "harmful," and "senseless" and were nondogmatic. Even more, it was the non-authoritarians who found the war to be dishonorable and illegal and who were more likely to have participated in antiwar activism. In contrast, those who scored high on scales of dogmatism and authoritarianism found the war to be necessary, worthwhile, or justified.[36]

Psychologists by the late 1960s found that those committed to significant political change possessed the central virtues that *The Authoritarian Personality* had once identified as characterizing the democratic mind. Those committed to social change and critical of the government were more flexible or "tolerant of ambiguity." On the other hand, those with narrow and "rigid" mentalities were more likely to support the use of force in international affairs.[37] As social psychologist Kenneth Keniston put it, members of the New Left were creative, open, flexible, nondogmatic, and anti-ideological.[38]

In fact, tolerance of ambiguity became more than a diagnostic category for professional psychologists. Not only did psychologists start finding tolerance of ambiguity among members of the Left—it also became an explicit value among leftists. It was so important to the movement that its members attacked one another for lacking this trait. One historian has noted that such internecine attacks carried enough weight that they contributed to the collapse of the movement.[39]

This marked a shift in the identification, by social scientists, of which individuals and which community possessed the qualities of character most needed in America. Such a change in the account of who was most psychologically virtuous was aided by the fact that leading members of the New Left, such as Richard Flacks and Paul Potter, were trained in social psychology and conducted seminal studies that compared the psychology of protestors with war supporters and those who supported American foreign policy.[40]

Sexism in the Academy, Racism in America

While left-wing social thought attacked liberal centrists for their narrow mindedness, ideology, and mere pretense of objectivity and expertise, there was a second and perhaps more fundamental development during the period that fractured the functions of open-mindedness in America. This was a sustained discussion about how academics used sexist and racist criteria for evaluating the intelligence and merit of other people. The charge of bias extended from the intellectuals' embrace of prejudice in everyday evaluations of their students and colleagues, to the embedding of prejudice in the very heart of scientific and purportedly objective measures of intelligence. If these charges were correct, then much of the psychological profession, the academy—and by extension, America— was not, as liberal centrist intellectuals claimed, a meritocracy. If this charge was correct, then those who held the most esteemed positions in the academic and policy worlds did not necessarily achieve their posts through their intelligence, merit, or expertise. Instead, their positions were, in part at least, achieved through their connections, social status, or the specific white and male bodies they inhabited. Further, those people who were in positions of power would be responsible for perpetuating a prejudiced, closed-minded system.

One of the most concentrated loci of criticism about academic sexism was in the social and human sciences, especially where these intersected with the second-wave feminist movement. Feminist psychologists deployed their scientific training to underline ways in which sexist criteria often accompanied judgments of whether individual women or men were qualified to pursue work in the academy.

In many instances, however, awareness of institutional sexism began not with a controlled scientific study but with direct personal experience. In some cases department chairs and deans directly expressed their opinion that women were incapable of the intellectual work required of academics. In other cases, the question was whether women belonged in graduate school; or whether women could teach, conduct research without the help of men, or could write well enough to get their work published.[41]

Often, it was only after discussion and the sharing of experiences that women initially understood as strictly personal that women began to form political and institutional critiques of the academy. Subsequently, they would develop and teach social scientific studies that would lend empirical weight to their personal experiences and political positions. This pattern held for psychologists including Rose Olver, Rhoda Unger, and Naomi Weisstein.[42] In Naomi Weisstein's case, the experience of sexism at the University of Chicago became a point of contact with other women in Chicago's Westside Group. Through discussion of the sexism that permeated her professional experiences—which had been ignored by male members of the Left—Weisstein received the support needed to continue her work as a cognitive scientist. At the urging of other members of the group, Weisstein wrote a widely read analysis of sexism inherent in psychology's representation of women.[43]

In an autobiographical essay, Weisstein critiqued the ways men had judged the potential merit of women as academics generally and, more specifically, as scientists. It was this set of informal judgments and prejudices that served as the basis of the differential treatment of women and men in psychology and in the sciences more generally. It was on such informal bases, Weisstein recalled, that advisors at Harvard denied her use of laboratory equipment, with the claim that she would break it.[44] (She subsequently gathered data for her dissertation at Yale, where she was allowed access to laboratory equipment.) Later, a related set of judgments led a job interviewer at the University of Chicago to ask Weisstein, "Who did your research for you?"[45]

First-hand accounts like this by Weisstein and other feminists pointed to a potential problem for the academy. They suggested that the academy's hierarchy was based not on true merit but on prejudice. Thus, while intellectuals from political scientist Zbigniew Brzezinski to historian of science Gerald Holton contended that the academy had been spectacularly successful in using "ingenious social devices" to recruit a sort of "meritocratic elite," Weisstein contended that these devices were simply mechanisms for choosing white men and excluding others from the profession.[46] As she put it, when people judged whether another person was an "excellent" scientist, they did not evaluate "imagination, originality, or even . . . intelligence." Instead, what they judged was only "mannerisms, external appearances, flashy skills, glibness" and "superficial fluency."[47] It was because of these superficial judgments of merit, Weisstein contended, that science was dominated by men.

Although Weisstein's claims about the problems of informal judgments of merit were themselves based on her own informal judgments and personal experiences, they were also mirrored by more formal social scientific studies on the topic. Social psychological studies found that evaluations of individuals and their work were consistently determined by the sex of the person being judged. Whether in hiring decisions or in opinions about the quality of a piece of scholarship, both men and women consistently rated women and their work lower than equivalent men and their work. Studies noted, for instance, that department chairs were more likely to evaluate a curriculum vitae with a man's name on it higher than the *exact same* curriculum vitae with a woman's name on it.[48] Similarly, a paper with a woman's name on it tended to be judged as inferior to the *exact same* work with a man's name on it. Such studies indicated how sexist assumptions permeated hiring decisions and judgments of merit in academia.[49]

Intellect, creativity, ingenuity, and merit are always at stake in decisions around grading students, as well as judgments over hiring, tenure, and peer review of grants and publications. Feminist social scientists argued that these judgments were systematically biased. In making this argument, they directly challenged the claim that intellectual life was conducted on an objective and meritocratic basis and even that the academy had been successful in recruiting and promoting the most excellent individuals.

Among such judgments were the ones most central to the academic community—the evaluation of whether individuals had intellectual merit and therefore belonged at all. Those decisions, typically made without the aid of systematic or scientific techniques, reflected a set of views about the criteria of merit and who it was who best fulfilled those criteria. While such modes of judgment received criticism for their sexism, left-wing scholars also criticized the more formal and systematic techniques that claimed to measure merit objectively: standardized tests, especially IQ tests.

In her most widely circulated political publication, Weisstein drew on the studies of psychologist Robert Rosenthal that showed that both people and non-human organisms in laboratories behave according to expectations. Rosenthal's results indicated a serious problem for psychological science. If the subjects of psychological experiments produce what the psychologists expect of them, then experimental studies are not necessarily evidence of nature or of anything that exists outside the laboratory. The experimental results would simply be a recording of what the scientists believed before the experiment had even started.[50]

Citing Rosenthal, Weisstein contended that standardized intelligence tests do not measure stable entities. She noted that psychologists had found that the expectations teachers had of their students produced a dramatic change in same students' intelligence as measured by nominally objective tests. Students who started out equal became unequal after their teachers were tricked into believing that some students were smarter than others. Weisstein noted that so powerful were expectations that not only humans but even laboratory rats would have their learning abilities affected by what people thought about them. Rather like the students, rats that were initially equal ended up unequal when experimenters falsely believed that some were "maze-bright" and some "maze-dull." That is, both human children and rats behave according to the expectations that other people have of them. They do so even when the expectations are completely unfounded. From these findings Weisstein argued that, contrary to the assumptions of psychology, humans do not have a particular personality or character, but generally confirm what is expected of them. She concluded that psychological tests that show differences between males and females simply indicated the differing social conditions in which males and females found themselves.[51] Weisstein added the contention that when psychologists did find sex differences, what they were measuring was their own sexist assumptions and not something preexisting in their research subjects. To support this claim Weisstein noted the lack of reliable results on sex differences even in the areas of biological function. If expectations could produce variability in rodents' learning abilities, then reports on sex differences were a product of the scientist's own preconceptions.[52]

While feminist commentary and research indicated that the ranks of university faculty had been populated on the basis of subjective, sexist criteria, a separate line of criticism suggested that even nominally objective and scientific measures of intelligence were biased. Because it related to the precollege and collegiate population, this second line of critique raised the possibility that the entire population of the academy—not only the faculty but even the students—was filled not on a reasonable basis having to do with true merit, but on ones having to do with racism. Even more, this critique contended that psychologists who pursued studies of IQ differences were not open-minded, but unscientific racists.

Articles published by educational psychologist Arthur Jensen and behavioristic psychologist Richard Herrnstein offered the occasion for the critique of standardized measures of intelligence. In 1969, Jensen noted that African Americans consistently scored lower than whites on IQ tests despite the fact that it had been more than a decade since the Supreme Court had declared segregated schools unconstitutional. Jensen concluded that "compensatory education" could not change IQ and that IQ tests accurately reflected inherent abilities. Therefore, according to Jensen, racial differences in average scores were a consequence not of problems with the tests or of environmental differences but a reflection of heredity. Jensen concluded that the American educational system should be modified to aim at the intellectual skills and capabilities to which blacks are biologically suited.[53]

Jensen's arguments received wide circulation. For instance William Shockley, a Nobel Prize–winning physicist, circulated copies of the article to all the members of the National Academy of Sciences.[54] Subsequently, Richard Herrnstein, a professor of psychology at Harvard, published a popular companion to Jensen's piece in the *Atlantic*. Herrnstein's contribution to the discussion was to speculate that not only intelligence but also socioeconomic factors were related to intelligence and heredity.[55]

Jensen and Herrnsein's articles elicited a heated debate. Critics attacked their methods, their arguments that intelligence tests actually measured intelligence, their claims that intelligence is based more on heredity than on environment, their assertions of racial difference, and their conclusions of the pointlessness of compensatory education. Many critiques centered on pointing to scientific errors that lay at the root of Jensen's and Herrnstein's claims.[56]

However, the debate ran on a separate level that concerned connections between political ideology and the positions that individuals took on IQ. This aspect of the debate centered on whether it was scientific and apolitical to study race differences in IQ. On the one hand, Jensen and Herrnstein contended that investigation of the question of racial differences in IQ was legitimate. They held that their opponents were anti-scientific reactionaries so committed to environmental determinism and racial equality that they were willing to shut off all rational investigation of the subject. Indeed, there were well-known cases where proponents of the hereditarian position were harassed.[57]

On the other hand, there was possibly a strategic component to the assertion that it was only opponents of the hereditarian view who sought to shut down debate. By adopting this position the hereditarians could mark themselves as open-minded and their opponents as ideologues who were closed to evidence that contradicted their views. Indeed, Richard Herrnstein found criticism of his work so intolerable that he refused to appear in public when he could not be assured in advance that his audience would either be entirely friendly or not express disagreement with him.[58] This sequence of events suggests that Herrnstein's pose as open-minded was less based on his willingness to engage in debate and more calculated for public consumption.

On the other side, critics of the hereditarian position noted that Jensen's research and associated IQ studies themselves depended on a set of dogmatic ideological assumptions, racist or otherwise. The argument was that the work of people like Jensen and Herrnstein was not scientifically viable. Therefore, because this research lacked a scientific basis, it would be interesting only to racists.

This was the position staked out by the journal *Cognition*, one of the first scholarly journals devoted to cognitive psychology. Articles in *Cognition* that attacked the hereditartian view of IQ adopted what was by then a standard critique of the behavioristic and operationist psychology used by Herrnstein and Jensen. As astrophysicist David Layzer put it, Jensen and Herrnstein misunderstood statistical methods, based their arguments on an obsolete strict empiricist scientific method invented by Francis Bacon in the sixteenth century, and depended on a metaphysical commitment to the inability of individual intelligence to change with experience.[59]

Noam Chomsky offered an even harsher critique in the first article ever published in *Cognition*. This article was actually a joint review of Herrnstein's *Atlantic* article and B. F. Skinner's recent best seller, *Beyond Freedom and Dignity*.[60] In this book Skinner had argued that it would be best to stop treating humans as if they were autonomous. Doing so would, according to Skinner, better enable the engineering of the good society.

Chomsky used his review to argue that both Skinner and Herrnstein held visions of human nature that were not grounded in science. Even more, Chomsky contended, their works were "dogmatic," thoroughly illogical, polluted by politics, and unsupported by evidence.[61] When discussing Skinner's book, Chomsky remarked that Skinner's behaviorist research program failed to produce a single nontrivial result. According to Chomsky this failure had stimulated Skinner to produce increasingly grandiose claims about his own success. When focusing on Herrnstein, Chomsky noted that the only reason that some people found racial dif-

ferences in IQ interesting was because America is a racist society. Without pervasive racism, he contended, questions about IQ differences between the races would be no more compelling than questions about IQ
differences according to height.

Here Chomsky added that studies of IQ by race are not scientifically
significant. They contribute nothing to our understanding of either mental processes or brain function. He noted that, lacking such clear scientific relevance, it was the responsibility of researchers such as Herrnstein
to show how they were not racist, to show how their studies would revo-
lutionize our understanding of the physiological relationship of the mind
to the body, and to demonstrate how their findings would not have dele-
terious social effects. Chomsky added that, without such evidence, these
IQ studies were no more defensible or objective than nineteenth-century
anthropological defenses of racial hierarchy or examinations in Nazi
Germany of the biological propensity of Jews to engage in usury. The
point was that Jensen's and Herrnstein's studies were not politically neutral, as they had claimed. Even asking the questions they had indicated a
person's racism.

Chomsky concluded his review by noting that both Skinner and
Herrnstein assumed, entirely without evidence, that people are only motivated by material reward (such as money) or punishment. Without the
assumption that people are motivated only by material reward, the link
Herrnstein presumed between intelligence and wealth would have no basis. For even if intelligence were hereditary, Chomsky contended, there
was no reason to assume that the intelligent people would apply themselves to the accumulation of wealth rather than to some other less well-
paying pursuit.

In making this remark, Chomsky advanced what was by then a quite
common criticism of behaviorism. Psychologists had long contended
that both humans and other organisms seek nonmaterial rewards. For
instance, Harry Harlow had critiqued behaviorism by noting that primates have an intrinsic need for love. Similarly Jerome Bruner had published studies noting the natural curiosity and drive for learning in animals from rats to people.[62]

As Chomsky saw it, this ungrounded focus on only material rewards
and punishments led both Skinner and Herrnstein both to scientific errors and to problematic moral positions. For instance, Chomsky contended that Skinner's vision of the good society did not have a way of re-
jecting National Socialism, concentration camps in which inmates are

frightened into following rules by screams and smoking ovens in the distance (but no direct physical punishment), or universities that promote only those professors who spout officially sanctioned dogma. This was because the sanctions in those situations were entirely mental and harm was done was only to individual autonomy—categories of analysis that Skinner rejected.

If Skinner's behaviorism and Herrnstein's work on IQ were both unscientific and politically pernicious, the path was clear: advance a vision of human nature that was grounded in valid science and proper method. A proper understanding of human nature unpolluted by either racist assumptions or by commitment to the dogma that humans only work or learn with material reward would not only be scientific, it would also open the possibility of a society that enhanced freedom. Chomsky noted that these problems could be resolved by acknowledging that humans are creative, autonomous beings.

Chomsky, then, aligned science with a leftist social vision and an account of humanity defined by its creativity and autonomy. He, along with fellow cognitive scientist Naomi Weisstein and other feminist scholars, underlined that the ways academics evaluated other people were often racist or sexist. Such criticisms undermined the field of psychology and its established measures of gender and intelligence. If such criticisms pointed to bias within the psychological sciences, they also highlighted ways in which the day-to-day operation of the academy was not open-minded but grounded in a sexist bias. The answer they offered was rooting out unscientific studies of human difference and recognition that humans are, by nature, creative, autonomous, and largely equal when not impeded by unjust social conditions. With recognition of creativity and autonomy as the basis of human nature, we would also have access both to the good society and to a coherent and truly scientific view of human nature.

Splitting the University Internally and from Society

As we have seen, many intellectuals of the 1950s took the academic community to be a model for how America could resolve the problems of modern mass society. Likewise, students of the 1960s believed the university could be the starting place for making America a nation that fostered real community and individual autonomy. The difference was that students believed that the university itself needed reform and was actually an instance of or even the cause of mass society. The conflict between students and their elders over whether the university was already an instance of the good society would have dramatic results. It would not only divide liberals into warring camps but also lead social thinkers to emphasize the distance between the academy the rest of America.

Although they valued creativity, autonomous thinking, and community, SDS members and left-wing intellectuals were not so convinced that these traits existed in great measure in America or even in the university. Nonetheless, many students and members of the second-wave feminist movement saw in the university the potential for genuine community, a place where true humanity could flourish and, therefore, serve as a model for the rest of society.[63] But they agreed that the university as it stood in the 1960s had not fulfilled that promise. It was neither a refuge nor a place for nurturing democratic culture by fostering autonomous thought and humane learning.[64] Instead, students saw the university as an impersonal "factory" that treated them as mechanical, interchangeable parts, that promoted "mindless conformity," that indoctrinated students in "middle class morality," and that made them into "twentieth century slaves."[65] While Clark Kerr saw the future in universities' becoming mini-societies, for students who experienced this transformation, the word that Kerr had used to characterize the change—the "multiversity"—became an epithet associated with social fragmentation and the very same problems that social theorists had pinned on "mass society."

The academy's role in promoting the formation of mass society offered members of the student movement an opportunity to work on social change in their own backyard. Students, like intellectuals from Clark Kerr to Daniel Bell, saw the university as a reflection of, the center of, or the engine for modern life.[66] Just as social critics of the 1940s and 1950s treated the academy as a model of society and worried that academic specialization would lead to social fragmentation, so, too, did SDS move easily between noting how the university mirrored society to commenting on how specialization of knowledge led to "parochial" views.[67]

Members of the student movement offered themselves and their own community as the answer to the problems that the nation faced.[68] For SDS, the answer to these problems was "participatory democracy," a vision of political life in which individuals would be directly involved in the making of decisions that affected them. Such involvement would not

only be the basis for a more authentic democracy, it would also be the locus of a true community that would address the alienation of individuals
from one another and even from themselves.[69] Thus, like the intellectuals and policy makers who had found in their own salons, dinner parties,
and conferences both community and the hope for a transformed America, so, too, did the New Left locate the answer to mass society in building communities, both their own and others'.[70]

According to this vision, the kind of community that would stand
against mass society and be held together by autonomous, authentic, creative selfhood would be populated by a specific kind of person. It would
no longer be the intellectuals who populated university common rooms.
The hope for reforming American character and political culture would
no longer reside with groups of elite intellectuals and policy makers
meeting at private retreats. In this new vision, the truly open-minded
people who would save America from the horrors of mass society were
those most critical of the status quo. Accordingly, men of the student
Left noted how direct protest actions produced a cure for many of the
ills that plagued the university, society, and individuals.

The philosophy of participatory democracy and the assignment of
centrality to the academy found fruition in a series of protest actions
throughout the second half of the 1960s. Among these were demonstrations, strikes, and sit-ins at universities from Berkeley to Columbia,
Harvard, the University of Chicago, and the University of Wisconsin.
In these cases, participants argued that universities were limiting free
speech, were perpetuating or exacerbating racism, were complicit in
American war efforts in Vietnam, or supporting an imperialist foreign
policy.

Participating in political action—whether community organizing,
protesting the war, engaging in strikes, or taking over university buildings—led both to true community and to a regeneration of the individual psyches for those involved. These activities not only cured alienation
and social fragmentation but made the participants feel more creative
and more human.[71]

Although movement activities answered the problems of mass society for men, the relationship between political involvement and community was more complicated for many women. They found that the class
and racial oppression the New Left opposed was actually reproduced as
sexual oppression within the movement itself.[72] Their subsequent efforts
to address these concerns were famously met with hostility. For instance,

cognitive psychologist Naomi Weisstein recalls that when her friends Jane Adams and Heather Booth tried to form a women's caucus within SDS, one member of SDS responded by exposing himself. The response of the assembled group was to laugh at Booth and Adams.[73] Such experiences led directly to forming groups devoted to women's liberation. These groups and the communities they founded both produced and were a consequence of growing political consciousness.[74] Ultimately, like male intellectuals of an older generation and like the men of the student Left, women involved in the second-wave feminist moment found their full potential as creative, autonomous beings in tight conjunction with their simultaneous work in political action and community formation.[75]

That students sought to solve the problems of modern society by protesting at universities was threatening to many in the academy. To those who maintained a commitment to the status quo—with respect to, at the minimum, debate—it seemed that members of the black power movement, SDS members, and other antiwar protestors had engaged in the cardinal sin of shutting down rational discussion. To Margaret Mead, the students' demands were simply incoherent. She contended that whereas Russians and domestic Communists had clearly defined goals that could be overcome or managed, the younger generation's mode of social interaction was a "social metastasis," a cancer, designed to be disorganized and disruptive. The younger generation could not be communicated with. They were much like the Viet Cong in that they could not be understood and explained by the set of tools Mead and fellow social critics had developed to analyze enemy cultures from Japan to the Soviet Union.[76]

While the analytic failure of her own mode of ethnography led Mead to believe that students and the Viet Cong had no rational goals, other professors such as Richard Hofstadter and David Riesman contended that the protests, strikes, university takeovers, and very language of the movement bore more than a passing resemblance to precisely the kinds of activities, both irrational and authoritarian, that characterized the radical Right, Communists, Nazis, and McCarthyites.[77]

This disjunction between the younger generation and their elders was not simply a matter of political views. Both opponents and supporters of the war took protest actions by antiwar students to be authoritarian. For instance, Irving Howe, a left-wing intellectual and founder of the journal *Dissent*, took SDS leader Tom Hayden to be a fanatic, rigid authoritarian when they first met.[78] Even a war opponent and supporter of the Berkeley Free Speech Movement like philosopher John Searle remarked: "To accuse a professor of doing secret war research for the Defense Department nowadays has the same delicious impact that accusations of secret Communist party membership did a decade ago."[79] Likewise, David Riesman was a longtime antinuclear activist and opponent of American foreign policy but nevertheless took student protestors to be "closed minded ideologues."[80]

George Kennan, also a critic of the war's prosecution, charged student protestors with the mental deficits that psychologists and centrist social scientists of the 1940s and 1950s had found among Communists and members of the radical Right. As evidence he detailed how the students of the "radical Left" were subverting the true nature of the university. For Kennan the university, properly understood, was defined by Woodrow Wilson at Princeton's Sesquicentennial. This vision of the university involved retreating from the "rough ways of democracy" to a place where one could "hear the truth about the past and hold debate about the affairs of the present, with knowledge and without passion." Kennan claimed that it was clear there was "a dreadful incongruity between [Wilson's] vision and the state of mind—and behavior—of the radical left on the American campus today." He added: "Instead of these affairs being discussed with knowledge and without passion, we find them treated with transports of passion and with a minimum, I fear, of knowledge."

Kennan's charge was more specific than saying that students had let emotion trump reason. He contended that their minds were clouded by "hatred and intolerance"; that, rather like the Communists about whom Kennan was an expert, the students were corrupted by an "extraordinary degree of certainty" about the "correctness of [their] own" "rectitude" and "certainty of the accuracy and profundity of [their] own [analyses] of the problems of contemporary society."[81]

Unmentioned in Kennan's paean to Wilson was that SDS was not simply assured of its own correctness. SDS was opposed to precisely the kind of university and kind of exclusionary politics that Wilson had promoted. During his tenure as president of Princeton, Wilson had successfully reinforced the university's status as a refuge from democracy's "rough ways" by making it the preeminent segregated bastion of WASP privilege, surpassing even Yale and Harvard in racial and religious exclusivity. On his subsequent election to president of the United States, Wilson made similar moves by expanding segregation in the federal government and in Washington, D.C.[82]

While Wilson had made both Princeton and America more segregated, many members of SDS and FSM had been involved in the Freedom Summer of 1964 and the movement to extend the franchise to African Americans in Mississippi. To these students, continued operation of the system as it stood—or even calmly discussing changes to the system—was immoral. Students had become politicized precisely because they had adopted opposition to the view of the university promulgated by Woodrow Wilson and George Kennan. In fact, a spark that set off the FSM protests against the University of California came when the university sought to restrict the efforts of students organizing for civil rights and against racial discrimination.[83] Berkeley students did not, as Wilson did, see the university as a refuge from democracy. Rather, they sought to make the university newly and actually democratic by noting that the university was a space for the exchange of ideas and that, as a consequence, it occupied a central place in the American social and political system.

In seeing the academy as a place for effecting social and political change, the student Left adopted a mode of social analysis that was, by that point, entirely conventional. As we have seen, scholars, social critics, educational reformers, and policy makers had made a regular practice of looking at the university as a microcosm of America. And they had looked to reform of the academy as a way to resolve national problems.

However, that students took the university as central to American life and politics provided centrist intellectuals the opportunity to attack the students in a variety of ways. While members of the Berkeley Free Speech movement contended that they were motivated by the University of California's decision to apply rules selectively that prohibited canvassing for off-campus political activities, Nathan Glazer, a professor of sociology at Berkeley, contended that the movement was not about free speech at all, that the protests were misplaced, and that the university was a mere substitute target for the real problems in America. Glazer's colleague Louis Feuer adopted the strategy of pathologizing opponents by arguing that the protests were simply a form of adolescent rebellion.[84]

Zbigniew Brzezinski implicitly agreed with the students that the university was a microcosm of the nation and suggested a political response. After the protests at Columbia University, he penned an article in the *New Republic* whose very title, "Revolution and Counterrevolution (but Not Necessarily About Columbia)," suggested that the efforts developed by experts such as himself could be used to handle student protests at

universities. In the piece, Brzezinski examined the various ways that institutions of authority could effectively respond to challenges. Whether
they were national leaders facing revolutionary opposition or university
presidents facing student protests, authorities might consider actions including having revolutionary leaders "physically liquidated."[85] Since, to
his mind, the majority of the movement was not very committed, the
movement would wither without its head. This analysis suggested that
the very claims of those in movements on campus or around the world
were more associated with following a charismatic leader than connected with serious or legitimate political commitments.

Daniel Bell, himself a Columbia sociologist, and Glazer, writing in a
journal co-founded and edited by Bell, found Brzezinski's article a bril-
liant analysis. His article fit with their understanding that the protest at
Columbia was led by a cohort of SDS rabble rousers who sought, like the
revolutionaries that Brzezinski discussed, not improvement of the situation but only to inflame conflict with the end of destroying rather than
changing the institutions of authority—whether Columbia University
or some, unspecified, authoritarian government. In so doing, Bell and
Glazer aligned reason with ongoing policies of established institutions
and sought to mark as irrational those who aimed for significant change
in the status quo.[86]

However, there was one point on which Bell and Glazer parted ways
with Brzezinski. Their response to the protests was not to see the university as comparable to the nation but to criticize SDS for making this
linkage. Glazer held that students made a mistake when they saw the
university as a microcosm of America or argued that social change could
happen through direct action within the university itself.[87] If Glazer was
critical, Bell's approach to SDS's views of the place of the university in
America was positively caustic. Bell argued that the conflation of the university with America was a delusion so deep that Columbia students had
become consumed by an ideology like the crowds described by Gustav
Le Bon.[88] While Bell invoked Le Bon only in passing, the reference itself
was telling. Aside from the strikers of late-nineteenth-century France
that Le Bon described, the best known Le Bon–style crowds were those
hypnotized by the lies of Benito Mussolini and Adolf Hitler.[89] Le Bon
had argued that "crowds" were unthinking, irrational, subhuman, primitive, and feminine homogenous masses that could only form when exposed to a contagious and hypnotic lie. Thus, Bell's reference to Le Bon
was a suggestion that the students' "ideology" was no mere cognitive

map of the world but rather a dangerous fiction that reduced their rationality, masculinity, and whiteness.

Bell and Glazer thus condemned SDS (and, in Bell's case, mildly scolded Brzezinski) for doing precisely what they and others had done over the course of the 1950s and early 1960s: seeing the university as a microcosm of America. The difference, of course, was that from the 1950s through the middle 1960s, intellectuals conceived of the university as a place for evolutionary improvement of the features of America that they supported. By the late 1960s, SDS and Brzezinski saw the university and America as sites for revolutionary or, in the case of Brzezinski, counter-revolutionary violence.

Although by 1968 Bell and like-minded scholars were careful to distinguish America from the academy, they apparently had forgotten their own habit of treating the academy as a model for America. At one time, they had seen the university as a place for generating a genuine pluralist and democratic community and also as the place for molding right-minded citizens for the modern and complex world that America entered after World War II. Such a vision was increasingly difficult to maintain as time passed. Once the virtues of the open mind had been claimed by their political opponents and once the university was to be a place for organizing change in America, the university no longer seemed to be the nation's center.

The Right-Wing Reaction

While members of the Left challenged the political system using the very analytical tools and values that had been professed by centrists, there was a parallel, though inverse challenge brewing on the Right. Just as the centrist vision of human nature provided a means for the Left to organize and critique the status quo, so too did that vision give members of the Right a platform, fundraising opportunities, and occasion for building their own political movement. The difference was that, rather than adopting the view that people are by nature creative, conservatives charged that this account was a tool of a liberal, secular plot to corrupt the nation and weaken it so as to provide for the takeover by international forces such as communism or the United Nations.

The storm would break in the early 1970s after the middle school social studies curriculum Man: A Course of Study (MACOS) was introduced outside of its initial trial locations. Many of the features which had made MACOS original—from its reliance on cognitive theories of mind, to its concern with learning by discovery, its foundation on the equality of all people, and its focus on making students more human by teaching them to think like scientists about the world—were precisely what disturbed MACOS's conservative critics.

Controversies in local school boards in Florida, Arizona, Texas, California, Georgia, Vermont, Washington, and Massachusetts flared when small but energized groups of conservatives raised objections to MACOS. The problems with MACOS included both its specific content and its methods of instruction. Although debate over MACOS began at the level of local school boards, it quickly attracted support from the Right's professional political figures as well conspiracy theorists, segregationists, and culture warriors including the Heritage Foundation, the National Coalition for Children, Pat Robertson, James J. Kilpatrick, and Richard Viguerie.[90]

The attack on MACOS operated as a political training ground. It crystallized a series of conservative views and organizational techniques that became characteristic of a newly political Christian Right. MACOS was significant enough that right-wing politicians like Jesse Helms would use newly developed techniques of direct mail to raise fears about the course in order to raise money for their own electoral campaigns. Later, during the 1980 campaign, Ronald Reagan used MACOS as a cudgel against secularism and liberalism.[91] However, national conservative leaders only followed and benefited from the organizing conducted at the grassroots level.

One conservative activist who first joined the fight against MACOS was John Steinbacher. Before focusing on educational matters, Steinbacher authored a pamphlet that explained that Robert Kennedy's assassination was the plot of an international conspiracy of occultists and the Illuminati.[92] After this publication, Steinbacher became a national figure on the right-wing lecture circuit by working on curricular issues, specifically to make public education religious and to eliminate sex education classes from schools even when they were supported by the majority of parents. His books *The Conspirators: Man Against God* and *The Child Seducers* and an accompanying record suggested that sex education was "sexploitation" designed to forcefully indoctrinate children into subservience to communism.[93]

As Steinbacher saw it, proper understanding of the previous five centuries of history could be best achieved by understanding them as a long conflict between two philosophical perspectives on human nature. On the one hand, there was what was, to Steinbacher, the proper view established two thousand years ago: "man is born *evil* and . . . he must be redeemed from this fallen state by faith in Christ." On the other hand, there was the view promoted by Communism, Nazism, the United Nations, the Illuminati, Satanists, and international bankers such as the Rothschilds. According to that "humanist" view, "man is a neutral being who can be manipulated by changing his environment. Since he is not a child of God he is a child of the state."[94]

When MACOS became an issue in Burlington, Vermont, Steinbacher arrived on the scene with the charge that MACOS taught "Deweyism, pragmatism, behaviorism, psychic manipulation, and humanism, and argued that such programs were paving the way for a communist takeover and the destruction of the religious faith of the younger generation."[95] Steinbacher was joined in Vermont by Norma Gabler, a Texas-based professional textbook censor who organized opposition to MACOS in Houston and who, during the 1960s, had orchestrated attacks on evolution in biology textbooks including the NSF-sponsored Biological Sciences Curriculum Study (BSCS).[96]

As the activities of Gabler and Steinbacher indicate, the opposition that MACOS attracted in local school districts drew both on regional concern as well as a coordinated national network of conservative activists. For instance, the opposition to MACOS in Houston relied on the prior work of conservatives in Phoenix.[97] Further, Gabler made her attacks on MACOS a mobile and national affair by showing up in numerous school districts that had adopted the curriculum.[98]

For conservatives, MACOS was archetypal of endemic problems in the public school system. The Baptist reverend Don Glenn claimed that "the course advocated sex education, evolution, a 'hippie-yippie philosophy,' pornography, gun control, and communism."[99] As Steinbacher put it, public schools had become a site for promoting the "Humanist revolt against God." These schools were "destroying the souls of an entire generation" by using the techniques of mind control that the behavioral sciences had developed in mental hospitals. On this account, schools that were seeking to cultivate "self-actualizing" people by teaching to the "whole child" were actually training children to accept "the concept of a socialized One World totalitarian state without question or struggle."[100] In a different fashion than critics on the Left, those on the Right argued that ideology was everywhere. These views linking MACOS, secularism, Jewish bankers, John Dewey, the Illuminati, and Satanic establishment of a New World Order would later find a more prominent platform in the works of Pat Robertson, the founder of the Christian Coalition and Christian Broadcasting Network.[101]

The evidence for these claims was various. Conservatives in Phoenix, Arizona; Bellevue, Washington; and Naples, Florida, were incensed by the way that MACOS treated humans as biological creatures—that is, as having a body or being a member of the animal kingdom. On this account, MACOS's mere activity of considering human physiology or contrasting humans with salmon, herring gulls, or baboons was illegitimate. Although MACOS course designers contended that they had not included any discussion of species evolution or the mechanisms of natural selection that account for the similarities and differences between humans and other organisms—and specifically noted that fact in the course materials—conservatives took MACOS's attention to physiology and life cycle as corrupting children because "Darwin's theory of evolution is taught as fact." In Florida, a citizens' committee reviewed both positions and the MACOS course materials themselves. These citizens concluded that the claim by conservatives that the curriculum included discussion of evolution was baseless and deceptive.[102]

Beyond its supposed inclusion of evolution, MACOS came under attack for its ethnographic attention to the Netsilik. As Steinbacher saw it, MACOS was an instance of how public schooling "attacks most of the civilizing influences in our society" by focusing on a "totally obscure Eskimo tribe" that had not "contributed" anything to civilization.[103] In Quincy, Massachusetts, the South Shore Citizens against Forced Busing objected to MACOS on the grounds that it exposed children to "unsavory and barbaric Eskimo practices."[104]

Other conservatives focused their attacks on MACOS less on the presumed backward nature of Netsilik civilization and more on what they characterized as its immoral culture. Right-wing critics in Houston noted that Netsilik culture included "cannibalism, infanticide, genocide, senilicide . . . stabbing, wife-swapping, animal beating, bloodletting and mating with all kinds of animals."[105] Conservatives in Bellevue had the same criticisms while also objecting to MACOS's recognition that divorce occurred in Netsilik culture.[106]

Conservatives were also enraged by MACOS's use of ethnographic attention to mythology. This complaint had two aspects. On the one hand,

MACOS exposed students to accounts of activities of mythological characters in Netsilik culture. It was these characters who engaged in many of the activities (e.g., cannibalism) that had so incensed conservatives. In the eyes of the course designers, mythology was one of the essential features of human nature along with such others as toolmaking and language. Mythology was simply a form of meaning-making that is a part of the human drive to make sense of the world, whether through narrative accounts or through formal scientific inquiry.[107] On the other hand, for some parents merely mentioning such fictional activities was tantamount to condoning them or even to encouraging students to engage in them. This was, however, a distinctly minority view.

The majority of parents did not agree with conservatives and understood that the discussion of Netsilik culture was not the same thing as approving of its actual activities or of its myths. They added that conservative critics confused MACOS's discussion of Netsilik culture with endorsing it. These parents further remarked that the conservative claim that MACOS approved of such things as infanticide was dishonest, for the course quite clearly depicted the Netsilik themselves deploring such activities. Majorities suggested that if conservatives were upset about stories that depicted killing babies, then they should ban the Bible due to its discussion of King Herod's order of infanticide.[108] Ultimately, however, the representation of MACOS as supporting rather than simply discussing features of Netsilik mythology was a persistent analytic frame repeated by conservatives that hopped from one local school district to another before reaching and being echoed in the national press.[109]

Conservatives' critique of MACOS's attention to mythology also had a religious component. They contended that MACOS invited its students to view Christianity just as it treated Netsilik cosmology—as mere myth rather than as fact.[110] For conservatives, this pedagogical position was problematic not only because it denigrated Christianity by reducing it to the level of Netsilik faith. For some, the problem was also that the course had not looked to the Bible for answers to what makes people human. By adopting a social scientific approach to culture, MACOS had failed to do what it should have: instruct students on human nature through the Christian tradition.[111]

Conservatives contended that the result of federal participation in education was promotion of leftist moral systems. They argued that the federal government, the NSF, and the curriculum planners were unacceptably unaccountable to the wishes of parents and local communities.

This disengagement, they felt, had led the NSF to miss the fact that the curricula funded with its money were strongly value-laden. MACOS's effort to treat the Netsilik as fully human was anathema to conservative activists, to politicians from Florida to Arizona, and to members of the conservative intelligentsia. For each group, MACOS's treatment of the Netsilik as human was tantamount to indoctrination of children in antireligious secular humanistic doctrine. As an expert at the Heritage Foundation put it, when MACOS sought to show how "one set of values is just as human as another" it had "in place of God, erected the god, Humanism."[112] Notably, this critique of MACOS did not center on the immorality of Netsilik values or on the superiority of Christian values, but instead relied on the assumption that the only values that can be classified as human are those that are Christian.

The Arizona state commissioner of education argued that, by the criteria given by liberal philosopher Sidney Hook, secular humanism was a religion and MACOS was proselytizing in that faith. He then threatened a lawsuit against schools that had adopted MACOS on the grounds that they had violated the establishment clause of the first amendment.[113] This argument circulated through the conservative education movement. It was adopted by Onalle McGraw, the curriculum director of Citizens United for Responsible Education. McGraw contended that MACOS was unconstitutional. In a column for the *Bethesda-Chevy Chase Tribune* she argued that teaching a curriculum that adopted a pluralist approach to religion was actually tantamount to the establishment of a specific religious faith: Secular Humanism.[114]

Others conservatives found separate reasons why MACOS was unconstitutional. James Kilpatrick took MACOS to be illegitimate simply because government support of curriculum design recalled Stalinism. This was, perhaps, an extension of a view of the Constitution that Kilpatrick had developed as a leader in the fight to keep Virginia schools segregated. According to his constitutional philosophy, *Brown v. Board of Education* was an illegitimate expression of federal power. Kilpatrick consequently had played a major role in organizing Virginia's massive resistance to school desegregation.[115] He later devoted efforts to attacking the Voting Rights Act of 1965 on the grounds that it involved an unconstitutional federal reach into an area where the rights of states should prevail.[116]

In general, objections of this sort did not reflect the position of most parents or teachers in the districts. Indeed, when polls were conducted, the overwhelming majority of parents approved of MACOS. Further when parents, school boards, city papers, or local task forces were appointed to investigate the charges conservatives leveled against MACOS, they found that these attacks were either without merit or were based on taking course materials so out of context as to misrepresent it.[117]

Available data indicates that opposition to MACOS was a minority position. A nationwide poll conducted by the superintendent in Montgomery County, Maryland, found that 78 percent of users found MACOS "excellent" and the remaining 22 percent thought it "good."[118] In Phoenix, the vast majority of parents approved of MACOS. In a survey conducted by parents in the district, only 8 percent of the parents were opposed to MACOS while 84 percent were in favor of it. The remainder had no opinion. This survey included a space for free-form commentary. In this regard, one parent remarked that attacks on the school by course opponents and their efforts to have the state override the local board were "more detrimental to my family's American prerogatives than anything this course covers. I resent outside agitation disrupting school communications and feel our children's progressive future needs progressive education." Another typical remark noted that attacks on MACOS were "hysterically emotional reactions" and a "smokescreen for an extremist attempt to de-authorize our Board of Trustees."[119] The last comment was directed toward a successful effort by the conservatives to have the state of Arizona or the superintendent of education block local rule and prevent the adoption of MACOS in new districts or schools.[120] Indeed, some in Arizona saw the MACOS controversy as nothing more than a "bible-thump opportunity" for fundamentalists to "impose" their views on others.[121] One teacher in Boston made a related point. He argued that opponents of MACOS were interested in using federal power to prevent local communities from making their own choices about whether to adopt the curriculum.[122] According to this perspective, the conservative position on federalism had nothing to do with blunting the power of the national government, but simply adopted a rhetoric of local control to achieve other goals.

At the surface, such disputes were framed in political terms, but they were energized by psychological analysis. Disputes over MACOS were thus about two issues at once. First, would America, citizenship, and humanity be defined with reference to their concept of the Christian tradition, as the conservative minority demanded? Was it unconstitutional to treat religions as equal rather than treating Christianity as the moral yardstick? Or would America, citizenship, and humanity be framed in the pluralist terms desired by the majority of parents?

Second, what kind of mental attributes should be cultivated in the classroom and, therefore, what kinds of citizens—and, ultimately, what kind of America—should these classrooms produce? This second question arose from more than the issue of whether social studies curricula would study humans as defined by one or another denomination of Christianity or through the social scientific lens of ethnography. These disagreements ultimately came down to the very different ways that MACOS proponents and critics understood human thinking and learning.

One of MACOS's central pedagogical methods was placing more emphasis on the development of thinking skills and cognitive abilities than on the learning of specific information. This approach had drawn on epistemology of science common in the natural and social sciences. According to this approach, the most important thing that students needed to learn in science classes was the modes of thought that professional scientists themselves used in their research. On this ground, memorization of facts was actually antithetical to the work of real science. The work of cognitive scientists like Jerome Bruner contributed the perspective that children learn and remember more if they do not learn by rote but by discovery, and, in the process, develop stronger mental abilities.

While the improvement in students' thinking skills was a primary route that MACOS saw to achieving its goal of making people more human, this pedagogical approach incensed MACOS's opponents. Indeed, for the Heritage Foundation, an attack on MACOS served as a useful means of overturning discovery-based education in all curricula, whatever the topic. As the foundation's Susan Marshner put it,

> Because of the wide acceptance of Bruner's theories, there is much more at stake in MACOS than 12 years and several million dollars. This single curriculum is important in determining the success or failure of the whole discovery method. MACOS represents discovery education at its most sophisticated level.[123]

While conservatives took discovery-based education as a liberal political project, it was also the case that from the 1950s into the 1960s learning through discovery had been for curriculum designers a way to make classrooms less authoritarian, to fashion a more liberal society, and to foster creativity, flexibility, and autonomy in students. Indeed, Jerrold

Zacharias had envisioned proper education in physics as a tool to fight McCarthyism. It was because of this democratic "social conscience" that he entered the curriculum design process in the first place.[124] When confronted with his pedagogical method, conservatives were, quite naturally, incensed.

They argued that the attention to thinking skills was illegitimate since the point of education was not leaning to think or how to ask useful questions, but knowledge and the memorization of facts.[125] The attention that MACOS devoted to developing children's thinking abilities was, in the eyes of course opponents, an effort at changing how students think and therefore nothing better than brainwashing. Where the course designers saw learning as an open, experimental process, conservatives from local school districts to congressional committees contended that MACOS's use of open-ended group discussions was either an illegal clinical experiment conducted upon children without the consent of their parents or an effort to undermine the country by drawing the schools into "Socialism and Communism—which is treason."[126]

This kind of criticism percolated upward into policy documents circulated by the conservative think tanks. For instance, Susan Marshner contended: "in MACOS, the enactive, iconic and symbolic ways of learning, which might respectively translate into games, films, and group discussions, are all enlisted for the purpose of making children more conscious of the interior workings of their own minds."[127] For Marshner this kind of self-awareness was a part of the MACOS program not of discovery-based learning but of "manipulation" or "psychoanalysis."[128] On the other hand, for Bruner and colleagues the enactive, iconic, and symbolic modes were fundamental and inescapable modes of representing the world that were characteristic of how humans from infancy to adulthood know and understand.[129] The reflective aspect of MACOS was conditioned by Bruner's view that the special power of symbolic modes of knowing—e.g. language or mathematics—as compared to enactive or physical modes of knowing was that symbolic knowledge can operate on or transform itself and therefore make new knowledge.[130]

In attacking MACOS as a mind-control device, critics repeatedly connected Bruner's work to B. F. Skinner's call for social engineering and ending attention to freedom and dignity. This conclusion was based on the observation that both had once been employed in the same department and that Bruner had once cited Skinner's views. Unmentioned in these attacks was any acknowledgment that their views of human nature and freedom were radically divergent.[131] Conservative intellectuals conflated MACOS and other NSF-sponsored science curriculum projects with behaviorism, with Deweyism, and/or with "life-skills" pedagogy that these curricula had explicitly avoided.[132]

If classroom discussion was problematic for conservatives, MACOS's stance toward its subject material was even more enraging. Specifically, where Bruner had wanted students to use "culture" in a scientific way by analyzing the structures that unified Netsilik society, conservatives saw a left-wing plot to use the national government to inculcate cultural relativism in children too young to have absorbed "American values and traditions."[133]

In 1975, after such protests had blossomed in several schools districts and become a rallying point for right-wing politics at the national level, representative John Conlan (R-Az.) made MACOS a focus of congressional politics. He proposed amending the appropriation for the NSF by striking funds for the dissemination of MACOS and by requiring congressional review of all NSF curriculum projects, including those in the natural sciences. NSF's congressional supporters charged that Conlan was proposing censorship and that it was the job of local school districts and not Congress to decide on the virtues of any particular curriculum. This was not acceptable to Conlan, perhaps because of the strong support that MACOS had in school districts.

When Conlan's amendment came to a vote, it failed by 215 to 196. However, Robert Bauman (R-Md.) introduced an amendment that was much more far-reaching. Under Bauman's amendment, the NSF would lose the right of peer review not only for curricula developed under its umbrella but for every one of its grants. This particular amendment to the NSF's authorization bill required the agency to submit all of its grant proposals to Congress for approval every thirty days. Baumann's amendment passed 212 to 199.[134]

Although Bauman's amendment was struck out of the final NSF bill in conference committee, this episode signifies a remarkable shift in the national political culture on the reciprocal roles of science and politics in American life. It marked the organizational and political skills of an energized conservative minority. It would also have ended the NSF's primary mode of operation since its inception in 1950. And although the NSF escaped the loss of its privilege of internal review of its projects, it lost the $9.2 million it had requested for fiscal 1976 for implementation of its precollege curricula.[135] This move ended the scientific establishment's twenty-year involvement in primary and secondary education and its effort to fight McCarthyism by making the next generation of Americans more human.

Conclusion

Over the course of the decade between 1965 and 1975, the once-strong connections between political centrism and the human/social sciences had completely dissolved. By the middle 1960s, it was no longer clear that academics, democratic citizens, and humans were interchangeable. The virtuous traits—autonomy, creativity, rationality—that had once united these types came to divide them. The politics associated with rationality were no longer so clear. Nor was it exactly clear who would best stand as an example of the creative, autonomous, creative American. With shifts in political culture, failures of U.S. foreign policy, and changes in understanding of who best exemplified valued character traits, the expertise of social scientists was itself undermined as their objectivity, rationality, and open-mindedness came into question.

Before this transformation in discourse about who would count as possessing true creativity, creative individuals were not members of the student movement. The canonical creative individuals were recipients of grants from the Atomic Energy Commission, members of the American Academy of Arts and Sciences, producers of art championed by the CIA, or military strategists at the RAND Corporation. These were the "Whiz Kids," the very people McNamara had recruited to bring creative, rational, and scientific management to the Department of Defense.[136]

For the first half of that decade, members of the Left, including those with social scientific training like Richard Flacks and cognitive scientists like Noam Chomsky and Naomi Weisstein, sought to promulgate the personal, epistemic, social, and political virtues of objectivity, universalism, and open-mindedness that centrist liberals had once successfully claimed for themselves, their own science, and their own political views. Once members of the New Left had successfully turned these very same virtues to their own ends, staunch liberals like Daniel Bell would cease arguing, as they had as recently as 1965, that the university was the center of American society. They would, as well, depart from liberalism to form a new and emergent branch of American political culture: neoconservatism.

For the second half of that decade, conservatives mounted a series of attacks on the liberal centrist view of self, politics, and human nature. Unlike feminists and members of the Left who had co-opted the scientific view of humans as, by nature, creative, flexible, and autonomous beings to further their own goals, members of the Right argued that such views were inherently bound up with a secular humanist, materialist, and potentially communist worldview that they opposed. In both instances, however, whether originating from the Right or the Left, movements of this decade broke what had seemed to many the unproblematic connections among academic merit, American character, and the scientific description of human nature.

The ties that had once existed between Cold War centrism and social science not only frayed, but provided both the Left and the Right organizational strategies. On the one hand, the Left attacked centrists' pretensions to objectivity. Feminists critiqued the academy as failing to live up to its meritocratic ideals. On the other, the Right built itself as a political movement by organizing around attacks on NSF-sponsored science curricula. Where the Left argued that scientists were tools of the American political machine or represented a sexist or racist ideology, the Right made the claim that scientists were instruments of an anti-American conspiracy. In both cases, the scientific establishment so thoroughly lost its claim to independent, objective, expert knowledge that the House of Representatives did not trust it enough to conduct peer review.

The History of the Open Mind

The open mind was a defining feature of Cold War culture. It framed the political virtues of the democratic character, marked the central virtues of scientists and intellectuals, and functioned as a criterion of human nature. Each of these three aspects of the open mind was tightly bound to the others during the early years of the Cold War.

More than anything, what held together the political, the academic, and the scientific visions of open-mindedness was that these aspects were not defined through a set of abstract or logical descriptions but by reference to real people whose psychological profiles served as exemplars for the category. Thus, creativity was operationally defined by developing a measuring tool that could separate individuals already known to be innovative from those who were already known to be only average. Human cognition was modeled on the mode of reasoning that anti-positivist scientists adopted. Within Cold War salons, political rationality was defined by reference to the views of the individuals who populated those same salons. However, these were not three prototypical individuals, but members of a community. This community and its members defined open-minded tolerance as the criterion of the good society, the human, and the intellectual all at once. More specifically, this community took itself and its members as exemplars for correct politics in America, for right thinking in the life of the mind, and even as the defining criterion of human thought itself.

The work of that community had significant effects. It developed, cultivated, and propagated tools and modes of analysis that policed the boundaries of reasoned discourse; it reshaped the academy by channeling resources to open-minded, interdisciplinary endeavors; it developed curricula to make Americans more broad-minded; and, through the work of cognitive science, it effectively redefined human nature as autonomous rather than purely subject to environmental influence.

Because articulations of open-mindedness occurred in multiple places at once and operated on all registers at once, this book is not a political history, a cultural history, an intellectual history, or a disciplinary history. For the authors of NSC-68, open-mindedness showed why Communism was evil and unnatural. For scientists such as Robert Oppenheimer, it defined the epistemic norm, the scientific spirit, and the hope for cultural coherence and community in an increasingly complex world. For cognitive scientists, it defined human nature. Recognition of how open-minded reason had multiple meanings and was, at the same time, defined by the attributes of specific individuals opens the possibility for reframing histories that have been treated as separate phenomena.

The Human Mind

Consider first the history of psychology and the emergence of cognitive science. In the time period covered by this study, neither psychology nor the other disciplines had sole claim to expertise on how thinking processes worked. Even more, cognitive scientists established their own credentials as researchers not by reinforcing disciplinary boundaries but by strategic boundary crossing.[1]

The struggle between behaviorists and their opponents was not simply a struggle over which group had access to the better set of scientific methods. While there has been disagreement in historical treatments as to whether or not the emergence of cognitive science constituted a "true" scientific revolution, these studies agree that these events involved the replacement of neo-behaviorism's stimulus-response model of human nature with an information-processing approach based on analogies between the human and the computer.[2]

I have shown that the development of cognitive psychology involved much more than adopting a computer model of human nature. In fact, researchers, including behaviorists, had wide latitude for the conclusions they could draw about people by comparing them to computers. By keeping focused on the wider intellectual community, this book shows that what was involved in the conflict between behaviorists and their opponents was a struggle over human nature that made use of language, rhetoric, and conceptual tools drawn from a multidisciplinary discussion of subjectivity.

Ultimately, then, no direct line can be drawn from the invention of computers to the rise of cognitive science or to behaviorism's loss of hegemony. Here is the heart of the issue: There are numerous ways of interpreting the meaning and significance of scientific models. And the computer is the most protean of scientific tools. What distinguishes the computer from other scientific instruments is precisely its plasticity. It is, after all, a general purpose, programmable tool.

This history indicates that we cannot write the history of subjectivity by treating the human self as something that can be read off the dials of scientific instruments. After all, these instruments led to multiple, ambiguous, and conflicting readings. This book thus shows that, unlike cases in the physical and biological sciences,[3] scientific tools and instruments, at least in their material form, did little to structure the community of researchers, determine the outcome of a scientific dispute, or even drive the early development of this field. Even more, telling the history of early cognitive science by giving agency to the computer and knowledge of its processes would get the arrow of causation backward. If anything, the human and psychological sciences did more to drive computer science than the other way around.

If it was not caused by the machine, how can we account for the cognitive revolution? The answer to this question comes from the recognition that the debate over human nature was at once also about the academic self and the democratic self. The resolution of debates over whether humans are creative, autonomous beings or not was not achieved according to the rules and methods of the single discipline of psychology. Instead, it occurred within the context of a multidisciplinary community of intellectuals, patrons, and administrators whose members simultaneously performed, discussed, and studied human reason. It is for this reason that this book has focused on salons as the primary spaces for the production of knowledge about human nature rather than such locations as laboratories, theaters, workshops, studies, "the field," or imperial courts that have inspired recent histories of science.

The Academic Mind

Many social scientists (including psychologists) did not place much stake in disciplinary studies. I have therefore not only examined the forms of disciplinary collaboration in which social scientists engaged but have also detailed how interdisciplinarity was equated with open-minded, creative, and virtuous thinking in this community. This was because interdisciplinary social scientists had the same character virtues as democratic citizens; both exhibited open-minded, flexible, and pluralist thinking.

The consequences of this approach to interdisciplinarity were numerous. On the one hand, this approach shaped the relative fortunes of numerous fields of inquiry. These included cognitive science and behaviorism. That the former was taken as interdisciplinary and the latter as disciplinary had effects on their relative attractiveness to patrons. On the other hand, this view of the academic open mind made interdisciplinarity into a seemingly inherent virtue shaping careers, departments, journals, and research fields. This perspective changed the contours of postwar intellectual life and helped to ratify cross-cutting endeavors ranging from the behavioral sciences to cybernetics and the systems sciences, both as basic and practical at the same time. This double vision of interdisciplinarity helped give the sciences of the Cold War period their distinctive character.

Considering the historical connections between the academic, human, and political forms of open-mindedness does more than reframe the history of scientific ideas of human nature. It also has underscored how powerfully the analogy between the academy and the political world operated. Recognizing that academics as a community tackled the problems of thinking restores to view how interconnected academic thinking, democratic thinking, and human thinking became for Cold War intellectuals. By paying attention to the interactions within this community, this book demonstrates that intellectuals' opinions about how they or their colleagues thought were thoroughly intertwined with their opinions about both human nature and democratic character.

When social and natural scientists attacked positivism, operationism, and behaviorism with the argument that science requires judgment and creative thinking, they engaged in a call for epistemic virtues akin to the "trained judgment" that Lorraine Daston and Peter Galison have examined.[4] However much the preference for a creative scientific self was an intellectual value, it was also much more. It was, at the same instant, an engagement in the politics of disciplines, a statement about the nature of human nature, and a move in domestic and international Cold War politics. The call to creativity operated in all of these ways precisely because scientists and other intellectuals were not insulated from but deeply engaged with questions of human nature, with institutional politics, and with domestic and international politics.

The political nature of the intellectual's self arose from the way that it was defined in the context of experiences of day-to-day life in work and leisure. Commitment to a pluralist political vision informed intellectuals' efforts to shape their salons, research centers, disciplines, and universities. Academics often treated their own disciplines, communities, and institutions as equivalent to the class and religious interest groups whose concerns would be recognized and balanced by American democracy. This is one reason why intellectuals so easily read interdisciplinarity as pluralism and disciplinarity as insular parochialism. Ultimately, what was important was being the kind of open-minded person who could recognize, understand, respect, and communicate with true experts from foreign disciplines. This ability would enable the functioning of common rooms, dinner clubs, interdisciplinarity centers, and, ultimately, America. Conversely, failure to achieve this understanding marked individuals as mentally handicapped authoritarians and even threatened the coherence of American culture. A consequence of this mode of character evaluation was exclusion of classes of individuals such as women or behaviorists who were regarded as insufficiently creative and open-minded. Through these events, we have seen the process by which evaluations of academic merit were attached to and justified by pluralist political sentiments—a mode of analysis that remains in effect today.[5]

The American Mind

While the analogy that linked the academy to the wider world helped intellectuals, administrators, and patrons organize their work, it also helped them make sense of the nation and of modern society. From the committee that wrote *General Education in a Free Society* to the participants of the Dedham conference on American character, elite intellectuals often imagined America as a diverse group of experts whose differences could be overcome through intimate dinners or cocktail parties. Administrators, descriptive social critics, theoretical sociologists, and specialists in small-group studies converged in seeing the academy as, variously, a model of the wider society or the center of society. It was because of this analogy that the very real distress intellectuals felt over disciplinary proliferation and the fracturing of the academy, a phenomenon they could not easily escape, would again and again become translated into a vision of a divided America where social order and democracy were under threat. At the same time, the solutions they offered for healing the academy would also resolve national problems.

Recognition of the role of personal experience in setting the frame for centrist social thought helps illuminate why postwar culture seemed to be characterized by its consensus and, then, after the late 1960s, by fracture, disunity, or unraveling. Consensus seemed present because social critics used themselves as models of the reasonable open mind. They developed tools of social and political analysis that ratified their own views and marginalized non-centrist political positions as illogical, rigid, narrow, authoritarian, and even subhuman. In so doing, they used the tools of science and social criticism to enact a form of identity politics.

However, despite their connection to a particular historical moment and their framing with respect to specific individuals, these scientific tools and modes of social criticism were not ultimately reducible to a single political or cultural orientation. Indeed, the tools developed in the era of the Cold War were sufficiently robust to function not only outside their original context but also when put to use for ends that conflicted with the aims of those who originally developed and deployed them. It would become clear by the late 1960s that although the liberal consensus was built upon a particular and ideological vision of human nature, that same vision could help undo it.

While Cold War centrist culture was attacked with the social scientific tools that had once supported it, the late 1960s and early 1970s brought a more fundamental challenge to this culture: a sustained critique of its vision of human nature. These charges had significant effects beyond destabilizing an account of generic human nature. Feminists, men on the Left, and also members of a newly powerful Right charged that human nature as depicted by the social sciences was inaccurate and represented pernicious political ideology. Such debates over human nature were deeply political because at stake in this debate was the question of who, precisely, best represented American values and who, precisely, could properly speak for what America should be in the future.

In this sense, the liberal framework built by midcentury intellectuals would also set the terms for its critics. By the middle 1960s, it was no longer clear that academics, democratic citizens, and humans were interchangeable. The virtuous traits—autonomy, creativity, rationality—

that had once united these types came to divide them. While centrists and leftists struggled over who was most open-minded, for conservatives it was not clear that open-minded flexibility was something that either characterized humans or was desirable. Thus, no longer was there a single kind of person who could stand as an exemplar for the American mind, the academic mind, and the human mind at once. Society seemed to fracture because it was not so clear who would best stand as an example of the creative, autonomous, open-minded American. Indeed, from the perspective of some conservatives, those traits were, themselves, un-American.

*   *   *

Although culture, politics, intellectual life, and human nature have lost what was once a unified exemplar of the open-minded person, the postwar cultural pattern linking the human mind, intellectual life, and politics remains. We continue to find that statements on one register have operated on the other registers as well. We can find that such persistent linkages inform cultural criticism through Christopher Lasch's contention that national decline could be found in NSF sponsored science curricula, in the New Math, and in social studies efforts to promote creative cognition.[6] Debates over curricula based on skills versus content, on distribution requirements, and on modes of post-structuralist or feminist academic analysis of the 1980s and 1990s drew on and found their energy in preexisting connections among politics, the academy, and cognition.[7] In the case of reader-response theory, claims about the effect of experience and preconscious cognitive process in the interpretation of texts would become polarizing political issues both inside and outside of the academy.[8] Connections between politics, curriculum, and cognition also appeared in Allen Bloom's jeremiad against inner-direction, the substitution of "creativity" for "virtue," and the abandonment of curriculum requirements in his *Closing of the American Mind*.[9] In the Christian Coalition's support of phonics-based language instruction, national politics has infused issues of both pedagogy and questions of human cognitive process.[10]

Recently, public discourse on political aspects of cognition has recovered the frameworks that were developed during the Cold War. Reflecting the debate over discovery-based instruction, while schools at all levels seek to promote active thinking to make their students better learners and citizens, the far right of the Republican party characterized college education as indoctrination and called for schools to abandon efforts to teach critical thinking.[11] Once again, discussion of the mental failings of authoritarianism has found a place in the public sphere. However, it has done so in a way that rejects the political orientation of those earlier studies. Recently some scholars have claimed that Cold War political psychology was itself an ideological product of centrism.[12] Other recent scholarship has argued that the deficits associated with the closed mind are a phenomenon that can be found essentially only in members of the right wing.[13] Professional historians and public intellectuals have adopted these perspectives in seeking to explain the mental orientation of the Tea Party and of the 22 percent of the population who supported George W. Bush at the end of his presidency.[14] These explanations have drawn on studies in cognitive science, neuroscience, and political psychology and have resuscitated the psychological diagnostic categories central to *The Authoritarian Personality* and the social criticism of Richard Hofstadter and Daniel Bell.[15] In their most recent form, such traits are associated not only with a mode of antidemocratic political thought but also with a range of medical pathologies, including sociopathy and autism.

According to behavioral economics—the most recent popular incarnation of cognitive science—it is not only conservatives, but all humans, who are not fully rational.[16] This view has been called on to support a centrist politics.[17] Academic policy makers have adopted behavioral economics as support for specific social policies based on the scientific demonstration that humans are irrational. With the appointment of these intellectuals to leading positions in his administration and the adoption of their policy prescriptions, Barack Obama brought behavioral economics and its methods into the center of government.

Despite such continuing connections between the norms of the academy, of politics, and of human nature, we have not seen the reemergence of the kind of coherent intellectual, cultural, and political system that centrists believed was characteristic of the postwar period. It is the very success of the open mind that makes the apparent cultural coherence of the postwar era a thing of the past. Once its social and political power was recognized by the Left and rejected by the Right, the open mind could no longer serve as a form of social-political cohesion. The emergence of such cohesion is unlikely to happen unless one contemporary group succeeds as thoroughly as mid-twentieth-century intellectuals did in making itself the model of intellectual values, political norms, and human nature.

Acknowledgments

In the years since I started this project I have been fortunate to receive support from the National Science Foundation, the Agence Nationale de la Recherche, the Center for Humanities at Wesleyan University, the Fishbein Center for the History of Science at the University of Chicago, and the Charles Warren Center for American History at Harvard.

I am grateful to George Miller and Jerome Bruner for sharing their time and insights, for allowing me access to their personal papers, and for commenting on early drafts of my work. I would also like to thank Doris Aaronson, Susan Carey, Noam Chomsky, Patricia Greenfield, Charles Gross, Daniel Kahneman, George Mandler, Jean Mandler, Alaistair Mundy-Castle, Donald Norman, Rose Olver, and Mary Potter for sharing their memories on the intellectual culture of psychology and cognitive science in the 1950s and 1960s.

This book draws on my earlier work. Portions of chapters 2 and 4 have been previously published as "The Creative American: Cold War Salons, Social Science, and the Cure for Modern Society," *Isis* 100, no. 2 (2009): 219–62. Portions of chapter 5 have previously appeared in "The Reflexivity of Cognitive Science: The Scientist as Model of Human Nature," *History of the Human Sciences* 18, no. 4 (2005): 107–39. Chapter 6 appeared as "Instituting the Science of Mind: Intellectual Economies and Disciplinary Exchange at Harvard's Center for Cognitive Studies," *British Journal for the History of Science* 40, no. 4 (2007): 567–97. My thanks to the University of Chicago Press, Cambridge University Press, and Sage Publications for permission to republish these materials.

This book is concerned with how thinking happens in communities. Its formation is itself a product of extended conversations with friends and colleagues at Princeton, the University of Chicago, Wesleyan, Yale, and Harvard. New colleagues in the American Studies Department at George Washington University have offered encouragement and support as I completed final revisions to the book.

From these communities and others I have been blessed over the years to have as readers, conversation partners, and mentors a group of individuals whose questions and suggestions have sharpened this work: John Agnew, Eric Ash, Peter Galison, Anne Harrington, Joel Isaac, Sheila Jasanoff, Andy Jewett, Adrian Johns, Bruce Kuklick, Mike Mahoney, Ole Molvig, Jill Morawski, Robert Richards, Julie Reuben, Suman Seth, Simon Schaffer, Mark Solovey, William Wimsatt, and Alison Winter. My thanks to Linn Cohen-Cole, Ada Muellner, and Richard Allen for copyediting. I owe special thanks to Elizabeth Lunbeck, Sarah Igo, and Roger Backhouse, who each read and critiqued the entire manuscript once or more during its development.

To my mother Linn and father Steve, brother Ethan and sister Anna, thank you for your encouragement from the beginning.

Most of all, thanks to my wife Eugenia and daughter Naomi for keeping me company while I worked and giving me the best of reasons to stop. In the last year of this project, Naomi watched final edits from her bouncer while I typed with one hand and rocked her with the other. To my wife Eugenia I owe a huge debt. During the days, months, and years I worked on this book, her involvement made the project both enjoyable and possible. She supported me materially, emotionally, and intellectually over the time this book was in formation. She helped me clarify what I hoped to argue, saw where my argument had holes, and understood where and why individual chapters and even the entire manuscript could use a new structure. This book is better on every level because of her.

Notes

Introduction

1. Carl E. Schorske, "A Life of Learning," in *Recasting America: Culture and Politics in the Age of Cold War*, ed. Lary May (Chicago: University of Chicago Press, 1989), 101.

2. Daniel Bell, *The End of Ideology: On the Exhaustion of Political Ideas in the Fifties* (New York: Free Press, 1960), 97. For examination of the midcentury move away from analysis of market forces, see Howard Brick, "Displacement of Economy in an Age of Plenty," in *Transcending Capitalism: Visions of a New Society in Modern American Thought* (Ithaca, N.Y.: Cornell University Press, 2006).

3. Samuel Lubell, "That 'Generation Gap,'" *Public Interest*, no. 13 (1968): 52–60, on 54.

4. "NSC-68: United States Objectives and Programs for National Security" (1950).

5. On the earlier removal of the soul from psychological work, see Edward Reed, *From Soul to Mind: The Emergence of Psychology from Erasmus Darwin to William James* (New Haven, Conn.: Yale University Press, 1997).

6. See, for instance, the way that behaviorists eradicated mental phenomena by "translating" them into the stimulus–response framework. William S. Verplanck, "A Glossary of Some Terms Used in the Objective Science of Behavior," *Psychological Review* 64, no. 6, part 2 (1957): i–vi, 1–42. For discussion of the mistranslation of the Freudian psyche into behaviorism, see Rebecca M. Lemov, *World as Laboratory: Experiments with Mice, Mazes, and Men* (New York: Hill and Wang, 2005), 125–44.

7. George Mandler, "Origins of the Cognitive (R)Evolution," *Journal of the History of the Behavioral Sciences* 38, no. 4 (2002): 339–53; Howard Gardner, *The Mind's New Science: A History of the Cognitive Revolution* (New York: Basic Books, 1985); Jean-Pierre Dupuy, *The Mechanization of the Mind: On the Origins of Cognitive Science* (Princeton, N.J.: Princeton University Press, 2000);

John D. Greenwood, "Understanding the 'Cognitive Revolution' in Psychology," *Journal of the History of the Behavioral Sciences* 35, no. 1 (1999): 1–22; Hunter Crowther-Heyck, *Herbert A. Simon: The Bounds of Reason in Modern America* (Baltimore, Md.: Johns Hopkins University Press, 2005); Bernard J. Baars, *The Cognitive Revolution in Psychology* (New York: Guilford Press, 1986); William Bechtel, Adele Abrahamsen, and George Graham, "The Life of Cognitive Science," in *A Companion to Cognitive Science*, ed. William Bechtel and William Graebner (Malden, Mass.: Blackwell, 1998); Gerd Gigerenzer, "Mind as Computer: The Social Origins of a Metaphor," in *Adaptive Thinking: Rationality in the Real World* (New York: Oxford University Press, 2000). Recent work on cybernetics and computer science discusses human subjectivity but focuses less on the psychological issues that are the concern of this project. Lily E. Kay, "From Logical Neurons to Poetic Embodiments of Mind: Warren S. McCulloch's Project in Neuroscience," *Science in Context* 4, no. 4 (2001): 591–614; Peter Galison, "The Ontology of the Enemy: Norbert Wiener and the Cybernetic Vision," *Critical Inquiry* 21 (1994): 228–68; N. Katherine Hayles, *How We Became Posthuman: Virtual Bodies in Cybernetics, Literature, and Informatics* (Chicago: University of Chicago Press, 1999); Steve J. Heims, *The Cybernetics Group: Constructing a Social Science for Postwar America* (Cambridge, Mass.: MIT Press, 1991).

8. Margaret A. Boden, *Mind as Machine: A History of Cognitive Science* (Oxford and New York: Oxford University Press, 2006).

9. Paul N. Edwards, *The Closed World: Computers and the Politics of Discourse in Cold War America* (Cambridge, Mass.: MIT Press, 1996); S. M. Amadae, *Rationalizing Capitalist Democracy: The Cold War Origins of Rational Choice Liberalism* (Chicago: University of Chicago Press, 2003); Philip Mirowski, *Machine Dreams: Economics Becomes a Cyborg Science* (Cambridge: Cambridge University Press, 2002).

10. For instance, see Lisa McGirr, *Suburban Warriors: The Origins of the New American Right* (Princeton, N.J.: Princeton University Press, 2001).

Chapter One

1. Wendy Wall, *Inventing the American Way: The Politics of Consensus from the New Deal to the Civil Rights Movement* (Oxford and New York: Oxford University Press, 2008); Benjamin Leontief Alpers, *Dictators, Democracy, and American Public Culture: Envisioning the Totalitarian Enemy, 1920s–1950s*, Cultural Studies of the United States (Chapel Hill: University of North Carolina Press, 2003); James H. Capshew, *Psychologists on the March: Science, Practice, and Professional Identity in America, 1929–1969* (Cambridge and New York: Cambridge University Press, 1999), chapter 2.

2. David E. Lilienthal, "The Unification of Specialized Knowledge in Practical Affairs," in *Science, Philosophy, and Religion: Third Symposium* (New York: Conference on Science, Philosophy and Religion in their Relation to the Democratic Way of Life, Inc., 1943), 237–39.

3. E.g., Mortimer J. Adler, "God and the Professors," in *Science, Philosophy, and Religion: A Symposium* (New York: Conference on Science, Philosophy and Religion in their Relation to the Democratic Way of Life, Inc., 1941).

4. Educational Policies Commission, *The Education of Free Men in American Democracy* (Washington, D.C.: Education Policies Commission, National Education Association of the United States, and the American Association of School Administrators, 1941), 71.

5. On the reach of the EPC, see Andrew Hartman, *Education and the Cold War: The Battle for the American School* (New York: Palgrave Macmillan, 2008), 59.

6. On the connection between the classical curriculum and disciplining mental faculties, see Laurence R. Veysey, *The Emergence of the American University* (Chicago: University of Chicago Press, 1965), 36.

7. Fredrick Rudolph, *Curriculum: A History of the American Undergraduate Course of Study since 1636* (San Francisco: Jossey-Bass, 1977), 117.

8. For an instance of the nonemphasis on the Morrill Acts, see Robert M. Hutchins, "What Is a General Education?" *Harpers*, November 1936; "The Confusion in Higher Education," *Harpers*, October 1936. For a contention that the crisis in education began in 1900, with the rise of electives, see Mortimer J. Adler, "This Pre-War Generation," *Harpers*, October 1940, 524–34.

9. Alvin C. Eurich, "A Renewed Emphasis Upon General Education," in *General Education in the American College: Thirty-Eighth Yearbook of the National Society for the Study of Education*, ed. Guy Montrose Whipple (Bloomington, Ill.: Public School Publishing Co., 1939), 7.

10. E.g., Hutchins, "What Is a General Education?" Although in this article Hutchins advocated "general education," the programs he advocated here and elsewhere were typically identified as "liberal" rather than general.

11. B. Lamar Johnson, "General Education Changes the College," *Journal of Higher Education* 9 (1938): 18–22.

12. Ibid.

13. James B. Conant, *My Several Lives: Memoirs of a Social Inventor* (New York: Harper and Row, 1970), 364.

14. The report's authors were Paul Buck, John Finley, Raphael Demos, Leigh Hoadley, Byron Hollinshead, Wilbur Jordan, I. A. Richards, Philip Rulon, Arthur Schlesinger Sr., Robert Ulich, George Wald, and Benjamin Wright. John Dunlop, John Gauss, Howard Jones, Alfred Simpson, and Howard Wilson were committee members for part of the time. See *General Education in a Free Society* (Cambridge, Mass.: Harvard University Press, 1945), xix.

15. Serial #31, October 28, 1943; Serial #41, November 4, 1943; Serial #60, November 30, 1943; Robert Ulich at the Meeting of the Committee, February 23, 1943, Records of the Committee on General Education in a Free Society, UAI 10.528.10, Harvard University Archives. Hereafter all serial numbers in this chapter refer to the records of this committee.

16. Serial #10, September 7, 1943; Serial #339, November 2, 1943; Serial #49, November 12, 1943; Serial #56, November 18, 1943; Serial #60, November 30, 1943.

17. The other figures discussed included Plato, Aristotle, Plutarch, Quintilian, John Amos Comenius, William Petty, Samuel Hartlib, John Locke, Benjamin Franklin, Thomas Jefferson, Rousseau, Pestallozzi, Herbart, Froebel, and Emerson. Robert Ulich, February 16 and 23, 1943.

18. See *General Education in a Free Society*, 40.

19. Raphael Demos, "Philosophical Aspects of the Recent Harvard Report on Education," *Philosophy and Phenomenological Research* 7, no. 2 (1946): 187–213, on 193; *General Education in a Free Society*, 41–46.

20. *General Education in a Free Society*, 41.

21. Ibid., 59, 64–66, 79–80, 190.

22. Ibid., 65.

23. Serial #29, October 28, 1943.

24. Hutchins, "What Is a General Education?"

25. Serial #39, November 2, 1943; Serial #68, December 9, 1943, p. 5.

26. Hutchins, "What Is a General Education?" 606.

27. *General Education in a Free Society*, 100, 91.

28. Serial #4, July 27, 1943.

29. *General Education in a Free Society*, 38, 53.

30. Serial #4, July 27, 1943, pp. 4–5.

31. *General Education in a Free Society*, 79–80. See also 106.

32. Serial #56, November 18, 1943.

33. ". . . encyclopedism is not enough . . ." *General Education in a Free Society*, 98.

34. Ibid., 110–11.

35. Educational Policies Commission, *Education for All American Youth* (Washington, D.C.: Educational Policies Commission, National Education Association of the United States, and the American Association of School Administrators, 1944), 51–52.

36. Wilford Merton Aikin, *The Story of the Eight-Year Study: With Conclusions and Recommendations* (New York: Harper and Brothers, 1942), part 4.

37. *General Education in a Free Society*, 54.

38. Ibid.

39. Ibid.

40. Demos, supp. 18, pp. 7–8. Records of the Committee on General Education in a Free Society, Harvard University Archives.

41. *General Education in a Free Society*, 54.

42. Joel Isaac, "The Harvard Pareto Circle Revisited." Paper presented at the Cambridge Seminars in Political Thought and Intellectual History, October 27, 2008.

43. Secondary literature suggests that James Conant founded the Shop Club in the 1920s when he was a member of Harvard's Chemistry Department. However, archival records indicate that there was already a Shop Club of Harvard faculty that dated to at least 1911. HUD 3787.505, Harvard University Archives.

44. One instance of inclusion of nondisciplinary specialists in ad hoc committees is discussed in James Conant to Paul Buck, 2/20/1951, papers of the Dean of the Faculty of Arts and Sciences, 1950–51, Courses of Instr. Changes since Final Edition. Folder: Ad Hoc Committee on Social Relations 1950–51, UA III, 5.55.26, Harvard University Archives. Discussion of the establishment of the up-or-out and ad hoc systems can be found in Morton Keller and Phyllis Keller, *Making Harvard Modern: The Rise of America's University* (New York: Oxford University Press, 2001), 64–71.

45. George C. Homans, *Coming to My Senses: The Autobiography of a Sociologist* (New Brunswick, N.J.: Transaction Books, 1984), 121, emphasis added. The same anecdote appears in Crane Brinton and George Caspar Homans, *The Society of Fellows* (Cambridge, Mass.: Society of Fellows of Harvard University, distributed by Harvard University Press, 1959), 25.

46. Brinton and Homans, *The Society of Fellows*, 29–31.

47. Ibid., 5–22. Brinton and Homans discuss how, although the house system was implemented before the founding of the Society, the idea of the Society preceded the concrete plan for houses by four years.

48. Demos, Supplement 18A, p. 22. See note 40 above.

49. Educational Policies Commission, *Education for All American Youth*, 135–37.

50. James B. Conant, *On Understanding Science* (New York: New American Library, 1951), 3–4, 6.

51. James B. Conant, "Foreword," in Thomas Kuhn, *The Copernican Revolution* (Cambridge, Mass.: Harvard University Press, 1957), xiii.

52. Ibid., xv.

53. Keller and Keller, *Making Harvard Modern*, 44.

54. See, for instance, the editorial that inaugurated the journal. Earl J. McGrath, "The General Education Movement," *Journal of General Education* 1, no. 1 (1946): 3–8.

55. Reuben Frodin, "Editorial Comment," *Journal of General Education* 3, no. 3 (1949): 161–62.

56. Harold Taylor, "The Philosophical Foundations of General Education," in *Fifty-First Year Book of the National Society for the Study of Education*, ed. Nelson B. Henry (Chicago: National Society for the Study of Education, 1952),

30–35; Daniel Bell, *The Reforming of General Education: The Columbia College Experience in Its National Setting* (New York: Columbia University Press, 1966), 38.

57. Nicholas Lemann, *The Big Test: The Secret History of the American Meritocracy* (New York: Farrar, Straus and Giroux, 1999).

58. David Eli Lilienthal, *TVA: Democracy on the March* (Chicago: Quadrangle Books, 1966), 77–92.

59. James B. Conant, "Some Aspects of Modern Harvard," *Journal of General Education* 4, no. 3 (1950): 175–83, on 180.

Chapter Two

1. John W. Gardner, *Self-Renewal: The Individual and the Innovative Society* (New York: Harper and Row, 1963), 32.

2. On the conformity of traditional societies and a celebration of the modern mind, see Alex Inkeles and David Horton Smith, *Becoming Modern: Individual Change in Six Developing Countries* (Cambridge, Mass.: Harvard University Press, 1974). For discussion of the place of this work in the field of modernization theory, see Nils Gilman, *Mandarins of the Future: Modernization Theory in Cold War America* (Baltimore, Md.: Johns Hopkins University Press, 2003).

3. The call for unity in diversity can be found in Alain Locke, "Pluralism and Intellectual Democracy," in *Science, Philosophy and Religion* (New York: Conference on Science, Philosophy and Religion, 2nd Symposium, 1942), 97; Daniel J. Boorstin, *The Genius of American Politics* (Chicago: University of Chicago Press, 1953), 28; J. Robert Oppenheimer, "Theory Versus Practice in American Values and Performance," in *The American Style: Essays in Value and Performance*, ed. Elting E. Morison (New York: Harper and Brothers, 1958); James B. Conant, *Education in a Divided World: The Function of the Public Schools in Our Unique Society* (Cambridge, Mass.: Harvard University Press, 1948), 179; Lyman Bryson, *Science and Freedom* (New York: Columbia University Press, 1948); Margaret Mead, *And Keep Your Powder Dry: An Anthropologist Looks at America* (New York: William Morrow, 1942); Lawrence K. Frank, "Psychology and Social Order," in *The Human Meaning of the Social Sciences*, ed. Daniel Lerner (New York: Meridian Books, 1959), 239.

4. That the 1950s represented a period of consensus and political and social quiescence was long accepted by both social critics and historians. For discussion of how social critics blinded themselves to divisions within American society, see T. J. Jackson Lears, "A Matter of Taste: Corporate Cultural Hegemony in a Mass-Consumption Society," in *Recasting America: Culture and Politics in the Age of Cold War*, ed. Lary May (Chicago: University of Chicago Press, 1989). For a more recent study of long-ignored divisions in Cold War America, see

Alan Brinkley, "The Illusion of Unity in Cold War Culture," in *Rethinking Cold War Culture*, ed. Peter J. Kuznick and James Gilbert (Washington and London: Smithsonian Institution Press, 2001).

5. See for instance, Daniel Bell, "America as a Mass Society: *A Critique*," in *The End of Ideology: On the Exhaustion of Political Ideas in the Fifties* (New York: The Free Press, 1962), 37; Talcott Parsons, "Evolutionary Universals in Society," *American Sociological Review* 29, no. 3 (1964): 339–57, on 355; Margaret Mead, *Coming of Age in Samoa: A Psychological Study of Primitive Youth for Western Civilisation* (New York: HarperCollins, 2001), 8; J. T. Dunlop et al., "Toward a Common Language for the Area of the Social Sciences" (unpublished white paper, Harvard University, 1941); George F. Kennan, "America's Administrative Response to Its World Problems," in *The American Style: Essays in Value and Performance*, ed. Elting E. Morison (New York: Harper and Brothers, 1958).

6. Elting E. Morison, "The Course of the Discussion," in *The American Style*, 396. C Wright Mills articulated a similar concern. For him the problem was not simply the fracturing of national community, but the obliteration of "small-scale discussion" in small communities, and the "leisurely human interchange of opinion." C. Wright Mills, "Mass Society and Liberal Education," in *Power, Politics, and People: The Collected Essays of C. Wright Mills*, ed. Irving Louis Horowitz (London and New York: Oxford University Press, 1967 [1954]), 362.

7. Seymour Martin Lipset and Leo Lowenthal, "Preface," in *Culture and Social Character: The Work of David Riesman Reviewed*, ed. Seymour Martin Lipset and Leo Lowenthal (Glencoe, Ill.: The Free Press, 1961), ix–x.

8. C. P. Snow, *The Two Cultures and the Scientific Revolution* (Cambridge: Cambridge University Press, 1959).

9. Richard Yates, *Revolutionary Road* (Boston: Little Brown, 1961); Sloan Wilson, *The Man in the Gray Flannel Suit* (New York: Simon and Schuster, 1955); Mills, "Mass Society and Liberal Education"; David Riesman, *The Lonely Crowd: A Study of the Changing American Character* (New Haven, Conn.: Yale University Press, 1950); William H. Whyte Jr., *The Organization Man* (New York: Simon and Schuster, 1956).

10. Bell, "America as a Mass Society: *A Critique*," 35.

11. On the survey see Clyde Kluckhohn, "Shifts in American Values: Review of Max Lerner, *America as a Civilization: Life and Thought in the United States*," *World Politics* 11, no. 2 (1959): 251–61. For anxieties about conformity, see Bryson, *Science and Freedom*; Conant, *Education in a Divided World*, 149.

12. Although intellectuals lacked unanimity as to whether the United States was becoming a mass society, there was more general agreement as to the linkage of conformity, mass society, and authoritarianism. Generally, through the early 1960s, liberals such as Daniel Bell did not believe that the United Sates was a mass society. See Bell, "America as a Mass Society: *A Critique*"; Sey-

mour Martin Lipset, *Political Man: The Social Bases of Politics* (Garden City, N.Y.: Doubleday, 1960). Leftists, including Irvin Howe, and European émigrés, including members of the Frankfurt School, were more likely to diagnose the United States as a mass society.

13. X [George F. Kennan], "The Sources of Soviet Conduct," *Foreign Affairs* 25 no. 4 July (1947): 566–82; "NSC-68: United States Objectives and Programs for National Security." The Truman Library has made a scan of one of the originals of NSC-68 available online: http://www.trumanlibrary.org/whistlestop/study_collections/korea/large/week2/nsc68_1.htm.

14. Educational Policies Commission, *American Education and International Tensions* (Washington: 1949), 4.

15. Allen L. Edwards, "Unlabeled Fascist Attitudes," *Journal of Abnormal and Social Psychology* 36 (1941): 575–82; Allen L. Edwards, "The Signs of Incipient Fascism," *Journal of Abnormal and Social Psychology* 39, no. 3 (1944): 301–16; Abraham H. Maslow, "The Authoritarian Character Structure," *Journal of Social Psychology* 18 (1943): 401–11. These built on earlier studies. See H. D. Lasswell, *Psychopathology and Politics* (Chicago: University of Chicago Press, 1930); Erich Fromm, *Escape from Freedom* (New York: Farrar and Rinehart, 1941); Ross Stanger, "Fascist Attitudes: Their Determining Conditions," *Journal of Social Psychology* 7, no. 4 (1936): 438–54; Ross Stanger, "Fascist Attitudes: An Exploratory Study," *Journal of Social Psychology* 7, no. 3 (1936): 309–19.

16. The chronology, history, and prehistory of the project may be found in R. Nevitt Sanford, "A Personal Account of the Study of Authoritarianism: Comment on Samelson," *Journal of Social Issues* 42 (1986): 209–14; William F. Stone, Gerda Lederer, and Richard Christie, "The Status of Authoritarianism," in *Strength and Weakness: The Authoritarian Personality Today*, ed. William F. Stone, Gerda Lederer, and Richard Christie (New York: Springer-Verlag, 1993). For discussion of the earlier research on psychology and fascism by Erich Fromm and other members of the Frankfurt School, see Franz Samelson, "The Authoritarian Character from Berlin to Berkeley and Beyond: The Odyssey of a Problem," in *Strength and Weakness*; Franz Samelson, "Authoritarianism from Berlin to Berkeley: On Social Psychology and History," *Journal of Social Issues* 42, no. 1 (1986): 191–208; Martin Jay, *The Dialectical Imagination: A History of the Frankfurt School and the Institute of Social Research, 1923–1950* (Boston: Little, Brown and Company, 1973).

17. Theodor W. Adorno et al., *The Authoritarian Personality* (New York: Harper and Brothers, 1950), 1.

18. Ibid., 100, 147–49, 464, 480.

19. See the discussion in Else Frenkel-Brunswik, "Sex, People and Self as Seen through the Interviews," in *The Authoritarian Personality*, 441.

20. Else Frenkel-Brunswik, "Comprehensive Scores and Summary of Interview Results," in *The Authoritarian Personality*, 480.

21. Else Frenkel-Brunswik, "Tolerance toward Ambiguity as a Personality Variable," *American Psychologist* 3 (1948): 268.

22. Herbert H. Hyman and Paul B. Sheatsley, *"The Authoritarian Personality*—a Methodological Critique," in *Studies in the Scope and Method of the Authoritarian Personality*, ed. Richard Christie and Marie Jahoda (Glencoe, Ill.: The Free Press, 1954).

23. Samelson, "The Authoritarian Character"; M. Brewster Smith, *"The Authoritarian Personality*: A Re-Review 46 Years Later," *Political Psychology* 18 (1997): 159–63.

24. Jerome Fisher, "The Memory Process and Certain Psychosocial Attitudes, with Special Reference to the Law of Prägnanz," *Journal of Personality* 19, no. 4 (1951): 406–20.

25. Gunnar Myrdal, Richard Mauritz, Edvard Sterner, and Arnold Rose, *An American Dilemma: The Negro Problem and Modern Democracy*, 2 vols. (New York: Harper and Brothers, 1944); Gordon W. Allport, *The Nature of Prejudice* (Cambridge, Mass.: Addison-Wesley Pub. Co., 1954).

26. Daniel J. Levinson, "The Study of Ethnocentric Ideology," in *The Authoritarian Personality*, 149–50.

27. Else Frenkel-Brunswik, "Dynamic and Cognitive Personality Organization as Seen through the Interviews," in *The Authoritarian Personality*, 454, 467.

28. S. E. Asch, "Effects of Group Pressure Upon the Modification and Distortion of Judgments," in *Groups, Leadership and Men: Research in Human Relations*, ed. Harold Guetzkow (Pittsburgh, Pa.: Carnegie Press, 1951); Richard S. Crutchfield, "Conformity and Character," *American Psychologist* 10, no. 5 (1955): 191–98.

29. Lindsay R. Harmon, "The Development of a Criterion of Scientific Competence," in *Scientific Creativity: Its Recognition and Development*, ed. Calvin W. Taylor and Frank Barron (New York: John Wiley and Sons, 1963), 45.

30. Sanford, "A Personal Account"; Barron, *Creativity and Psychological Health*, 12.

31. For results based on freeform and projective tests, see Anne Roe, "A Psychological Study of Physical Scientists," *Psychological Monographs* 43 (1951): 121–239; Anne Roe, *The Making of a Scientist* (New York: Dodd, Mead, 1953).

32. Richard S. Crutchfield, "Conformity and Creative Thinking," in *Contemporary Approaches to Creative Thinking*, ed. Howard E. Gruber, Glenn Terrell, and Michael Wertheimer (New York: Atherton Press, 1962), 130.

33. Ibid. On Crutchfield's error, see Alice H. Eagly, "Sex Differences in Influenceability," *Psychological Bulletin* 85, no. 1 (1978): 86–116, on 92. Eagly noted that the textbook had erroneously used a famous study on conformity by Solomon Asch to argue for sex differences in that trait and that Asch's study had been conducted on only male subjects. The textbook in question is David Krech,

Richard S. Crutchfield, and Edgerton L. Ballachey, *Individual in Society* (New York: McGraw-Hill, 1962).

34. Frank Barron, "Originality in Relation to Personality and Intellect," *Journal of Personality* 25 (1957): 730–42. This remark and a thorough analysis of gender bias in creativity and conformity studies may be found recorded in a piece by another IPAR researcher: Ravenna Helson, "Creativity in Women," in *The Psychology of Women: Future Directions in Research*, ed. Julie A. Sherman and Florence L. Denmark (New York: Psychological Dimensions, 1978).

35. Richard Christie, "Authoritarianism Reexamined," in *Studies in the Scope and Method of the Authoritarian Personality*, 168–70; Sanford, "A Personal Account," 213.

36. Sanford, "A Personal Account," 213.

37. Hyman and Sheatsley, "*The Authoritarian Personality*—a Methodological Critique," 94.

38. Adorno et al., *The Authoritarian Personality*, 95. Quotations from *TAP* are reproduced in and criticisms thereof may be found in Hyman and Sheatsley, "*The Authoritarian Personality*—a Methodological Critique," 95.

39. Else Frenkel-Brunswik, "Further Explorations by a Contributor to *the Authoritarian Personality*," in *Studies in the Scope and Method of the Authoritarian Personality*, 233.

40. Richard Christie, "Some Experimental Approaches to Authoritarianism: I. A Retrospective Perspective on the *Einstellung* (Rigidity?) Paradigm," in *Strength and Weakness: The Authoritarian Personality Today*, 88.

41. Lipset, *Political Man*.

42. Seymour Martin Lipset, "Three Decades of the Radical Right: Coughlinites, McCarthyites, and Birchers," in *The Radical Right*, ed. Daniel Bell (Garden City, N.Y.: Doubleday and Company, 1964); Michael Paul Rogin, *The Intellectuals and McCarthy: The Radical Specter* (Cambridge, Mass.: MIT Press, 1967).

43. See, for instance, M. Brewster Smith, "Review of T. W. Adorno, E. Frenkel-Brunswik, D. J. Levinson and R. Nevitt Sanford, *The Authoritarian Personality*," *Journal of Abnormal and Social Psychology* 45 (1950): 775–79; Lipset, *Political Man*. For a historical analysis that drew on similar models, see Richard Hofstadter and Wilson Smith, eds., *American Higher Education: A Documentary History* (Chicago: University of Chicago Press, 1961).

44. Samelson, "Authoritarianism from Berlin to Berkeley."

45. Edward Shils, "Authoritarianism: 'Right' and 'Left,'" in *Studies in the Scope and Method of the Authoritarian Personality*. For linkages of communism and fascism, see Seymour Martin Lipset, "The Sources of the 'Radical Right,'" in *The Radical Right*, 327. For general discussion of this phenomenon, see Les K. Adler and Thomas G. Paterson, "Red Fascism: The Merger of Nazi Germany and Soviet Russia in the American Image of Totalitarianism, 1930s–1950s," *American Historical Review* 75 (1970): 1046–64.

46. Milton Rokeach, *The Open and Closed Mind: Investigations into the Nature of Belief Systems and Personality Systems* (New York: Basic Books, 1960), 76, 77, 80.

47. Ibid., 76.

48. Ibid., 83–84.

49. Lipset, *Political Man*.

50. Giuseppe Di Palma and Herbert McClosky, "Personality and Conformity: The Learning of Political Attitudes," *The American Political Science Review* 64, no. 4 (1970): 1054–73, on 1060.

51. The connection between the CIA and Hofstadter are discussed in David S. Brown, *Richard Hofstadter: An Intellectual Biography* (Chicago: University of Chicago Press, 2006), 150–52. Lipset's work initially appeared as a Fund for the Republic pamphlet.

52. Lionel Trilling, *The Liberal Imagination: Essays on Literature and Society* (New York: Viking Press, 1950).

53. The lack of concerted attention given to conservatism is discussed in Alan Brinkley, "The Problem of American Conservatism," *American Historical Review* 99, no. 2 (1994): 409–29.

54. Educational Policies Commission, *American Education and International Tensions*, 16.

55. See, for instance, Bell, *The End of Ideology*; Boorstin, *The Genius of American Politics*; Arthur M. Schlesinger Jr., *The Vital Center: The Politics of Freedom* (Boston: Houghton-Mifflin, 1949).

56. "Ideology vs. Democracy." Text of a speech by Arthur Schlesinger Jr. before the Indian Council on World Affairs, New Delhi, February 15, 1962. Papers of J. Robert Oppenheimer, Box 65, Folder: Schlesinger, Arthur. Manuscript Division, Library of Congress.

57. For discussion of Hook's and Lovejoy's views that Communists could not think for themselves, see Ellen Schrecker, *No Ivory Tower: McCarthyism and the Universities* (New York: Oxford University Press, 1986), 105–6.

58. Educational Policies Commission, *American Education and International Tensions*, 40.

59. Testimony delivered to the Subcommittee on the Reorganization of the Committee on Government Operations of the U.S. Senate, March 15, 1955.

60. One such individual, because of his having supported Henry Wallace's 1948 presidential campaign, was Everett Mendelsohn, then a graduate student in History of Science. Schrecker, *No Ivory Tower*, 260.

61. Raymond B. Allen, "Communists Should Not Teach in American Colleges," *Educational Forum* 13, no. 4 (1949): 433–40. Significantly, Allen held that people who believed in Marxism could be teachers, but that members of the Communist Party could not.

62. For discussion of the University of Washington see Ellen Schrecker, *The*

Lost Soul of Higher Education: Corporatization, the Assault on Academic Freedom, and the End of the American University (New York: The New Press, 2010), 35–38. Schrecker argues that the Washington case is critical because it set the pattern for how universities would deal with accused Communists during the McCarthy period.

63. For an expression of this sentiment, see John W. Gardner, *Excellence: Can We Be Equal and Excellent Too?* (New York: Harper and Row, 1961), 35.

64. J. G. Schimek, "Creative Originality: Its Evaluation by Use of Free-Expression Tests" (Ph.D. diss., University of California, Berkeley, 1954), 44. Cited in Donald W. MacKinnon, "IPAR's Contribution to the Conceptualization and Study of Creativity," in *Perspectives in Creativity*, ed. Irving A. Taylor and Jacob W. Getzels (Chicago: Aldine Publishing Company, 1975), 64–65. Frank Barron, *Creativity and Psychological Health: Origins of Personal Vitality and Creative Freedom* (Princeton, N.J.: Van Nostrand, 1963); Frenkel-Brunswik, "Tolerance toward Ambiguity"; Lawrence S. Kubie, "Blocks to Creativity," in *Explorations in Creativity*, ed. Ross L. Mooney and Taher A. Razik (New York: Harper and Row, 1967); Abraham H. Maslow, "Creativity in Self-Actualizing People," in *Creativity and Its Cultivation*, ed. Harold H. Anderson (New York: Harper and Row, 1959); Rokeach, *The Open and Closed Mind*, 58.

65. Levinson, "Ethnocentric Ideology," 150; Else Frenkel-Brunswik, "Comprehensive Scores and Summary of Interview Results," in *The Authoritarian Personality*, 466. See also Frenkel-Brunswik, "Tolerance toward Ambiguity."

66. See, for instance, Herbert Gutman, "The Biological Roots of Creativity," in *Explorations in Creativity*, ed. Ross L. Mooney and Taher A. Razik (New York: Harper and Row, 1967), 3. Originally published as Herbert Gutman, "The Biological Roots of Creativity," *Genetic Psychology Monographs* (1961): 419–58.

67. Arthur Koestler, *The Act of Creation* (New York: Macmillan, 1964); Arthur Koestler, *The Sleepwalkers: A History of Man's Changing Vision of the Universe* (London: Hutchinson, 1959); Arthur Koestler, *Darkness at Noon* (New York: The Macmillan Company, 1941).

68. This particular individual disagreed with the connection between creativity and democracy, arguing that the Soviet Union's success in producing innovative science was a product of its investment in education, research, and development. Situated as he was in the Department of Defense's Advanced Research Projects Agency [DARPA], this commentator had a personal stake in connecting creative ideas not to democracy but to monetary investments. N. E. Golvin, "The Creative Person in Science," in *Scientific Creativity: Its Recognition and Development*, 20.

69. For discussion of the defense of bourgeois society by formerly radical intellectuals, see Neil Jumonville, *Critical Crossings: The New York Intellectuals in Postwar America* (Berkeley and Los Angeles: University of California Press,

1991), 221–29. A continuation of this defense of bourgeois values and the associated charge that cultural radicals lacked true creativity can be found in Daniel
Bell, "The Sensibility of the Sixties," in *The Cultural Contradictions of Capitalism* (New York: Basic Books, 1976), 120–43.

70. B. Clark, "You Can Learn to Think Creatively," *Readers Digest* 77 (1960):
66–70.

71. C. Burt, "Eugenics: Intelligence and Genius," *British Medical Bulletin*
6 (1949): 78–79; E. Kretschmer, "The Breeding of the Mental Endowments of
Genius," *Psychiatric Quarterly* 4 (1930): 74–80; Lewis M. Terman, ed., *Genetic
Studies of Genius*, 4 vols. (Stanford: Stanford University Press, 1925, 1926, 1930,
1957). There were, of course, exceptions to this trend of American studies on
the environmental etiology of creativity. The British psychologist Charles Spearman, for instance, saw creativity as genetic. See Spearman, *Creative Mind* (New
York: D. Appleton, 1931).

72. Lawrence S. Kubie, *Neurotic Distortions of the Creative Processes* (Lawrence: University of Kansas Press, 1958); Jeannette Marks, *Genius and Disaster:
Studies in Drugs and Genius* (New York: Adelphi, 1925); J. F. Nisbet, *The Insanity of Genius* (New York: Scribner's, 1912); R. K. White, "Note on the Psychopathology of Genius," *Journal of Social Psychology* 1 (1930): 311–15; Xavier
Francotte, *La génie et la folie* (Brussels: Société belge de librairie, 1890); P. Funatoli, *Il Genio e La Follia* (Siena: Tip. dell'Ancora, 1885); Barron, *Creativity
and Psychological Health*; Donald W. MacKinnon, "The Nature and Nurture of
Creative Talent," *American Psychologist* 17, no. 7 (1962): 484–95, on 488.

73. For discussion of the historical connections between isolation and insightful thinking, see Steven Shapin, "'The Mind Is Its Own Place': Science and Solitude in 17th-Century England," *Science in Context* 4 (1991): 191–218; Martin
Kusch, "Recluse, Interlocutor, Interrogator: Natural and Social Order in Turn-
of-the-Century Psychological Research Schools," *Isis* 86 (1995): 419–39.

74. This result is recounted in Ross L. Mooney and Taher A. Razik, "Preface," in *Explorations in Creativity*, ed. Ross L. Mooney and Taher A. Razik
(New York: Harper and Row, 1967).

75. Taher A. Razik, *Bibliography of Creativity Studies and Related Areas*
(Buffalo: State University of New York at Buffalo, 1965).

76. H. W. Gabriel, "On Teaching Creative Engineering," *Journal of En-
gineering Education* 45 (1955): 794–801; Howard Gammon, "Some Practical
Techniques for Increasing Creativity in Engineering," *Plant Maintenance and
Engineering* 21 (1960): 25–27; G. Brown, "An Experiment in the Teaching of
Creativity," *School Review* 72, no. 4 (1964): 437–50.

77. Golvin, "The Creative Person in Science"; John C. Flanagan, "The Definition and Measurement of Ingenuity," in *Scientific Creativity: Its Recognition
and Development*, edited by Calvin W. Taylor and Frank Barron (New York and
London: John Wiley and Sons, 1963); Jack D. Summerfield, Lorlyn Thatcher,

and with commentary by Lyman Bryson, eds., *The Creative Mind and Method: Exploring the Nature of Creativeness in American Arts, Sciences, and Professions* (New York: Russell and Russell, 1964).

78. L. R. Bittel, "Brainstorming: Better Way to Solve Plant Problems," *Factory Management* 114, no. 5 (1956): 98–107.

79. H. R. Wallace, "Creative Thinking: A Factor in Sales Productivity," *Vocational Guidance Quarterly* 9 (1961): 223–26.

80. C. H. Greenwalt, *The Uncommon Man: The Individual in the Organization* (New York: McGraw-Hill, 1959). For examination of industrial research practices, see Steven Shapin, *The Scientific Life: A Moral History of a Late Modern Vocation* (Chicago: University of Chicago Press, 2008).

81. Donald W. Taylor, *Thinking* (New Haven: Yale University Department of Psychology, 1962); D. Cohen, J. W. Whitmyre, and W. H. Funk, "Effect of Group Cohesiveness and Training on Creative Thinking," *Journal of Applied Psychology* 44 (1960): 319–22; Jerome S. Bruner, "The Conditions of Creativity," in *Contemporary Approaches to Creative Thinking*, ed. Howard E. Gruber, Glenn Terrell, and Michael Wertheimer (New York: Atherton Press, 1962).

82. Summerfield, Thatcher, and Bryson, eds., *The Creative Mind and Method*. Other members of the intellectual establishment expressed similar reactions to the Beats. Among them were Daniel Bell and the contributors to *Commentary* and *Partisan Review*. See Bell, "America as a Mass Society: *A Critique*," 35–36; Richard Pells, *The Liberal Mind in a Conservative Age: American Intellectuals in the 1940s and 1950s* (New York: Harper and Row, 1985), 377–80; Jumonville, *Critical Crossings*, 186–93.

83. Bryson, *Science and Freedom*, x.

84. Summerfield and Thatcher, "Preface," in *The Creative Mind and Method*, ix.

85. Riesman, *The Lonely Crowd*, 246; Crutchfield, "Conformity and Creative Thinking," 126; Paul Goodman, *Growing Up Absurd: Problems of Youth in the Organized System* (New York: Random House, 1960); Daniel Horowitz, *Betty Friedan and the Making of the Feminine Mystique: The American Left, the Cold War, and Modern Feminism* (Amherst: University of Massachusetts Press, 1998), 172–76. For a similar argument about the counter-conformity of juvenile delinquents, see Albert Kircidel Cohen, *Delinquent Boys: The Culture of the Gang* (Glencoe, Ill.: The Free Press, 1955).

86. Barron, "Complexity-Simplicity as a Personality Dimension."

87. Frances Stonor Saunders, *Who Paid the Piper? The CIA and the Cultural Cold War* (London: Granta Books, 1999), 252–78. For its part, the State Department sponsored international tours of jazz musicians. Penny M. Von Eschen, *Satchmo Blows Up the World: Jazz Ambassadors Play the Cold War* (Cambridge, Mass.: Harvard University Press, 2006).

88. Serge Guilbaut, *How New York Stole the Idea of Modern Art: Abstract*

Expressionism, Freedom, and the Cold War (Chicago: University of Chicago Press, 1983); Erika Doss, "The Art of Cultural Politics: From Regionalism to Abstract Expressionism," in *Recasting America: Culture and Politics in the Age of Cold War,* ed. Lary May (Chicago: University of Chicago Press, 1989).

89. Sarah Schrank, "The Art of the City: Modernism, Censorship, and the Emergence of Los Angeles's Postwar Art Scene," *American Quarterly* 56, no. 3 (2004): 663–91, on 670; William S. Schlamm, "The Self-Importance of Picasso," *National Review,* July 13, 1957; William S. Schlamm, "150 Drawings—But Out of This World," *National Review,* May 23, 1956. See also Mike Davis, *City of Quartz: Excavating the Future in Los Angeles* (New York: Vintage Books, 1992), 63.

90. Samuel A. Stouffer, *Communism, Conformity, and Civil Liberties: A Cross Section of the Nation Speaks Its Mind* (New York: Doubleday, 1955).

91. William F. Buckley and Brent Bozell, *McCarthy and His Enemies: The Record and Its Meaning* (Chicago: H. Regnery Co., 1954); William F. Buckley, *God and Man at Yale; The Superstitions of Academic Freedom* (Chicago: Regnery, 1951).

92. For the review, see McGeorge Bundy, "The Attack on Yale," *The Atlantic,* November 1951, 50–52. An undated memo from the Yale Office of University Development to members of the University Committee on Endowments and Gifts calls Bundy's review (copies of which it circulated) a "complete and convincing answer" to Buckley: a copy of the memo is in the McGeorge Bundy Personal Papers, Box 24, Folder: "God and Man at Yale" (1 of 2), John F. Kennedy Library, Boston, Massachusetts.

93. McGeorge Bundy, "McGeorge Bundy Replies," *The Atlantic,* December 1951, 84.

94. Bundy, Testimony delivered to the Subcommittee on the Reorganization of the Committee on Government Operations of the U.S. Senate, March 15, 1955; Allen, "Communists Should Not Teach in American Colleges."

95. See chapter 1 for discussion of this point. More broadly, on the division between an empirical and secular approach to learning and culture as opposed to one based on religious values—and the shift of the American academy to favor the former, see David A. Hollinger, *Science, Jews, and Secular Culture: Studies in Mid-Twentieth Century Intellectual History* (Princeton, N.J.: Princeton University Press, 1996).

96. Sidney Hook, *Reason, Social Myths and Democracy* (1940; reprint, Buffalo, N.Y.: Prometheus Books, 1991), 76, 79; John T. McGreevy, "Thinking on One's Own: Catholicism in the American Intellectual Imagination, 1928–1960," *Journal of American History* 84, no. 1 (1997): 97–131; David A. Hollinger, "Science as a Weapon in *Kulturkämpfe* in the United States during and after World War II," *Isis* 86 (1995): 440–54.

97. As John Carson points out, assessing the validity of a specific intelligence

test requires comparing it to results gathered by other means. John Carson, *The Measure of Merit: Talents, Intelligence, and Inequality in the French and American Republics, 1750–1940* (Princeton, N.J.: Princeton University Press, 2007), 227.

98. On Terman and gender norms, see Jill G. Morawski, "Impossible Experiments and Practical Constructions: The Social Bases of Psychologists' Work," in *The Rise of Experimentation in American Psychology*, ed. Jill G. Morawski (New Haven, Conn., and London: Yale University Press, 1988); Peter Hegarty, *Gentlemen's Disagreement: Alfred Kinsey, Lewis Terman, and the Sexual Politics of Smart Men* (Chicago: University of Chicago Press, 2013). On the extent of testing for authoritarianism, see Jos D. Meloen, "The F Scale as a Predictor of Fascism: An Overview of 40 Years of Authoritarianism," in *Strength and Weakness: The Authoritarian Personality Today*, 49; Christie, "Authoritarianism Reexamined," 154.

99. On merit, see Jerome Karabel, *The Chosen: The Hidden History of Admission and Exclusion at Harvard, Yale, and Princeton* (Boston: Houghton Mifflin, 2005).

100. Frank Barron, *Creativity and Personal Freedom* (Princeton, N.J.: Van Nostrand, 1968); MacKinnon, "IPAR's Contribution," 77; Ross L. Mooney, "Groundwork for Creative Research," *American Psychologist* 9, no. 9 (1954): 544–48.

Chapter Three

1. Martin Landau, Harold Proshansky, and William Ittelson, "The Interdisciplinary Approach and the Concept of Behavioral Sciences," in *Decisions, Values and Groups*, ed. Norman Washburne (Oxford: Pergamon, 1962), 15.

2. For further discussion of this point see Robert K. Merton, "The Mosaic of the Behavioral Sciences," in *The Behavioral Sciences Today*, ed. Bernard Berelson (New York: Basic Books, 1963).

3. John W. Bennett and Kurt H. Wolff, "Toward Communication between Sociology and Anthropology," *Yearbook of Anthropology* (1955): 329–51, on 333, 345.

4. An example of this appeared in Sidney J. Kaplan, "An Appraisal of an Interdisciplinary Social Science Course," *Journal of Educational Sociology* 34, no. 2 (1960): 70–77.

5. This point about the use of "cross-fertilization" rather than "cross-sterilization" follows from discussion in Margaret B. Luszki, *Interdisciplinary Team Research: Methods and Problems* (Washington, D.C.: National Training Laboratories, 1958).

6. Self Study Committee, *A Report on the Behavioral Sciences at the University of Chicago* (Chicago: University of Chicago, 1954), 24 (emphasis added).

7. Luszki, *Interdisciplinary Team Research*, 193.

8. George Peter Murdock, "The Conceptual Basis of Area Research," *World Politics* 2, no. 4 (1950): 571–78.

9. R. Richard Wohl, "Some Observations on the Social Organization of Interdisciplinary Social Science Research," *Social Forces* 33, no. 4 (1955): 374–83, on 375.

10. Ibid., 374.

11. *The Behavioral Sciences at Harvard* (Cambridge, Mass.: Harvard University, 1955), 12; University of North Carolina, S*urvey of the Behavioral Sciences, 1953–54: Report of the Visiting Committee* (1954), 15; Gordon W. Blackwell, "Multidisciplinary Team Research," *Social Forces* 33, no. 4 (1955): 367–74, on 370.

12. For instances of this phenomenon, see Roy R. Grinker, ed., *Toward a Unified Theory of Human Behavior* (New York: Basic Books, 1956); John Gillin, ed., *For a Science of Social Man: Convergences in Anthropology, Psychology, and Sociology* (New York: Macmillan, 1954).

13. A small sub-sample of the 107 major participants in this project included figures elsewhere discussed in this book, including Else Frenkel-Brunswik, Alex Inkeles, David McClelland, Margaret Mead, Talcott Parsons, R. Nevitt Sanford, and Edward Shils. Luszki, *Interdisciplinary Team Research*, xxi, xxv–xxvii.

14. Ibid., xxii.

15. Merton, "The Mosaic of the Behavioral Sciences," 254.

16. Wohl, "Some Observations," 374.

17. Lorraine Daston and Peter Galison, *Objectivity* (New York: Zone Press, 2007); Theodore M. Porter, *Trust in Numbers: The Pursuit of Objectivity in Science and Public Life* (Princeton, N.J.: Princeton University Press, 1995); M. Norton Wise, ed., *The Values of Precision* (Princeton, N.J.: Princeton University Press, 1995); Joan L. Richards, "Geometry in the Age of Reason: Euclid and the Enlightenment," paper presented at the History of Science Society, Kansas City, 1998.

18. On these developments, see Thomas L. Haskell, *The Emergence of Professional Social Science: The American Social Science Association and the Nineteenth Century Crisis of Authority* (Urbana: University of Illinois Press, 1977); Dorothy Ross, *The Origins of American Social Science* (Cambridge and New York: Cambridge University Press, 1991).

19. On such a moment, see David A. Hollinger, "Two NYUs and the 'Obligation of Universities to the Social Order' in the Great Depression," in *Science, Jews, and Secular Culture: Studies in Mid-Twentieth Century American Intellectual History* (Princeton, N.J.: Princeton University Press, 1996).

20. See "Abstractions in Social Inquiry," in Hook, *Reason, Social Myths and Democracy*, 12–32.

21. For an example of this division from a supporter of modernity and empiricism, see Joseph Jastrow, "Psychology," in *Encyclopaedia of the Social Sciences*,

ed. Edwin R. Seligman (New York: The Macmillan Company, 1934). For one instance of the conflict between metaphysics and science, see the debate between Hutchins on the one hand and John Dewey and Harry Gideonse on the other. Harry D. Gideonse, *The Higher Learning in a Democracy: A Reply to President Hutchins' Critique of the American University* (New York and Toronto: Farrar and Rinehart, 1937); Robert M. Hutchins, *The Higher Learning in America* (New Haven, Conn.: Yale University Press, 1936); John Dewey, "The Higher Learning in America" *Social Frontier* 3, no. 24 (March 1937): 167–69.

22. See for instance, Phillip Frank, "Science and Democracy," in *Science, Philosophy, and Religion: A Symposium* (New York: Conference on Science, Philosophy and Religion in their Relation to the Democratic Way of Life, Inc., 1941). Frank's analysis of the distinction between the rationalistic medieval mindset and the pragmatic modern and scientific way of thinking draws on Alfred N. Whitehead, *Science in the Modern World* (1925). While Whitehead made a similar distinction as Frank between the medieval rationalistic and the modern empiricist scientific method, Frank, unlike Whitehead, wholeheartedly praised the empiricist bent of modern science.

23. Kerry W. Buckley, *Mechanical Man: John Broadus Watson and the Beginnings of Behaviorism* (New York and London: Guilford Press, 1989); Franz Samelson, "Organizing for the Kingdom of Behavior: Academic Battles and Organizational Policies in the Twenties," *Journal of the History of the Behavioral Sciences* 21 (1985): 33–47.

24. John M. O'Donnell, *The Origins of Behaviorism: American Psychology, 1870–1920* (New York: New York University Press, 1985).

25. Edward Franklin Buchner, "Psychological Progress in 1910," *Psychological Bulletin* 8, no. 1 (1911): 1–10.

26. John B. Watson, "Psychology as the Behaviorist Views It," *Psychological Review* 20 (1913): 158–77.

27. John B. Watson, *Behaviorism* (New York: W. W. Norton and Company, 1925), 3.

28. Although this was a point Watson voiced early on, other behaviorists made similar arguments. See, for instance, Clark L. Hull, "Mind, Mechanism and Adaptive Behavior," *Psychological Review* 44 (1937): 1–32; Kenneth W. Spence, *Behavior Theory and Conditioning* (New Haven, Conn.: Yale University Press, 1956).

29. Jill G. Morawski, "Organizing Knowledge and Human Behavior at Yale's Institute of Human Relations," *Isis* 77 (1986): 219–42, on 236.

30. Gordon W. Allport, "The Psychologist's Frame of Reference," *Psychological Bulletin* 37, no. 1 (1940): 1–28, on 4; Jerome S. Bruner and Gordon W. Allport, "Fifty Years of Change in American Psychology," *Psychological Bulletin* 37, no. 10 (1940): 757–77, on 769.

31. Allport, "The Psychologist's Frame of Reference," 5; Bruner and Allport, "Fifty Years of Change," 765.

32. S. S. Stevens, "The Attributes of Tones," *Proceedings of the National Academy of Sciences* 20 (1934): 457–59; "The Relation of Pitch to Intensity," *Journal of the Acoustical Society of America* 6 (1935): 150–54; "Tonal Density," *Journal of Experimental Psychology* 17 (1934): 585–91.

33. S. S. Stevens, "A Scale for the Measurement of a Psychological Magnitude," *Psychological Review* 43, no. 5 (1936): 405–16.

34. S. S. Stevens, "Psychology, the Propaedeutic Science," *Philosophy of Science* 3 (1936): 90–103; "The Operational Definition of Psychological Concepts," *Psychological Review* 42 (1935): 512–27; "The Operational Basis of Psychology," *American Journal of Psychology* 43 (1935): 323–30; "Psychology and the Science of Science," *Psychological Bulletin* 36, no. 4 (1939): 221–63.

35. Gary L. Hardcastle, "S. S. Stevens and the Origins of Operationism," *Philosophy of Science* 62 (1995): 404–24. For analysis of the multiple approaches to operationism, see Joel Isaac, *Working Knowledge: Making the Human Sciences from Parsons to Kuhn* (Cambridge, Mass.: Harvard University Press, 2012).

36. Stevens, "Psychology and the Science of Science," 227–28.

37. Ibid., 227, 33.

38. Ibid., 221.

39. Ibid., 244.

40. Ibid., 236.

41. Ibid., 243.

42. Ibid., 236 (emphasis added).

43. Stevens, "A Scale for the Measurement of a Psychological Magnitude." Stevens produced the loudness scale based on a review article. B. G. Curcher, "A Loudness Scale for Industrial Noise Measurements," *Journal of the Acoustical Society of America* 6 (1935): 216–26. The research on guinea pigs was reported by H. Davis. Stevens does not give a reference for this work.

44. Stevens, "Psychology and the Science of Science," 250. See also pages 223, 240, and 248 for Stevens linking behavioristic psychology to operationism and logical positivism.

45. Stevens himself would not properly have been identified as a behaviorist and therefore would have fallen into the multitude that Spence identified as believing in behaviorism without identifying themselves as such.

46. C. C. Pratt, *The Logic of Modern Psychology* (New York: Macmillan, 1939); Edwin G. Boring, *The Physical Dimensions of Consciousness* (New York: The Century Co., 1933).

47. George Lundberg, "Quantitative Methods in Social Psychology," *American Sociological Review* 1 (1936): 38–54. For Lunberg's position relative to the colleagues at Columbia and his student Harry Alpert, see Mark Solovey and Jefferson D. Pooley, "The Price of Success: Sociologist Harry Alpert, the NSF's First Social Science Policy Architect," *Annals of Science* 68, no. 2 (2011): 229–60. More generally on Lundberg's views, see Jennifer Platt, *A History of Sociologi-*

cal Research Methods in America: 1920–1960, Ideas in Context 40 (New York: Cambridge University Press, 1996).

48. Bruce Kuklick, *The Rise of American Philosophy: Cambridge, Massachusetts, 1860–1930* (New Haven, Conn.: Yale University Press, 1977), 459–63.

49. Newman would subsequently join and eventually chair the Psychology Department at Harvard.

50. This account of the PRT draws on Ludy T. Benjamin, "The Psychological Roundtable: Revolution of 1936," *American Psychologist* 32, no. 7 (1977): 542–49; Gary L. Hardcastle, "The Cult of Experiment: The Psychological Round Table, 1936–1941," *History of Psychology* 3, no. 4 (2000): 334–70.

51. Department of Psychology, Meeting of the Permanent Members. Dated December 11, 1959 (perhaps 1958, based upon a pencil notation and content of the minutes). Department of Psychology and Social Relations, Permanent Member Meetings, Minutes, 1952–83, UAV 714.4, Harvard University Archives.

52. Charles G. Gross, "Psychological Round Table in the 1960s," *American Psychologist* 32, no. 12 (1977): 1120–21. Thanks to Peter Mandler for drawing my attention to this source.

53. On the sexist exclusionary practices of the SEP, see Laurel Furumoto, "Shared Knowledge: The Experimentalists, 1904–1929," in *The Rise of Experimentation in American Psychology*, ed. Jill G. Morawski (New Haven, Conn.: Yale University Press, 1988).

54. Gross, "Psychological Round Table in the 1960s."

55. S. S. Stevens, ed., *Handbook of Experimental Psychology* (New York: Wiley, 1951).

56. The ACLS was composed of the American Philosophical Society, American Academy of Arts and Sciences, American Antiquarian Society, American Oriental Society, American Philosophical Association, Archaeological Institute of America, Modern Language Association, American Philological Association, American Historical Association, American Economic Association, and the American Sociological Society. Charles E. Merriam, "Annual Report of the Social Science Research Council," *The American Political Science Review* 20, no. 1 (1926): 185–89, on 185; Charles E. Merriam, "Annual Report of the American Council of Learned Societies," *The American Political Science Review* 20, no. 1 (1926): 189–92, on 189; Charles E. Merriam, "Annual Report of the Social Science Research Council," *The American Political Science Review* 21, no. 1 (1927): 159–61, on 161.

57. Donald Fisher, *Fundamental Development of the Social Sciences: Rockefeller Philanthropy and the United States Social Science Research Council* (Ann Arbor: University of Michigan Press, 1993), 44.

58. In Louis Wirth, ed., *Eleven Twenty-Six: A Decade of Social Science Research* (Chicago: University of Chicago Press, 1940), 123, 139.

59. See Roundtable discussions in ibid.

60. Robert E. Kohler, "Warren Weaver and the Rockefeller Foundation Program in Molecular Biology, a Case Study in the Management of Science," in *The Sciences in the American Context: New Perspectives*, ed. Nathan Reingold (Washington, D.C.: Smithsonian Institution Press, 1979).

61. John Dewey, "Unity of Science as a Social Problem," in *International Encyclopedia of Unified Science*, ed. Otto Neurath, Rudolf Carnap, and Charles Morris (Chicago: University of Chicago Press, 1938), 33; George A. Reisch, *How the Cold War Transformed Philosophy of Science: To the Icy Slopes of Logic* (Cambridge and New York: Cambridge University Press, 2005), chapter 4.

62. Reisch, *How the Cold War Transformed Philosophy of Science*. Recently David Hollinger has noted that the Unity of Science movement was only a small and "sectarian" segment of a much larger and longer running endeavor that linked science to cultural pluralism and inclusiveness. David A. Hollinger, "The Unity of Knowledge and the Diversity of Knowers: Science as an Agent of Cultural Integration in the United States between the Two World Wars," *Pacific Historical Review* 80, no. 2 (2011): 211–30.

63. Alain Locke, "Pluralism and Intellectual Democracy," in *Science, Philosophy and Religion* (1942).

64. Allport, "The Psychologist's Frame of Reference," 10, 11, 16.

65. Ibid., 23.

66. The connection between operationism and Islam appears in a reference to its "carpet of prayer," ibid., 22.

67. Ibid., 26.

68. Ibid.

69. Ibid., 1. The speech from which the paper was taken was delivered on September 7, 1939.

70. Ibid., 26.

71. Ibid., 21.

72. Ibid., 18.

73. Ibid., 20.

74. J. T. Dunlop et al., "Toward a Common Language for the Area of the Social Sciences" (unpublished white paper, Harvard University, 1941), 2. For discussion of the significance of this document, see "Department and Laboratory of Social Relations: The First Decade, 1946–1956" (Cambridge, Mass.: Harvard University, 1956); Talcott Parsons, "Clyde Kluckhohn and the Integration of the Social Sciences," in *Culture and Life: Essays in Memory of Clyde Kluckhohn*, ed. Walter W. Taylor, John L. Fischer, and Evon Z. Vogt (Carbondale: Southern Illinois University Press, 1973). Although the report is unpublished, a copy may be found in the Tozzer Library at Harvard University. Portions also appear in Talcott Parsons, *Essays in Sociological Theory* (Glencoe, Ill.: The Free Press, 1949).

75. Talcott Parsons and Edward Shils, eds., *Toward a General Theory of Action* (Cambridge, Mass.: Harvard University Press, 1951).

76. For discussion of Harvard's leadership in theoretical sociology after World War II, see Henrika Kuklick, "A 'Scientific Revolution': Sociological Theory in the United States, 1930–1945," *Sociological Inquiry* 43 (1973): 3–22; Edward Shils, "Transition, Ecology, and Institution in the History of Sociology," *Daedalus* 4 (1970): 760–825.

77. Dunlop et al., "Toward a Common Language," 1.

78. Ibid., 11.

79. Ibid., 10 (emphasis added).

80. Talcott Parsons, *The Structure of Social Action: A Study in Social Theory with Special Reference to a Group of Recent European Writers* (New York: McGraw-Hill, 1937); Harry Alpert, *Emile Durkheim and His Sociology* (New York: Columbia University Press, 1939). This reading of Alpert's position on Durkheim depends on the analysis in Solovey and Pooley, "The Price of Success."

81. Dunlop et al., "Toward a Common Language," 22.

82. In fact, several years later, Clyde Kluckhohn equated "ethos" with *Zeitgeist*. Clyde Kluckhohn, *Mirror for Man: The Relation of Anthropology to Modern Life* (New York: McGraw-Hill, 1949), 34. While this work was not published until 1949, it originated in work Kluckhohn undertook in 1945–46.

83. Dunlop et al., "Toward a Common Language," 23.

84. Letter reprinted in G. W. Allport, C. K. M. Kluckhohn, O. H. Mowrer, H. A. Murray, and T. Parsons, "Confidential Memorandum on the Reorganization of the Social Sciences at Harvard." September 1, 1943. Papers of the Dean of the Faculty of Arts and Sciences, Letters to/from, 1945–46, Box: Social Science–Z; Folder: Social Sciences, UA III 5.55.26, Harvard University Archives.

85. Ibid., 3, 5; Wirth, ed., *Eleven Twenty-Six*.

86. Allport et al., "Confidential Memorandum," 10–11.

87. Ibid., 8.

88. On the fraught relationship between Parsons and economics, see Howard Brick, "Talcott Parsons's 'Shift Away from Economics,' 1937–1946," *Journal of American History* 87, no. 2 (2000): 490–514.

89. "Confidential Memorandum," 5 (emphasis added).

90. Ibid., 15.

91. Gordon Allport to E. G. Boring, March 23, 1944. E. G. Boring Correspondence, 1919–69, First File, 1919–56. Box Allen, L.K.–Amer Journal of Psychology. Box 2 of 83. Folder Allport, Gordon, 1944–45, HUG 4229.5, Harvard University Archives.

92. For expression of these debates see E. G. Boring to Gordon Allport, May 28, and June 7, 1945. E. G. Boring Papers, Correspondence, 1919–1969, First File 1919–56, Box 2 of 83, Folder 36, HUG 4229.5, Harvard University Archives.

93. On exams and departmental requirements, see Allport to Boring, April 29, 1941 and E. G. Boring to Gordon Allport, April 27, 1941, Department

of Psychology Papers, Correspondence etc., 1937–1943, Box 2, Birkhoff-Ca/Cn, Folder, Boring, E. G. UAV 714.12, Harvard University Archives.

94. Rodney G. Triplet, "Henry A. Murray and the Harvard Psychological Clinic, 1926–1938: A Struggle to Expand the Disciplinary Boundaries of Academic Psychology" (Ph.D. diss., University of New Hampshire, 1983).

95. Gordon Allport to Dean of the Faculty G. D. Birkhoff, March 30, 1938, Department of Psychology, Correspondence, etc.

96. Talcott Parsons to Paul Buck, May 1, 1945, Papers of the Dean of the Faculty of Arts and Sciences, Letters to/from, 1944–45, R-Sociology Folder: Sociology, UA III 5.55.26, Harvard University Archives.

97. Parsons to Buck, February 19, 1945 and September 11, 1945, Papers of Clyde K. M. Kluckhohn, Correspondence, ca. 1938–1947, AC, Folder: Buck, Paul, HUG 4490.3, Harvard University Archives.

98. Gordon Allport to Paul Buck, February 14, 1945, 2, Papers of Clyde K. M. Kluckhohn.

99. Talcott Parsons to Paul Buck, May 1, 1945, Papers of the Dean of the Faculty of Arts and Sciences, Letters to/from, 1944–45, R-Sociology, Folder: Sociology, UA III 5.55.26, Harvard University Archives.

100. The reasons for Buck's support of interdisciplinarity in the social sciences were probably multiple. As Barry Johnston has shown, Parsons had threatened to leave Harvard if its social sciences were not allowed to progress according to his vision. While that would certainly have been a significant consideration, it seems insufficient for explaining why Buck would have allowed the threat of a single, newly minted full professor to convince him to assent in the restructuring of several social science departments. Barry V. Johnston, "The Contemporary Crisis and the Social Relations Department at Harvard: A Case Study in Hegemony and Disintegration," *The American Sociologist* 29, no. 3 (1998): 26–42, on 3.

101. On the timing of Buck's joining the Parsonian project, see Jamie Cohen-Cole, "Thinking About Thinking in Cold War America" (Ph.D. diss., Princeton University, 2003).

102. Discussion of the suppression of opposition to these plans is recounted in Record of Interview, CD [Charles Dollard], Gordon Allport, Talcott Parsons, Thursday, January 10, 1946, Carnegie Corporation Grant Files, Series 1, Box 163, Folder Harvard University Laboratory of Social Relations. CUA.

103. Boring's opposition is made clear in Boring to Stevens and Newman, November 24, 1945; Boring to Buck, November 27, 1945, Department of Psychology Papers, Department of Psychology Chairman's Correspondence—Gordon W. Allport, 1943–46, A-D Box 1 of 4, Folder Buck re: Fission, UAV 714.12, Harvard University Archives; E. G. Boring to Robert Yerkes, December 3 and 7, 1945, E. G. Boring Papers, first file, Box 66 of 83, Folder #1513, HUG 4229.5, Harvard University Archives.

104. Administrative pressure is recounted in E. G. Boring to Robert Yerkes, December 7, 1945, E. G. Boring Papers.

105. Boring to Buck, November 27, 1945, Department of Psychology Papers, Harvard University Archives.

106. Robert Seidel, "The Origins of Lawrence Berkeley Laboratory," in *Big Science: The Growth of Large-Scale Research*, ed. Peter L. Galison and Bruce Hevly (Stanford, Calif.: Stanford University Press, 1992).

107. On Weaver's promotion of interdisciplinarity at Rockefeller and the transfer of these techniques to further war aims, see David A. Mindell, *Between Human and Machine: Feedback, Control, and Computing before Cybernetics* (Baltimore, Md.: Johns Hopkins University Press, 2002), 188–89.

108. Roundtable on "One Science or Many" in Wirth, ed., *Eleven Twenty-Six*.

109. On war work, see Peter Buck, "Adjusting to Military Life: The Social Sciences Go to War, 1941–1950," in *Military Enterprise and Technological Change: Perspectives on the American Experience*, ed. Merritt Roe Smith (Cambridge, Mass.: MIT Press, 1985); James H. Capshew, *Psychologists on the March: Science, Practice, and Professional Identity in America, 1929–1969*; Ellen Herman, *The Romance of American Psychology: Political Culture in an Age of Experts* (Berkeley and Los Angeles: University of California Press, 1995). The outcomes of these projects include major fixtures of Cold War social science. For instance: OSS Assessment Staff, *Assessment of Men: Selection of Personnel for the Office of Strategic Services* (New York: Reinhart, 1948); Samuel A. Stouffer et al., *The American Soldier* (Princeton, N.J.: Princeton University Press, 1949).

110. William Sewell, "Some Reflections on the Golden Age of Interdisciplinary Social Psychology," *Annual Review of Sociology* 15 (1989): 1–16, on 5.

111. Ibid.

112. Robin M. Williams Jr., "Some Observations on Sociological Research in Government During World War II," *American Sociological Review* 11, no. 5 (1946): 573–77, on 574.

113. Bruner to Allport, September 17, 1945, Papers of Jerome S. Bruner, HUG 4242.5, Harvard University Archives.

114. While he had the support of the primaries in psychology (E. G. Boring and Gordon Allport) and some support in sociology (Talcott Parsons, Gordon Allport, and Clyde Kluckhohn), Bruner's appointment to the Department of Sociology was blocked by Pitrim Sorokin and Carle Zimmerman, who saw Bruner's interests and methods as too far outside their field and interests. (Although their appointments were, respectively, in anthropology and psychology, Clyde Kluckhohn and Gordon Allport were voting members of the Department of Sociology in 1945.) Bruner's efforts to be appointed to the Department of Sociology are discussed in Jerome S. Bruner to Gordon Allport, May 24, 1945 and E G. Boring

to Paul Buck, September 6, 1945, Papers of Jerome S. Bruner, Correspondence 1944–47, A-E, Folder: Allport, G. W. 1943–44, HUG 4242.5, Harvard University Archives; Talcott Parsons to Paul Buck, May 1, 1945, Papers of the Dean of Faculty of Arts and Sciences, Letters to/from 1944–45, R-Sociology. Folder: Sociology, UAIII 5.55.26, Harvard University Archives.

115. Samuel A. Stouffer, "Research Techniques," in *The Social Sciences at Mid-Century* (Freeport, N.Y.: Books for Libraries Press, 1968 [1952]), 69.

116. Leonard S. Cottrell Jr., "Strengthening the Social Sciences in the Universities," in *The Social Sciences at Mid-Century*, 32.

117. Charles Dollard, "Strategy for Advancing the Social Sciences," in *The Social Sciences at Mid-Century*, 18. For discussion of Dollard and his relationship with Stouffer, see Ellen Condliffe Lagemann, *The Politics of Knowledge: The Carnegie Corporation, Philanthropy, and Public Policy* (Middletown, Conn.: Wesleyan University Press, 1989).

118. The Center for Advanced Study in the Behavioral Sciences, "Report of the Planning Group," and meeting minutes of CASBS, June 1952. Papers of Paul Herman Buck, Box: Center for Advanced Study in the Behavioral Sciences, 1952–late 1960s, Folder: Misc., HUG (FP) 9.20, Harvard University Archives.

119. The yearly expenditure figures given above are only approximate since the available data gives total project allocations. Most lasted over several years and the data does not specify the expenditure burn rate. Figures given above are thus calculated as a fraction of the total grant with the assumption that funds would be used at an equal rate year to year. The total figures for 1952 are: CASBS ($1,150,000) and "Improving Content and Methods" ($300,000). These are also only approximate figures, but are included in the available data. Summary Report of the Behavioral Sciences Division, 1952–53, Papers of Paul Herman Buck, Harvard University Archives.

120. John W. Gardner, "International Relations and Sociology: Discussion," *American Sociological Review* 13, no. 3 (1948): 273–75, on 273.

121. Carnegie Corporation of New York, *Annual Report* (1946), 27–28; Roger L. Geiger, "American Foundations and Academic Social Science, 1945–1960," *Minerva* 26 (1988): 315–41, on 319.

122. *The Behavioral Sciences at Harvard*, 12.

123. Lagemann, *The Politics of Knowledge*, 170.

124. Cohen-Cole, "Thinking About Thinking," 179–81.

125. John Gardner and Talcott Parsons, January 5, 1949, record of interview. Carnegie Corporation Grant Files. Talcott Parsons and Edward Shils, eds., *Toward a General Theory of Action* (Cambridge, Mass.: Harvard University Press, 1951).

126. For discussion of theory as tools in this project, see Joel Isaac, "Tool Shock: Technique and Epistemology in the Postwar Social Sciences," *History of*

Political Economy 42 (Annual Supplement 2010), 133-64; Joel Isaac, "Theorist at Work: Talcott Parsons and the Carnegie Project on Theory, 1949-1951," *Journal of the History of Ideas* 71, no. 2 (2010): 287-311.

127. A conversation with Dollard on May 13, 1946, had "fortified" Stouffer enough to "tell Conant bluntly that Harvard should not ask for any outside funds until the show was underway." Record of Interview, CD [Charles Dollard] and Samuel Stouffer, September 26, 1946. Carnegie Corporation Grant Files.

128. Samuel Stouffer to Paul Buck, May 14, 1946, Papers of the Dean of the Faculty of Arts and Sciences, Letters to/from, 1945-46, Romance Langs. and Lits.—Social Relations, Folder: Social Relations, UAV 5.55.26, Harvard University Archives.

129. One instance was a five-page document from Conant to the Carnegie Corporation arguing just how much Harvard had committed to Social Relations. "Memorandum to the Carnegie Corporation," James B. Conant. Receipt acknowledged by Charles Dollard July 11, 1946, Carnegie Corporation Grant Files, Series 1, Box 163, Folder: Harvard University Laboratory of Social Relations.

130. Record of Interview, CD [Charles Dollard] and Samuel Stouffer, September 26, 1946. Carnegie Corporation Grant Files. Series 1, Box 163, Folder: Harvard University Laboratory of Social Relations.

131. Samuel Stouffer to Paul Buck, May 14, 1946, Harvard University Archives.

132. These developments are discussed in the report of Harvard's self-study, *The Behavioral Sciences at Harvard* (1955), and in correspondence between Talcott Parsons and McGeorge Bundy, who was then dean of the faculty. Talcott Parsons to McGeorge Bundy, May 9, 1955, Papers of Talcott Parsons, Correspondence and Related Papers, ca. 1930-59, Box 23 of 28, Folder: Social Relations Dept, 1948-55, HUG(FP) 15.2, Harvard University Archives.

133. Discussion of these events may be found in Ethan Schrum, "Administering American Modernity: The Instrumental University in the Postwar United States" (Ph.D. diss., University of Pennsylvania, 2009), 200-201; Emily Hauptmann, "The Constitution of Behavioralism: The Influence of the Ford Foundation's Behavioral Science's Program on Political Science," paper presented at the Sixth Annual Workshop on the History of Economics as History of Science, École Normale Supérieure de Cachan, June 19, 2009.

134. The committee members were Thomas Carroll, John Gillin, Ernest Hilgard, F. F. Hill, Clark Kerr, Harold Lasswell, Paul Lazarsfeld, Robert K. Merton, Talcott Parsons, Edward Shils, Samuel Stouffer, Ralph Tyler, Donald Marquis, and Hans Speier.

135. People attending the meetings of the third committee were David McClelland, Irving Hallowell, Richard Hofstadter, Carl Hovland, Paul Lazarsfeld, Robert K. Merton, Henry Murray, Edward Shils, Joseph Spengler, Ralph Tyler, Allen Barton, Bernard Berelson, Thomas Cochran, Herbert Simon, and Robert

Dahl. Meeting Minutes of the Informal Planning Group on the Center for Advanced Study, Papers of Paul H. Buck, Center for Advanced Study in the Behavioral Sciences, 1952–late 1960s, Folder: board minutes, group minutes, HUG (FP) 9.20, Harvard University Archives.

136. Jefferson Pooley and Mark Solovey, "Marginal to the Revolution: The Curious Relationship between Economics and the Behavioral Sciences Movement in Mid-Twentieth-Century America," *History of Political Economy* 42, annual supplement (2010): 199–233, on 226.

137. John W. Gardner Interview, Carnegie Corporation Oral History Project, Columbia University Archives; Charles Thomas O'Connell, "Social Structure and Science: Soviet Studies at Harvard" (Ph.D. diss., UCLA, 1990).

138. Pooley and Solovey, "Marginal to the Revolution."

139. Hunter Crowther-Heyck, "Patrons of the Revolution: Ideals and Institutions in Postwar Behavioral Science," *Isis* 97, no. 3 (2006): 420–46; Hunter Crowther-Heyck, "Herbert Simon and the GSIA: Building an Interdisciplinary Community," *Journal of the History of the Behavioral Sciences* 42, no. 4 (2006): 311–34.

140. See, for instance, the discussion of political science in Gilman, *Mandarins of the Future.*

141. W. Allen Wallis, *The 1953–54 Program of University Surveys of the Behavioral Sciences* (Behavioral Sciences Division, Ford Foundation, 1955).

142. On the effect of Ford grants on Stanford, see Rebecca S. Lowen, *Creating the Cold War University: The Transformation of Stanford* (Berkeley and Los Angeles: University of California Press, 1997), chapter 7.

143. Sewell, "Some Reflections on the Golden Age of Interdisciplinary Social Psychology," 5; Wallis, *The 1953–54 Program*, 5.

144. James Bryant Conant, "Memorandum to the Carnegie Corporation," n.d. (receipt acknowledged July 11, 1946). Carnegie Corporation Grant Files.

145. "Some Considerations on the Integration of Research," Papers of the Department Social Relations, Correspondence, etc. 1948–194[?], A-K, UAV 801.2010, Harvard University Archives.

146. Carnegie Corporation Grant Files, Series 1, Box 163, Folder: Harvard University Laboratory of Social Relations.

147. Talcott Parsons to Paul Buck, October 10, 1947, S. S. Wilks Papers, Folder: Mosteller, Frederic C., B W665, American Philosophical Society Archives.

148. Samuel Stouffer to Parsons October 9, 1947, S. S. Wilks Papers (emphasis added).

149. Talcott Parsons to McGeorge Bundy, May 9, 1955; Jerome Bruner to McGeorge Bundy, May 10, 1955; Samuel Stouffer to McGeorge Bundy, May 9, 1955; Gordon Allport to McGeorge Bundy, May 9, 1955, Papers of Talcott Parsons, Harvard University Archives.

150. Harry Alpert, "The National Science Foundation and Social Science Re-

search," *American Sociological Review* 19, no. 2 (1954): 208–11, on 211; Thomas Gieryn, "The U.S. Congress Demarcates Natural Science and Social Science (Twice)," in *Cultural Boundaries of Science: Credibility on the Line* (Chicago: University of Chicago Press, 1999). This point about the nonscientific nature of the social sciences was not limited to reactionary congressmen. As Daniel Lerner reported, humanistically oriented academics like Nathan Glazer also held this position. See *"The American Soldier* and the Public," in *Continuities in Social Research: Studies in the Scope and Method of the American Soldier*, ed. Robert K. Merton and Paul F. Lazarsfeld (Glencoe, Ill.: The Free Press, 1950).

151. Bernard Barber, *Science and the Social Order* (Glencoe, Ill.: The Free Press, 1952), 248.

152. On the Kinsey reports and reactions to its methods, see Sarah E. Igo, *The Averaged American: Surveys, Citizens, and the Making of a Mass Public* (Cambridge, Mass.: Harvard University Press, 2007).

153. Vannevar Bush to Harlow Shapley on November 9, 1946, Papers of Vannevar Bush, Box 13, Folder: Bohman, Isaiah. Manuscript Division, Library of Congress. For further analysis of this incident and how Bush and the science establishment reacted, see Mark Solovey, "Riding the Natural Scientists' Coattails onto the Endless Frontier: The SSRC and the Quest for Scientific Legitimacy," *Journal of the History of the Behavioral Sciences* 40, no. 4 (2004): 393–422.

154. George A. Lundberg, "The Social Sciences in the Post-War Era," *Sociometry* 8, no. 2 (1945): 137–49.

155. Barber, *Science and the Social Order*, 238.

156. Solovey, "Riding the Natural Scientists' Coattails," 406.

157. Dollard, "Strategy for Advancing the Social Sciences," 12.

158. Gieryn, "The U.S. Congress Demarcates Natural Science and Social Science (Twice)."

159. See, for instance, Barber, *Science and the Social Order*, 12–16, 242. Proponents of interdisciplinarity tended to rely on Conant. See Self-Study Committee, *A Report on the Behavioral Sciences at the University of Chicago*, 6–7.

160. James B. Conant, *Modern Science and Modern Man* (New York: Doubleday, 1952), 106–7. Cited in Alpert, "The National Science Foundation and Social Science Research," 210.

161. Barber, *Science and the Social Order*, 14.

162. Luszki, *Interdisciplinary Team Research*, 168–69.

163. Harry Alpert, "The Social Sciences and the National Science Foundation: 1945–1955," *American Sociological Review* 20, no. 6 (1955): 653–61, on 656–57 (emphasis added).

164. Alpert, "The National Science Foundation and Social Science Research," 210. This is a direct quote from Conant, *Modern Science and Modern Man*, 106–7. The fit between Conant's philosophy of science and that of social scientists who pressed for interdisciplinarity is likely a product of the fact that

both borrowed from the epistemological views of L. J. Henderson. On Henderson and his influence on Conant and social sciences, see "Making a Case: The Harvard Pareto Circle," chapter 2 in Joel Isaac, *Working Knowledge: Making the Human Sciences from Parsons to Kuhn* (Cambridge, Mass.: Harvard University Press, 2012).

165. Alpert, "The National Science Foundation and Social Science Research," 210.

166. Cohen-Cole, "Thinking About Thinking," 168–72.

167. Harry H. Harman, *The Psychologist in Interdisciplinary Research* (Santa Monica, Calif.: RAND Corporation, 1955), 7.

168. Ibid.

169. Luszki, *Interdisciplinary Team Research*, 10.

170. Kenneth Roose, "Observations on Interdisciplinary Work in the Social Sciences," in *Interdisciplinary Relationships in the Social Sciences*, ed. Muzafer Sherif and Carolyn Sherif (Chicago: Aldine, 1969), 323–24.

171. Raymond B. Cattell, "The Personality and Motivation of the Researcher from Measurements of Contemporaries and from Biography," in *Scientific Creativity: Its Recognition and Development* (New York and London: John Wiley and Sons, 1963); Crutchfield, "Conformity and Creative Thinking"; Gutman, "The Biological Roots of Creativity"; Lawrence S. Kubie, "Blocks to Creativity," in *Explorations in Creativity*, ed. Ross L. Mooney and Taher A. Razik (New York: Harper and Row, 1967), 36; Frank Barron, "The Needs for Order and for Disorder as Motives in Creative Activity," in *Scientific Creativity: Its Recognition and Development*, ed. Calvin W. Taylor and Frank Barron (New York and London: John Wiley and Sons, 1963), 155.

172. Sharon Ghamari-Tabrizi, "Simulating the Unthinkable: Gaming Future War in the 1950s and 1960s," *Social Studies of Science* 30, no. 2 (2000): 163–223, on 170.

173. Luszki, *Interdisciplinary Team Research*, 62, 175, 204, 205, 212, 234.

174. Ibid., 205. For another call for tolerance and flexibility in interdisciplinary endeavors, see Marshall K. Powers, "Area Studies," *Journal of Higher Education* 26, no. 2 (1955): 82–89, 113, on 113.

175. Blackwell, "Multidisciplinary Team Research," 371.

176. Mooney, "Groundwork for Creative Research."

177. Frenkel-Brunswik, "Tolerance Toward Ambiguity"; Rokeach, *The Open and Closed Mind*, 16.

178. Charles N. Cofer and George Horsley Smith, "A Simple Inexpensive Classroom Demonstration of Concept Formation," *American Psychologist* 2, no. 11 (1947): 521–22, on 522; Milton Rokeach, "'Narrow-Mindedness' and Personality," *Journal of Personality* 30, no. 2 (1951): 234–51.

179. My thanks to Eugenia Lao for this observation.

180. Donald T. Campbell, "Ethnocentrism of Disciplines and the Fish-Scale

Model of Omniscience," in *Interdisciplinary Relationships in the Social Sciences*, ed. Muzafer Sherif and Carolyn Sherif (Chicago: Aldine, 1969).

Chapter Four

1. Clark Kerr, *The Uses of the University*, 4th ed. (Cambridge, Mass.: Harvard University Press, 1995). Originally published in 1963.

2. Daniel Bell, *The Reforming of General Education: The Columbia College Experience in Its National Setting* (New York: Columbia University Press, 1966). For a similar analysis that situated the academy as alternately the center or the microcosm of society, see Kerr, *The Uses of the University*, 15, 27, 31.

3. Daniel Bell, "Notes on the Post-Industrial Society," *The Public Interest*, no. 6 (1967): 24–35, on 30. This work found later publication in book form. Daniel Bell, *The Coming of Post-Industrial Society: A Venture in Social Forecasting* (New York: Basic Books, 1973).

4. A more recent characterization of science as a paper chase may be found in Bruno Latour and Steven Woolgar, *Laboratory Life: The Construction of Scientific Facts* (Princeton, N.J.: Princeton University Press, 1986).

5. On the efforts of academics to separate their work from corporate research, see Shapin, *The Scientific Life*.

6. On the dramatic growth of physics during World War II and the Cold War, see Daniel J. Kevles, *The Physicists: The History of a Scientific Community in Modern America* (Cambridge, Mass.: Harvard University Press, 1995); David Kaiser, "Cold War Requisitions, Scientific Manpower, and the Production of American Physicists after World War II," *Historical Studies in the Physical and Biological Sciences* 33, no. 1 (2002): 131–59; Peter L. Galison, *Image and Logic: A Material Culture of Microphysics* (Chicago: University of Chicago Press, 1997); Paul Forman, "Behind Quantum Electronics: National Security as a Basis for Physical Research in the United States, 1940–1960," *Historical Studies in the Physical and Biological Sciences* 18, no. 1 (1987): 149–229. On the use of the idioms of social criticism within the physics community, see David Kaiser, "The Postwar Suburbanization of American Physics," *American Quarterly* 56 (2004): 851–88. For the expansion of science and of the university more generally during this period, see Peter L. Galison and Bruce Hevly, eds., *Big Science: The Growth of Large-Scale Research* (Stanford, Calif.: Stanford University Press, 1992); David Hounshell, "The Cold War, RAND, and the Generation of Knowledge," *Historical Studies in the Physical Sciences* 27 (1997): 236–67; Stuart W. Leslie, *The Cold War and American Science: The Military-Industrial-Academic Complex at M.I.T. and Stanford* (New York: Columbia University Press, 1993); Capshew, *Psychologists on the March*; Roger L. Geiger, *Research and Relevant Knowledge: American Research Universities since World War II* (New York and

Oxford: Oxford University Press, 1993); R. C. Lewontin, "The Cold War and the Transformation of the Academy," in *The Cold War and the University*, ed. Noam Chomsky et al. (New York: The New Press, 1997).

7. For typical expressions of these concerns, see, for instance, Daniel Bell, "The Disjunction of Culture and Social Structure: Some Notes on the Meaning of Social Reality," *Daedalus* 94, no. 1 (1965): 208–22; Lyman Bryson, Louis Finkelstein, and Robert M. MacIver, eds., *Approaches to National Unity: Seventh Symposium, Conference on Science, Philosophy and Religion* (New York: Harper and Brothers, 1945); James B. Conant, "Unity and Diversity in Secondary Education," *Vital Speeches* 18, no. 15 (1952): 463–65.

8. Peter L. Galison, Bruce Hevly, and Rebecca Lowen, "Controlling the Monster: Stanford and the Growth of Physics Research, 1935–1962," in *Big Science*; Mirowski, *Machine Dreams*; Edwards, *The Closed World: Computers and the Politics of Discourse in Cold War America*.

9. On speaking properly as a marker of cultural capital, see Pierre Bourdieu, "The Production and Reproduction of Legitimate Language," in *Language and Symbolic Power: The Economy of Linguistic Exchanges* (Cambridge, Mass.: Harvard University Press, 1991). For effects of genteel discourse and self-presentation on the production, communication, and stabilization of scientific work, see Mario Biagioli, *Galileo, Courtier: The Practice of Science in the Culture of Absolutism* (Chicago: University of Chicago Press, 1993); Steven Shapin, *A Social History of Truth: Civility and Science in Seventeenth-Century England* (Chicago: University of Chicago Press, 1994).

10. For insightful analysis of the "subculture" as a category of analysis in intellectual history, see Joel Isaac, "Tangled Loops: Theory, History and the Human Sciences in Modern America," *Modern Intellectual History* 6 (2009): 397–424.

11. Thomas C. Schelling, *The Strategy of Conflict* (Cambridge, Mass.: Harvard University Press, 1980 [1960]), 12–13. Schelling's framing was far from idiosyncratic. For discussion of how the U.S. press and foreign policy establishment portrayed Mossadeq as a child in need of discipline, see John Foran, "Discursive Subversions: *Time* Magazine, the CIA Overthrow of Mussaddiq, and the Installation of the Shah," in *Cold War Constructions: The Political Culture of United States Imperialism*, ed. Christian G. Appy (Amherst: University of Massachusetts Press).

12. Amadae, *Rationalizing Capitalist Democracy*; Nicholas Giuilhot, "Cyborg Pantocrator: International Relations Theory from Decisionism to Rational Choice," *Journal of the History of the Behavioral Sciences* 47, no. 3 (2011): 279–301.

13. For discussion of the content and context-free nature of small group studies as well as its rise and fall, see Joseph E. McGrath, "Small Group Research, That Once and Future Field: An Interpretation of the Past with an Eye to the Future," *Group Dynamics: Theory, Research, and Practice* 1, no. 1 (1997):

7–27; John D. Greenwood, *The Disappearance of the Social in American Social Psychology* (Cambridge and New York: Cambridge University Press, 2004), 185–215.

14. Margaret Mead and Paul Byers, *The Small Conference: An Innovation in Communication* (The Hague, Paris: Mouton, 1968), 10–13, 113–16.

15. For instance, Robert F. Bales, "How People Interact in Conferences," *Scientific American* 192 (1955): 31–55; D. G. Marquis, Harold Guetzkow, and R. W. Heyns, "A Social Psychological Study of the Decision-Making Conference," in *Groups, Leadership and Men: Research in Human Relations*, ed. Harold Guetzkow (Pittsburgh, Pa.: Carnegie Press, 1951).

16. See, for instance, Sidney Verba, *Small Groups and Political Behavior: A Study of Leadership* (Princeton, N.J.: Princeton University Press, 1961). One seminal study, completed in the 1950s and which circulated in draft form for two decades, argued that the nation is a network of individuals connected to one another. That is, it is not an anonymous, faceless mass. Even a "hermit in the Okefenokee Swamps" and another "hermit in the Northwest woods" would be linked to each other through a chain of only about seven individuals, each who knew at least one or two other members of the chain. Ithiel de Sola Pool and Manfred Kochen, "Contacts and Influence," *Social Networks* 1, no. 1 (1978): 5–51, on 18.

17. Mead and Byers, *The Small Conference*, 3, 9.

18. Ibid., 111.

19. Ibid., 108.

20. UNESCO, *The Technique of International Conferences: A Progress Report on Problems and Methods* (Paris: UNESCO/SS/3, 1951).

21. Brock Chisholm, preface to Mary Capes, *Communication or Conflict: Conferences: Their Nature, Dynamics, and Planning* (New York: Association Press, 1960), xi. General biographical and professional information on Chisholm may be found in Allan Irving, *Brock Chisholm: Doctor to the World* (Markham, Ont.: Fitzhenry and Whiteside, 1998); John Farley, *Brock Chisholm, the World Health Organization, and the Cold War* (Vancouver: University of British Columbia Press, 2008).

22. Techniques for studying conferences were supplied by the specialists in small group studies. These were the University of Michigan Research Center for Group Dynamics, Harvard Department of Social Relations, Tavistock Institute of Human Relations, and the University of Liverpool Department of Social Science. UNESCO, *The Technique of International Conferences*, 5. For discussion of the UN and its relationship to small group research, see Mead and Byers, *The Small Conference*.

23. Perrin Selcer, "The View from Everywhere: Disciplining Diversity in Post–World War II International Social Science," *Journal of the History of the Behavioral Sciences* 45, no. 4 (2009): 309–29, on 324.

24. The Macy conferences are detailed in Heims, *The Cybernetics Group*; Dupuy, *The Mechanization of the Mind*.

25. Margaret Mead, "Cybernetics of Cybernetics," in *Purposive Systems: Proceedings of the First Annual Symposium of the American Society for Cybernetics*, ed. Heinz von Foerster et al. (New York and Washington: Spartan Books, 1968), 9–10.

26. Margaret Mead, preface to Charles Darwin, *The Expression of the Emotions in Man and Animals. With Photographic and Other Illustrations. With a Preface by Margaret Mead* (New York: Philosophical Library, 1955), vi.

27. The conference proceedings, first published in early 1965, indicate that the conference took place about a year earlier. The participants were Daniel Bell, Frederick Burkhardt, Douglas Bush, Bruce Chalmers, Benjamin DeMott, Lillian Hellman, Hudson Hoagland, György Kepes, Leon Kirchner, Harold Lasswell, William Letwin, Herbert Marcuse, Margaret Mead, Robert K. Merton, Elting Morison, Robert Morison, F. S. C. Northrop, Talcott Parsons, Don K. Price, Edward Purcell, W. V. O. Quine, I. A. Richards, Walter Rosenblith, B. F. Skinner, Krister Stendahl, Julius Stratton, George Wald, and Harry Woolf. "Preface to the Issue," *Daedalus* 94, no. 1 (1965): iii–iv, on iii.

28. Daniel Bell, "The Disjunction of Culture and Social Structure: Some Notes on the Meaning of Social Reality," in *Science and Culture: A Study of Cohesive and Disjunctive Forces*, ed. Gerald Holton (Boston: Beacon Press, 1967), 244. Originally published as Bell, "The Disjunction of Culture and Social Structure: Some Notes on the Meaning of Social Reality," *Daedalus* 94, no. 1 (1965): 208–22.

29. James S. Ackerman, "On *Scientia*," in *Science and Culture: A Study of Cohesive and Disjunctive Forces*, ed. Gerald Holton (Boston: Beacon Press, 1967), 19.

30. Ackerman's sense of the creativity of interdisciplinarity also appeared in his analysis of the relative architectural skills of Michelangelo and Antonio da Sangallo the Younger. James S. Ackerman, *The Architecture of Michelangelo* (New York: Viking Press, 1961), xxxiii–xxxiv.

31. Gerald Holton, "The Thematic Imagination in Science," in *Science and Culture: A Study of Cohesive and Disjunctive Forces*, ed. Gerald Holton (Boston: Beacon Press, 1967); Gerald Holton, *Thematic Origins of Scientific Thought*, 2d ed. (Cambridge, Mass.: Harvard University Press, 1988).

32. W. W. Rostow, "The National Style," in *The American Style: Essays in Value and Performance*, ed. Elting E. Morison (New York: Harper and Brothers, 1958); George F. Kennan, "America's Administrative Response to Its World Problems," in *The American Style*; Richard M. Bissell Jr., "Comment on Kennan," in *The American Style*.

33. Robert S. McNamara, "Convocation Address at Millsaps College." Febru-

ary 24, 1967. Copy in Joshua Lederberg Papers, 1904–2003, Box 105, Folder 28. Modern Manuscripts Collection, History of Medicine Division, National Library of Medicine, Bethesda, Md.; MS C 552. *Unique Identifier*: BBGDMS. Copy: http://profiles.nlm.nih.gov/BB/G/D/M/S/.

34. On the elite theory of democracy in postwar political science, see Gilman, *Mandarins of the Future*, 47–56.

35. Mills, "Mass Society and Liberal Education," 357–58.

36. *General Education in a Free Society*, 54.

37. Jacob Bronowski, *Science and Human Values* (New York: Julian Messner, 1956).

38. For discussion of the pervasiveness of this practice, see Dennis Wrong, "The United States in Comparative Perspective: Max Lerner's America as a Civilization," *The American Journal of Sociology* 65, no. 5 (1960): 499–604. Wrong suggests that disciplines treating themselves as nations fulfill Emile Durkheim's hopes for the management of anomie in mass society.

39. OSS Assessment Staff, *Assessment of Men: Selection of Personnel for the Office of Strategic Services*.

40. For commentary on Chicago's social sciences, see Leonard D. White, "Introduction," in *The State of the Social Sciences*, ed. Leonard D. White (Chicago: University of Chicago Press, 1956), ix; Merton, "The Mosaic of the Behavioral Sciences."

41. Gardner, *Excellence: Can We Be Equal and Excellent Too?* 10.

42. John Gillin, "Grounds for a Science of Social Man," in *For a Science of Social Man*, ed. John Gillin (New York: Macmillan, 1954), 4.

43. On anti-Semitism in the fields of psychology and history, see Andrew S. Winston, "'As His Name Indicates': R. S. Woodworth's Letters of Reference and Employment for Jewish Psychologists in the 1930s," *Journal of the History of the Behavioral Sciences* 32, no. 1 (1996): 30–43; Andrew S. Winston, "The Defects of His Race: E. G. Boring and Antisemitism in American Psychology," *History of Psychology* 1, no. 1 (1998): 27–51; Peter Novick, *That Noble Dream: The "Objectivity Question" and the American Historical Profession* (Cambridge and New York: Cambridge University Press, 1988), 172–73.

44. Hollinger, "Jewish Intellectuals and the De-Christianization of American Public Culture in the Twentieth Century."

45. Gillin, "Grounds for a Science of Social Man," 5. For other instances of the equation of interdisciplinarity with social toleration, see Allport, "The Psychologist's Frame of Reference"; Locke, "Pluralism and Intellectual Democracy"; Powers, "Area Studies," 84.

46. Selcer, "The View from Everywhere."

47. Kerr, *The Uses of the University*, 15, 27, 31.

48. For an explicit instance of this sentiment, see Mead and Byers, *The Small Conference*, 5–6.

49. One study noted the inverse correlation between authoritarianism scores and several different measures of intelligence including mathematical operations, mathematical problems, scientific vocabulary, scientific facts, scientific experience, scientific interpretation, intellectual curiosity, and sophistication. Frank N. Jacobson and Salomon Rettig, "Authoritarianism and Intelligence," *Journal of Social Psychology* 50, no. 2 (1959): 213–19, on 215, 17.

50. For commentary by founders, see "Points of Discussion Regarding A Center for Advanced Study in the Behavioral Sciences," Papers of Paul H. Buck, Center for Advanced Study in the Behavioral Sciences, 1952–late 1960s, Folder: board minutes, group minutes. Subsequent discussion about the Center can be found in "Report to the Board of Trustees and Advisory Committee of a Committee on the Future of the Center Appointed by the Director," November 16, 1954, p. 12, Papers of Paul Hovland, Box 13, Folder 19, Yale University Archives.

51. Frank Fremont-Smith, "The Josiah Macy, Jr. Foundation Conference Program," in *Cybernetics: Circular, Causal, and Feedback Mechanisms in Biological and Social Systems, 10th Annual Meeting*, ed. Heinz Von Foerster (New York: Josiah Macy, Jr. Foundation, 1955), 11.

52. Among the authors in Anshen's several series were Marshall McLuhan, Jonas Salk, Werner Heisenberg, Paul Tillich, Konrad Adenauer, Erich Fromm, Kenneth Boulding, George Miller, H. Stuart Hughes, Raymond Aron, Howard Mumford Jones, Jacques Maritain, René Dubos, and Robert McIver.

53. Mary Catherine Bateson, "Notes on Varieties of Unification," n.d. (evidence in other documents indicates c. 1968), Papers of Ruth Nanda Anshen, Box 15, no folder, Columbia University Archives.

54. Crutchfield, "Conformity and Creative Thinking"; Crutchfield, "Conformity and Character"; McGreevy, "Thinking on One's Own: Catholicism in the American Intellectual Imagination, 1928–1960."

55. E. G. Boring to Paul Buck, September 9, 1946, Dean, Faculty of Arts and Sciences, Letters to/from, 1946–47, Philosophy—S Misc., UA III 5.55.26, Harvard University Archives.

56. Donald T. Campbell, "A Tribal Model of the Social System Vehicle Carrying Scientific Knowledge," *Knowledge* 1, no. 2 (1979): 181–201. These figures are based on the research reported in D. L. Krantz and L. Wiggins, "Personal and Impersonal Channels of Recruitment in the Growth of Theory," *Human Development* 16 (1973): 133–56. My thanks to William Wimsatt for bringing this source to my attention.

57. Morawski, "Organizing Knowledge," 237.

58. Charles Gross recounting his experiences with Skinner, personal communication.

59. Verplanck, "Glossary."

60. On Charles Hockett's unifying efforts at CASBS, see Gregory Radick,

The Simian Tongue: The Long Debate about Animal Language (Chicago: University of Chicago Press, 2007), 288–89.

61. Center for Advanced Study in the Behavioral Sciences, "List of Fellows," 1955–2008. My thanks to Melissa Johnson, director of Development and Communications for the Center, for making this information available.

62. George A. Miller, Eugene Galanter, and Karl H. Pribram, *Plans and the Structure of Behavior* (New York: Henry Holt and Company, 1960).

63. "CASBS Special Projects by year, 1956–2010, by final year," http://www .casbs.org/userfiles/Special_projects_1956–2010.pdf.

64. Herbert Simon Archive, "Consulting—1942–2000—Box 42—Administrative Record—Ford Foundation Center for Advanced Study in the Behavioral Sciences—Planning Group: December 1, 1952": 4. http://diva.library.cmu.edu/ webapp/simon/item.jsp?q=/box00042/fld03362/bd10005/doc0001/.

65. Meeting Minutes of the Informal Planning Group on the Center for Advanced Study, fourth meeting, December 20, 1952, p. 8, Papers of Paul H. Buck, Harvard University Archives.

66. Ibid., p. 23.

67. Ibid.

68. Sigmund Koch, "Epilogue: Some Trends of Study I," in *Psychology: A Study of a Science*, vol. 3: *Formulations of the Person and the Social Context*, ed. Sigmund Koch (New York: McGraw-Hill, 1959); Jerome S. Bruner, "Mechanism Riding High: Review of Kenneth W. Spence, *Behavior Theory and Conditioning*," *Contemporary Psychology* 2, no. 6 (1957): 155–57; Noam Chomsky, "Review of B. F. Skinner, *Verbal Behavior*," *Language* 35, no. 1 (1959): 26–58; Miller, Galanter, and Pribram, *Plans*.

69. O'Donnell, *The Origins of Behaviorism: American Psychology, 1870–1920*.

70. Morawski, "Organizing Knowledge."

71. Jerome S. Bruner to Talcott Parsons, March 30 (context indicates 1949), Papers of Talcott Parsons, Correspondence and Other Papers, 1935–1955, Box 6 of 19, Folder: Carnegie Project Correspondence, HUG 42.8.4, Harvard University Archives.

72. For an instance of pluralism's use as a marker of democratic character, see Lipset, *Political Man*.

73. Gerald Holton, "On the Educational Philosophy of the Project Physics Course," in *The Scientific Imagination: Case Studies* (Cambridge: Cambridge University Press, 1978), 184–85. For instances of the use of de Tocqueville as an authority on America and its character, see Henry Steele Commager, *The American Mind: An Interpretation of American Thought and Character since the 1880s* (New Haven, Conn.: Yale University Press, 1950); Max Lerner, *America as a Civilization: Life and Thought in the United States Today* (New York,

Conn.: Simon and Schuster, 1957), 72. Lerner would also go onto edit a 1965 edition of de Tocqueville's *Democracy in America*.

74. J. Robert Oppenheimer, "The Growth of Science and the Structure of Culture: Comments on Dr. Frank's Paper," in *Science and the Modern Mind*, ed. Gerald Holton (Cambridge, Mass.: Harvard University Press, 1958), 73. For further discussion of Oppenheimer's call for cultural integration by intimate discussion, see Charles Thorpe, *Oppenheimer: The Tragic Intellect* (Chicago: University of Chicago Press, 2006), 273–80.

75. Westbrook, *John Dewey and American Democracy*; Keith Michael Baker, *Condorcet: From Natural Philosophy to Social Mathematics* (Chicago: University of Chicago Press, 1975).

76. Amy Gutmann and Dennis F. Thompson, *Democracy and Disagreement* (Cambridge, Mass.: Belknap Press of Harvard University Press, 1996).

77. For instance, see Jürgen Habermas, *Communication and the Evolution of Society* (Boston: Beacon Press, 1979); Dennis F. Thompson, *The Democratic Citizen: Social Science and Democratic Theory in the Twentieth Century* (Cambridge: Cambridge University Press, 1970).

78. Testimony to the role of parties in the community and intellectual life in Cambridge can also be found in the diaries of Edmund Wilson and Arthur Schlesinger Jr.

79. Discussion of the value of community and interdisciplinary discussion at the American Academy of Arts and Sciences can be found in its bulletin and in its house organ, *Daedalus*. See, for instance, "Summary of the Discussion at the Meeting on February 13," *Bulletin of the American Academy of Arts and Sciences* 5, no. 6 (1952): 2; "Academy Conferences," *Bulletin of the American Academy of Arts and Sciences* 10, no. 1 (1956): 2–4. For discussion of the role of creativity and avant garde sensibility for the community at the RAND Corporation, see Sharon Ghamari-Tabrizi, *The Worlds of Herman Kahn: The Intuitive Science of Thermonuclear War* (Cambridge, Mass.: Harvard University Press, 2005).

80. On Georgetown society and its membership, see C. David Heymann, *The Georgetown Ladies' Social Club: Power, Passion, and Politics in the Nation's Capital* (New York: Atria Books, 2003). On the CIA's connections with the American press, see Carl Bernstein, "The CIA and the Media," *Rolling Stone*, October 20, 1977. On connections between the Bundys and journalists like Lippman and the Alsop brothers, see Kai Bird, *The Color of Truth: McGeorge Bundy and William Bundy, Brothers in Arms: A Biography* (New York: Simon and Schuster, 1998).

81. The membership of the supper club included Elting Morison, Kingman Brewster, Max Millikan, George Homans, Myron Gilmore, Robert Lee Wolff, McGeorge Bundy, Victor Weisskopf, Edward Purcell, Edwin Land, Will Haw-

thorne, Julius Stratton, Wassily Leonteif, and Jerome Bruner. Occasional visitors such as Robert Oppenheimer, Ernst Gombrich, Walter Lippmann, Alfred Kazin, and Carl Kaysen joined the club for dinner and conversation. Allan A. Nedell, "'Truth Is Our Weapon': Project Troy, Political Warfare, and Government-Academic Relations in the National Security State," *Diplomatic History*, 17, no. 3 (1993): 399–420; Allan A. Nedell, "Project Troy and the Cold War Annexation of the Social Sciences," in *Universities and Empire: Money and Politics in the Social Sciences During the Cold War*, ed. Christopher Simpson (New York: The New Press, 1998); Jerome S. Bruner, *In Search of Mind: Essays in Autobiography* (New York: Harper and Row, 1983), 210.

82. On government-military-academic interdisciplinary study groups, see Jack S. Goldstein, *A Different Sort of Time: The Life of Jerrold Zacharias* (Cambridge, Mass.: MIT Press, 1992).

83. On Los Alamos, see interviews in "The Day After Trinity"; on the Rad Lab, see Goldstein, *A Different Sort of Time*; on social science see Bruner, *In Search of Mind*.

84. On the relationships among Project Troy, CENIS, the State Department, the Ford Foundation, and the CIA see Nedell, "Project Troy"; Nedell, "'Truth Is Our Weapon.'"

85. Bird, *The Color of Truth*, 139–40; Bruce Kuklick, *Blind Oracles: Intellectuals and War from Kennan to Kissinger* (Princeton, N.J.: Princeton University Press, 2006), 82.

86. Christopher Rand, "Center of a New World, " part 3, *The New Yorker*, April 25, 1964, 92. The full series may be found in book form as Christopher Rand, *Cambridge, U.S.A.: Hub of a New World* (New York: Oxford University Press, 1964).

87. McGeorge Bundy's use of private meetings that screened critical debate continued after the 1950s and again after his tenure as National Security Advisor. In May 1971, Bundy held an off-the-record and invitation-only set of talks at the Council of Foreign Relations defending U.S. policy in Vietnam. Bird, *The Color of Truth*, 396.

88. Bruner, *In Search of Mind*, 225.

89. Ibid., 223.

90. Ibid., 219.

91. Victor Frederick Weisskopf, *The Joy of Insight: Passions of a Physicist* (New York: Basic Books, 1991), 161.

92. Bruner, *In Search of Mind*, 224.

93. Jerome Bruner interview, March 1952, p. 10. Anne Roe Papers, American Philosophical Society.

94. Weisskopf, *The Joy of Insight*, 161.

95. Richard Parker, *John Kenneth Galbraith: His Life, His Politics, His Economics* (New York: Farrar, Straus and Giroux, 2005), 251–52, 264–67.

96. Initial plans for the Center stipulated that it, like the Institute for Advanced Study, would be populated by a long-term faculty and shorter-term visitors. See Meeting Minutes of the Informal Planning Group on the Center for Advanced Study, Papers of Paul H. Buck, Harvard University Archives.

97. Morison, "The Course of the Discussion," 395–96.

98. Herbert Goldhamer as quoted in "Conference of Social Scientists, September 14–19, 1947." RAND Corporation, June 9, 1948. Report R-106, viii.

99. Mead and Byers, *The Small Conference*, 36.

100. Ibid., 35.

101. Transcripts of the UN-sponsored conferences on how to run a conference may be found in Capes, *Communication or Conflict*. On similar policies of the Macy Foundation in sponsoring conferences, see also the comments of Frank Fremont-Smith in Heinz von Foerster, ed., *Cybernetics*, 11–12.

102. Capes, *Communication or Conflict*, 61.

103. Ibid., 61–62.

104. Morison, "The Course of the Discussion," 413–14.

105. Henry B. Phillips, "Smoker in the Lounge," *Bulletin of the American Academy of Arts and Sciences* 6, no. 8 (1953): 1. Substantive analysis of this question can also be found in contemporary publications. See, for instance, Daniel Lerner, "The Social Sciences: Whence and Whither?" in *The Human Meaning of the Social Sciences*, ed. Daniel Lerner (New York: Meridian Books, 1959); Max Millikan, "Inquiry and Policy: The Relation of Knowledge to Action," in *The Human Meaning of the Social Sciences*.

106. Bird, *The Color of Truth*, 193.

107. "Ode to the Friday Evening on the Thursday Evening," April 28, 1960, Papers of Jerome Bruner, Correspondence, Box 3 of 4, Folder: Bundy, McGeorge, HUG 4242.9, Harvard University Archives.

108. For instance, McGeorge Bundy's brother William was married to Acheson's daughter Mary. Bundy's family friendship with luminaries of the legal world paved the way for an invitation to McGeorge to clerk for Supreme Court Justice Felix Frankfurter although he had no formal legal training. When he declined the post, it was offered to William who did have a law degree, but who also declined. Bird, *The Color of Truth*, 100, 55.

109. Bird, *The Color of Truth*, 150–51.

110. For Bundy's comparison of Cuban counterrevolutionaries to Harvard professors who had been given their walking papers, see David Halberstam, *The Best and the Brightest* (New York: Random House, 1972), 95. For discussion of Bundy's own role in the Bay of Pigs operation, in preventing views opposed to the operation from reaching the president, and the conclusion that he was not responsible for the operation's failure, see Bird, *The Color of Truth*, 193–99.

Chapter Five

1. On the Voice of America, see Alan L. Heil, *Voice of America: A History* (New York: Columbia University Press, 2003); Holly Cowan Shulman, *The Voice of America: Propaganda and Democracy, 1941–1945* (Madison: University of Wisconsin Press, 1990); David F. Krugler, *The Voice of America and the Domestic Propaganda Battles, 1945–1953* (Columbia: University of Missouri Press, 2000).

2. George Miller, "Thinking, Cognition, and Learning," in *The Behavioral Sciences Today*, ed. Bernard Berelson (New York: Basic Books, 1963), 150.

3. Graham Richards, "Of What Is History of Psychology a History?" *British Journal for the History of Science* 20 (1987): 201–11; Didier Anzieu, *Freud's Self-Analysis*, The International Psycho-Analytical Library, no. 118 (London: Hogarth Press and the Institute of Psycho-analysis, 1986); John E. Toews, "Historicizing Psychoanalysis: Freud in His Time and for Our Time," *Journal of Modern History* 63 (1991): 504–45; Carl E. Schorske, *Fin-De-Siècle Vienna: Politics and Culture* (New York: Vintage Books, 1981); Ian A. M. Nicholson, "Gordon Allport, Character, and the 'Culture of Personality,' 1897–1937," *History of Psychology* 1, no. 1 (1998): 52–68.

4. Laurence D. Smith, "Metaphors of Knowledge and Behavior in the Behaviorist Tradition," in *Metaphors in the History of Psychology*, ed. David E. Leary (Cambridge: Cambridge University Press, 1990), 237–66; Laurence Smith, *Behaviorism and Logical Positivism: A Reassessment of the Alliance* (Stanford, Calif.: Stanford University Press, 1986); Ernest R. Hilgard and Gordon H. Bower, *Theories of Learning*, 3d ed. (New York: Appleton-Century-Crofts, 1966), 104.

5. In an examination of a later period and somewhat different vision of mind than that considered in this chapter, Gerd Gigerenzer has shown how cognitive psychologists' own ways of thinking and working became the model for human cognition. Gigerenzer, "Discovery in Cognitive Psychology: New Tools Inspire New Theories," *Science in Context* 5, no. 2 (1992): 329–50; Gigerenzer, "From Tools to Theories: A Heuristic of Discovery in Cognitive Psychology," *Psychological Review* 98, no. 2 (1991): 254–67. For general discussion of the cognitive revolution, see Gardner, *The Mind's New Science*; Mandler, "Origins of the Cognitive (R)Evolution"; John D. Greenwood, "Understanding the 'Cognitive Revolution' in Psychology," *Journal of the History of the Behavioral Sciences* 35, no. 1 (1999): 1–22; Baars, *The Cognitive Revolution in Psychology*; Dupuy, *The Mechanization of the Mind*. Robins, Gosling, and Craik have provided an empirical study of how cognitive psychology supplanted behaviorism. See "An Empirical Analysis of Trends in Psychology," *American Psychologist* 54, no. 2 (1999): 117–28. Thomas Leahey has argued that the cognitive revolution was not actually a "real" revolution: see "The Mythical Revolutions of American Psychology," *American Psychologist* 47 (1992): 308–18. On the other hand, histori-

ans and psychologists continue to use the term. Without seeking to judge the validity of the term, this chapter follows the actors' categories in adopting the term
"cognitive revolution."

6. Some of Simon's works in this vein are "Cognitive Science: The Newest Science of the Artificial," *Cognitive Science* 4 (1980): 33–46, reprinted in *Mind and
Cosmos: Essays in Contemporary Science and Philosophy*, University of Pittsburgh Series in the Philosophy of Science, vol. University of Pittsburgh Series
in the Philosophy of Science, vol. 3 (Pittsburgh: University of Pittsburgh Press,
1966); "Thinking by Computers," in *Mind and Cosmos*; "Scientific Discovery
and the Psychology of Problem Solving," in *Mind and Cosmos*.

7. Theo Lentz, "The Attitudes of World Citizenship," *Journal of Social Psychology* 32 (1950): 207–14, on 213–14.

8. Archives of the Ford Foundation, Grant Files, Reel R-004, Grant 53–78,
Project Proposal, Jerome S. Bruner to Ford Foundation, 1/9/53.

9. Abraham H. Maslow, "Problem-Centering Versus Means-Centering in
Science," *Philosophy of Science* 13 (1946): 326–31; Abraham H. Maslow, "Cognition of the Particular and of the Generic," *Psychological Review* 55 (1948):
22–39; Hadley Cantril, "Toward a Scientific Morality," *The Journal of Psychology* 27 (1949): 363–76; Hadley Cantril, "An Inquiry Concerning the Characteristics of Man," *Journal of Abnormal and Social Psychology* 45, no. 3 (1950): 490–
503; Clyde Kluckhohn, "Universal Values and Anthropological Relativism," in
Modern Education and Human Values (Pittsburgh, Pa.: University of Pittsburgh
Press, 1952); Lawrence K. Frank, *Nature and Human Nature: Man's New Image
of Himself* (New Brunswick, N.J.: Rutgers University Press, 1951).

10. Rokeach, *The Open and Closed Mind*, 65. For a similar, but only slightly
less politically charged, argument, see Edward C. Tolman, "Cognitive Maps in
Rats and Men," *Psychological Review* 55, no. 4 (1948): 189–208, on 207–8.

11. Herbert Feigl, "The Philosophical Embarrassments of Psychology,"
American Psychologist 14, no. 3 (1959): 115–28, on 118.

12. The split was noted by many leaders of psychology. The social psychologist Gordon Allport parsed the division in psychology similarly. To Allport,
psychology was split between "Lockeans" and "Leibnizians." Allport's behavioristic Lockeans believed that humans could be completely explained (and
were determined by) their experiences and the stimuli that impinged upon their
senses. According to Allport, Leibnitzians believed that the human mind was, at
least in part, autonomous. Gordon W. Allport, *Becoming: Basic Considerations
for a Psychology of Personality* (New Haven, Conn.: Yale University Press,
1955). Others who saw a similar division in psychology included six presidents
of the APA (Tolman[1937], Hebb [1960], Rogers[1947], Hilgard[1949], Harlow
[1958], and Miller [1969]), the editor of *The Journal of Experimental Psychology*
(Melton), and the author of one of the most widely read textbooks in psychology
(Hilgard). See Miller, Galanter, and Pribram, *Plans and the Structure of Behav-*

ior, 7–8; D. O. Hebb, "The American Revolution," *American Psychologist* 15 (1960): 735–45; Harry F. Harlow, "Current and Future Advances in Physiological and Comparative Psychology," *American Psychologist* 11 (1956): 273–77, on 274; George A. Kelley, "I Itch Too," *American Psychologist* 10, no. 4 (1955): 172–73; Ernest R. Hilgard, *Theories of Learning* (New York: Appleton-Century-Crofts, 1948), 9; Theodore C. Kahn, "Clinically and Statistically Oriented Psychologists Split Our Profession," *American Psychologist* 10, no. 4 (1955): 171–72; Carl R. Rogers, "Persons or Science? A Philosophical Question," *American Psychologist* 10, no. 7: 267–78; Tolman, "Cognitive Maps"; Arthur W. Melton, "Present Accomplishments and Future Trends in Problem-Solving and Learning Theory," *American Psychologist* 11 (1956): 278–81. Hilgard's *Theories of Learning* is particularly significant due to its wide readership. In surveys conducted in 1953–54 and 1958–59, it was recommended for graduate education by over half of surveyed departments. Hilgard's text was one of only six books to rate so highly. See Norman D. Sundberg, "Basic Readings in Psychology," *American Psychologist* 15 (1960): 343–45.

13. David McClelland noted: "[P]sychologists used to be interested in what went on in people's heads. . . . [B]ut with the rise of modern scientific psychology we lost interest in ideas, by and large." David C. McClelland, "The Psychology of Mental Content Reconsidered," *Psychological Review* 62, no. 4 (1955): 297–302, on 297.

14. Positivist, operationist, and behavioristic psychology had such claim on being the "fundamental" part of psychology that it was a notable event when social and clinical psychologists characterized their own work as "fundamental." Sigmund Koch, "Epilogue: Some Trends of Study I," 5.

15. In his presidential address to the American Psychological Association, Donald Hebb remarked that learning had been "the fundamental issue in psychology" ever since the work of J. B. Watson and Edward Thorndike in the early twentieth century. Hebb, "The American Revolution," 736.

16. "Between 1944 and 1950, 70 percent of the articles on learning and motivation in the *Journal of Experimental Psychology* and the *Journal of Comparative and Physiological Psychology* cited Hull." Alexandra W. Logue, "The Growth of Behaviorism: Controversy and Diversity," in *Points of View in the Modern History of Psychology*, ed. Claude E. Buxton (Orlando: Academic Press, 1985), 180. Logue cites evidence collected in Kenneth W. Spence, "Clark Leonard Hull: 1884–1952," *American Journal of Psychology* 65 (1952): 639–46.

17. A useful overview of learning theory, including both perspectives, is Hilgard and Bower, *Theories of Learning*. Discussion of the marginality of Gestalt theory may be found in Michael Sokal, "The Gestalt Psychologists in Behaviorist America," *American Historical Review* 89, no. 5 (1984): 1240–63; Mitchell G. Ash, "Gestalt Psychology: Origins in Germany and Reception in the United States," in *Points of View in the Modern History of Psychology*, ed. Claude E.

Buxton (Orlando: Academic Press, 1985). On understanding and cognition, see I. Krechevesky, "The Genesis of 'Hypotheses' in Rats," *Psychological Review* 45 (1932): 107–33; I. Krechevesky, "'Hypothesis' vs. 'Chance' in the Pre-Solution Period in Sensory Discrimination-Learning," *University of California Publications in Psychology* 6 (1932): 27–44; Tolman, "Cognitive Maps."

18. In descending order of size of APA membership, the top ten psychological speciàlties in 1952 were clinical, educational, experimental, industrial, vocational, social, general, developmental, personality, and physiological. S. H. Sanford, "Annual Report of the Executive Secretary," *American Psychologist* 7 (1952): 686–96; cited in William A. Hunt, *The Clinical Psychologist* (Springfield, Ill.: Charles C. Thomas, 1956), 14. For excellent analysis of the changing composition of the discipline, see Capshew, *Psychologists on the March*.

19. Frederick L. McGuire, "On the Issue 'What Is Science?'" *American Psychologist* 11, no. 3 (1956): 152–53, on 153; George A. Kelley, "I Itch Too"; Carl R. Rogers, "Persons or Science? A Philosophical Question"; Hans H. Strupp et al., "Comments on Rogers' 'Persons or Science,'" *American Psychologist* 11, no. 3 (1956): 153–57; E. B. Skaggs, "Personalistic Psychology as Science," *Psychological Review* 52 (1945): 234–48; E. Parker Johnson, "On Readmitting the Mind," *American Psychologist* 11 (1956): 712–14, on 712.

20. Johnson, "On Readmitting the Mind," 712.

21. Lawrence N. Solomon, "The Paradox of the Experimental Clinician," *American Psychologist* 10, no. 4 (1955): 170–71, on 170.

22. Allport, "The Psychologist's Frame of Reference"; Harry F. Harlow, "The Nature of Love," *American Psychologist* 13, no. 12 (1958): 673–85, on 674; Sanford C. Ericson, "Unity in Psychology: A Survey of Some Opinions," *Psychological Review* 48, no. 1 (1941): 73–82, on 76.

23. Bruner, "Mechanism Riding High," 156; Daniel Brower, "The Problem of Quantification in Psychological Science," *Psychological Review* 56, no. 6 (1949): 325–33, on 326, 328, 330, and 332.

24. Howard E. Gruber, Kenneth R. Hammond, and Richard Jessor, "Foreword," in *Contemporary Approaches to Cognition* (Cambridge, Mass.: Harvard University Press, 1957), v.

25. B. F. Skinner, "Are Theories of Learning Necessary?" *Psychological Review* 57, no. 4 (1950): 193–216; B. F. Skinner, "A Case History in Scientific Method," *American Psychologist* 11, no. 5 (1956): 221–33.

26. On the scientific virtues of mechanical objectivity, "trained judgment," and the role of intuition in the latter, see Daston and Galison, *Objectivity*, 115–90, 358–61.

27. Else Frenkel-Brunswik, "Intolerance of Ambiguity as an Emotional and Perceptual Personality Variable," *Journal of Personality* 18, no. 1 (1949): 108–43; Frenkel-Brunswik, "Further Explorations by a Contributor to *the Authoritarian Personality*."

28. W. C. H. Prentice, "Operationism and Psychological Theory: A Note," *Psychological Review* 53, no. 4 (1946): 247–49, on 247.

29. Frederick C. Thorne, "Psychologists, Heal Thyselves!" *American Psychologist* 11, no. 3 (1956): 152.

30. For discussion about the inter-species applicability of laws of behavior, see John Dollard and Neal E. Miller, *Personality and Psychotherapy: An Analysis in Terms of Learning, Thinking, and Culture* (New York: McGraw-Hill, 1950), 63; Skinner, "A Case History in Scientific Method," 221–33.

31. M. E. Bitterman, "Toward Comparative Psychology of Learning," *American Psychologist* 15 (1960): 704–12, on 705, 11. For other attacks on this methodological assumption, see William F. Dukes, "The Snark Revisited," *American Psychologist* 15 (1960): 157; Frank Λ. Beach, "The Snark Was a Boojum," *American Psychologist* 5, no. 5 (1950): 115–24.

32. Donald Hebb, *The Organization of Behavior: A Neuropsychological Theory* (New York: John Wiley and Sons, 1949), 4. For other examples of psychologists noting the religious nature of behaviorism, see Harry F. Harlow, "Mice, Monkeys, Men, and Motives," *Psychological Review* 60, no. 1 (1953): 23–31, on 23; Jerome S. Bruner, "Neural Mechanisms in Perception," *Psychological Review* 64 (1957): 340–58, on 341; Hunt, *The Clinical Psychologist*, 7.

33. Chomsky, "Review of B. F. Skinner, *Verbal Behavior*," 32, 35, 39. On forgetting and ignoring inconvenient evidence see 40, nn. 23, 42.

34. Lucille C. Birnbaum, "Behaviorism in the 1920s," *American Quarterly* 7, no. 1 (1955): 15–30, on 15.

35. Ibid., 30.

36. Koestler, *The Act of Creation*, 560–61.

37. Koestler, *The Sleepwalkers*.

38. This image of religion and particularly Catholicism as dogmatic and antiscientific (and antidemocratic) appeared particularly strongly during World War II and afterward in the writings of Sidney Hook. Sidney Hook, "Democracy and Education: Introduction," in *The Authoritarian Attempt to Capture Education: Papers from the Second Conference on the Scientific Spirit and Democratic Faith* (New York: King's Crown Press, 1945); Hook, *Reason, Social Myths and Democracy*.

39. Sigmund Koch, "Epilogue: Some Trends of Study I," 732, 734–35, 69, 776–77, 786.

40. Ibid., 783.

41. Ibid.

42. Ibid., 748, 70. Koch also suggested a contrast between the "realities of science" and the "facile myth" advanced by the Age of Theory. See ibid., 786

43. Ibid., 769–70.

44. Ibid., 776–78.

45. Ibid., 785–86.

46. Ibid., 786.

47. Ibid., 770–85.

48. See P. McKellar, *Imagination and Thinking: A Psychological Analysis* (New York: Basic Books, 1957).

49. Koch, "Epilogue," 783.

50. Ibid., 748, 786; see also Hunt, *The Clinical Psychologist*, 6.

51. Koch, "Epilogue," 787–88.

52. However, according to Lawrence D. Smith in *Behaviorism and Logical Positivism*, logical positivism was not, as Koch argued, a phenomenon that arose outside of psychology and was subsequently adopted by it. Instead, Smith argues, behaviorists like Clark Hull had developed methodological commitments on their own, and only subsequently adopted positivist language to describe their prior programs. If Smith is correct and Koch was wrong about the origins of logical positivism in psychology, it only strengthens the argument I have been making. It suggests the particular strength of the cultural construct that explained rigidity and closed-mindedness as products of lack of autonomy and external influence.

53. Christopher D. Green, "Of Immortal Mythological Beasts: Operationism in Psychology," *Theory and Psychology* 2, no. 3 (1992): 291–320; Andrew S. Winston and Daniel J. Blais, "What Counts as an Experiment? A Transdisciplinary Analysis of Textbooks, 1930–1970," *American Journal of Psychology* 109, no. 4 (1996): 559–616.

54. For example, see Kenneth MacCorquodale and Paul E. Meehl, "On a Distinction between Hypothetical Constructs and Intervening Variables," *Psychological Review* 55 (1948): 95–107, on 96.

55. Herbert Feigl, "The Philosophical Embarrassments of Psychology," 115.

56. Bruner, "Mechanism Riding High," 155–57; Gustav Bergmann and Kenneth W. Spence, "Operationism and Theory in Psychology," *Psychological Review* 48, no. 1 (1941): 1–14; Kenneth W. Spence, "The Empirical Basis and the Theoretical Structure of Psychology," *Philosophy of Science* 24, no. 2 (1957): 97–108.

57. See Oppenheimer, "Theory Versus Practice"; James B. Conant, *Science and Common Sense* (New Haven, Conn.: Yale University Press, 1951); Conant, *Modern Science and Modern Man*; Conant, *On Understanding Science*; Warren Weaver, "Science and Faith," Delivered on Layman's Sunday in the Congregational Church of New Milford, Connecticut, May 16, 1954. Papers of Vannevar Bush, Box 117, Folder: Weaver, Warren, (1948–1954), Manuscript Division, Library of Congress.

58. Koestler, *The Act of Creation*; Bronowski, *Science and Human Values*.

59. Bernard Barber, *Science and the Social Order*. On Parsons's philosophy

of science, see Charles Camic, "Introduction: Talcott Parsons before *the Structure of Social Action*," in *Talcott Parsons: The Early Essays*, ed. and with an Introduction by Charles Camic (Chicago: University of Chicago Press, 1991).

60. Conant, *On Understanding Science*; Thomas S. Kuhn, *The Structure of Scientific Revolutions* (Chicago: University of Chicago Press, 1962). On the relationship between Conant and Kuhn, see Steve Fuller, *Thomas Kuhn: A Philosophical History for Our Times* (Chicago: University of Chicago Press, 2000).

61. Bronowski, *Science and Human Values*, 48–49.

62. Jacob Bronowski, "Analogy in Science," *American Psychologist* 11, no. 3 (1956): 127–35, on 130.

63. Ibid., 134

64. Ibid., 135.

65. A similar critique of narrowness and overconcern with rigor can be found in the work of psychologist Abraham Maslow; see his "Problem-Centering Versus Means-Centering in Science." Several years later, Maslow would focus this general critique of methodological fetishism in a discussion of the problem of means-centering in psychology. Maslow, "The Expressive Component of Behavior," *Psychological Review* 56, no. 5 (1949): 261–72, on 261.

66. "Bruner, "Mechanism Riding High"; Spence, *Behavior Theory and Conditioning.*

67. Koch, "Epilogue," 784–86.

68. Surveying the field in his 1960 presidential address to the APA, Donald Hebb remarked: "Though this opposition of aims may seem over-simplified, I believe it is fundamentally sound. How [should we] understand otherwise the learning theorist's [i.e., the behaviorist's] bland refusal even to discuss attention or purpose, or the cognitive psychologist's happy preference for phenomena he cannot explain—so long as the other cannot explain them either?" Hebb, "The American Revolution," 737.

69. Jerome S. Bruner, Jacqueline J. Goodnow, and George A. Austin, *A Study of Thinking* (London: John Wiley and Sons, 1956), 6.

70. Ibid., 233.

71. Ibid., 10, 14, 17, 19, 31, 37–38, 54, 56, 92, 233, 244, 246.

72. Ibid., 7, emphasis added.

73. Ibid., 19; Ernest Mayr, "Concepts of Classification and Nomenclature in Higher Organisms and Microorganisms," *Annals of the New York Academy of Science* 56 (1952): 391–97. Six years before publication of *A Study in Thinking*, the psychologist Hadley Cantril, with whom Bruner had worked during World War II, made similar points. He noted that "variables scientists use do not exist in their own right. They are only aspects abstracted out of the total situation by scientists as inquiring human beings endowed with the capacity to manipulate ideas." Citing Einstein and Infeld, Cantril also noted the "creative imagination" necessary in science. Cantril, "An Inquiry Concerning the Characteris-

tics of Man," 491; Albert Einstein and L. Infeld, *The Evolution of Physics*, 2d ed. (New York: Simon and Schuster, 1942), 95.

74. Bruner, Goodnow, and Austin, *A Study of Thinking*, 7.

75. Ibid.

76. Ibid., 8.

77. See Koch, "Epilogue," 768–69; Bruner, "Mechanism Riding High," 156; Bruner, "Neural Mechanisms in Perception," 340–58; Miller, Galanter, and Pribram, *Plans*, 21–25; D. O. Hebb, "The Problem of Consciousness and Introspection," in *Brain Mechanisms and Consciousness*, ed. E. D. Adrian (Oxford: Blackwell, 1954), 404.

78. "Cognition and the Limits of Scientific Inquiry." Unpublished paper read at the Institute for the Unity of Science at the American Academy of Arts and Sciences, 1951, Jerome S. Bruner Papers, HUG 4242.28, Harvard University Archives.

79. Bruner based this argument on a series of experiments he conducted on how expectation determines what people see. For historians of science, the most notable of these experiments is one that examines how individuals experience gestalt switches in their perception when viewing miscolored playing cards such as black hearts. Jerome S. Bruner and Leo J. Postman, "On the Perception of Incongruity: A Paradigm," *Journal of Personality* 18 (1949): 206–23. This paper was one of Thomas Kuhn's central examples of gestalt switching in *The Structure of Scientific Revolutions*.

80. Bruner, "Mechanism Riding High."

81. Eugene Galanter and Murray Gerstenhaber, "On Thought: The Extrinsic Theory," *Psychological Review* 63, no. 4 (1956): 218–27, on 219. Bruner later made a similar argument: "I am inclined to think of mental development as involving the construction of a model of the world in the child's head, an internalized set of structures for representing the world around him." Jerome S. Bruner, "On Learning Mathematics," in *On Knowing: Essays for the Left Hand* (Cambridge, Mass.: Harvard University Press, 1979 [1962]), 103.

82. An earlier, but uncited, form of this argument can be found in Kenneth James Williams Craik, *The Nature of Explanation* (Cambridge: Cambridge University Press, 1943). While Craik has been credited with inspiring cognitive psychology, remarkably the first citation of his work in *Psychological Review*, the leading theoretical journal of the American Psychological Association, occurred well after the cognitive project was off the ground. See Lawrence E. Arend, "Spatial Differential and Integral Operations in Human Vision," *Psychological Review* 80, no. 5 (1973): 374–95.

83. Bruner, Goodnow, and Austin, *A Study of Thinking*, 6–7; Galanter and Gerstenhaber, "On Thought"; Miller, Galanter, and Pribram, *Plans*; Noam Chomsky, "Review of B. F. Skinner, *Verbal Behavior*," in *The Structure of Language*, ed. J. A. Fodor and J. J. Katz (Englewood Cliffs, N.J.: Prentice-Hall,

1959); Herbert A. Simon, "Understanding Creativity," in *Creativity: Its Educational Implications*, ed. John Curtis Gowan, George D. Demos, and E. Paul Torrence (New York: John Wiley and Sons, 1967); Allen Newell, J. C. Shaw, and Herbert A. Simon, "The Processes of Creative Thinking," in *Contemporary Approaches to Creative Thinking*, ed. Howard E. Gruber, Glenn Terrell, and Michael Wertheimer (New York: Atherton Press, 1962); Simon, "Scientific Discovery and the Psychology of Problem Solving"; A. Newell, J. C. Shaw, and H. Simon, "Elements of a Theory of Human Problem Solving," *Psychological Review* 65 (1958): 151–66.

84. Noam Chomsky, "Three Models for the Description of Language," *IRE Transactions on Information Theory* IT-2, no. 3 (1956): 113–24, on 113, 116; Chomsky, *Syntactic Structures* (The Hague: Mouton Publishers, 1957); Chomsky, "Logical Structure of Linguistic Theory," 6 (typescript, 1955–56, available at http://alpha-leonis.lids.mit.edu/chomsky/).

85. Skinner, "Are Theories of Learning Necessary?"; Skinner, "A Case History in Scientific Method."

86. Chomsky, "Review of B. F. Skinner, *Verbal Behavior*," 56–57.

87. Roger W. Sperry, "On the Neural Basis of the Conditioned Response," *British Journal of Animal Behavior* 3 (1955): 41–44; cited in Chomsky, "Review of B. F. Skinner, *Verbal Behavior*," 44. This use of biological evidence to attack behaviorism appeared in the work of other cognitive scientists. See, for instance, Jerome Bruner's review of physiological evidence that indicated that behaviorists failed to understand reflexes—the very foundation of their own use of conditioned reflexes as the model of all behavior. Bruner, "Neural Mechanisms in Perception."

88. For discussion of S-R chains, see Hilgard and Bower, *Theories of Learning*.

89. The proof hinged on the point that embedded sentences can be of arbitrary length. However, that would require the machine which generated the sentence to have an arbitrary number of states—so that it would have to be an *in*finite-state machine.

90. Miller, Galanter, and Pribram, *Plans*; see, for instance, Galanter and Gerstenhaber, "On Thought." Noam Chomsky and George A. Miller, "Finite State Languages," *Information and Control* 1 (1958): 91–112; George A. Miller and Noam Chomsky, "Finitary Models of Language Users," in *Handbook of Mathematical Psychology*, vol. 2, ed. R. D. Luce, R. R. Bush, and Eugene Galanter (New York: Wiley, 1963); Noam Chomsky and George A. Miller, "Introduction to the Formal Analysis of Natural Languages," in *Handbook of Mathematical Psychology*, vol. 2.

91. Randy Allen Harris, "Chomsky's Other Revolution," in *Chomskyan (R) Evolutions*, ed. Douglas A. Kibbee (Philadelphia: John Benjamins Publishing Company, 2010).

92. Baars, *The Cognitive Revolution in Psychology*, 212.

93. Hebb, "The American Revolution," 740, emphasis added.

94. For discussion of this paradox, see Hunter Crowther-Heyck, "George A. Miller, Language, and the Computer Metaphor of Mind," *History of Psychology* 2 (1999): 37–64.

95. Miller, Galanter, and Pribram, *Plans*, 41.

96. B. F. Skinner, "Why I Am Not a Cognitive Psychologist," in *Reflections on Behaviorism and Society* (Englewood Cliffs, N.J.: Prentice Hall, 1978).

97. Herbert Simon and Allen Newell, "Computer Simulation of Human Thinking and Problem Solving," in *Management and the Computer of the Future*, ed. Martin Greenberger (New York: MIT Press and Wiley, 1962).

98. Newell, Shaw, and Simon, "Elements of a Theory of Human Problem Solving"; Newell, Shaw, and Simon, "The Processes of Creative Thinking"; Miller, Galanter, and Pribram, *Plans*.

99. Henri Poincaré, *Science and Method*, trans. F. Maitland (New York: Dover, 1952); G. Polya, *How to Solve It* (Princeton, N.J.: Princeton University Press, 1945); G. Polya, *Mathematics and Plausible Reasoning* (Princeton, N.J.: Princeton University Press, 1954); Michael Polyani, *Personal Knowledge* (Chicago: University of Chicago Press, 1958). Cited in Miller, Galanter, and Pribram, *Plans*, 87, 160, 167–69, 179, 180, 183, 191. For additional references to Poincaré and Polya see Newell, Shaw, and Simon, "The Processes of Creative Thinking."

100. "Bibliography—Miscellaneous Comments on Starred Items," Commentary—1958 [bundled] (Consulting—Social Science Research Council—Conferences and Seminars—Summer Research Training Institute on Simulation of Cognitive Processes—1958). Herbert Simon Digital Archive. http://diva.library .cmu.edu/Simon/index.html. See also Newell, Shaw, and Simon, "Elements," 153 n. 2; Allen Newell and Herbert Simon, *The Simulation of Human Thought* (Santa Monica, Calif.: The RAND Corporation, 1959), 10.

101. For instance, Philip N. Johnson-Laird, *The Computer and the Mind: An Introduction to Cognitive Science* (Cambridge, Mass.: Harvard University Press, 1988).

102. John von Neumann, "The General and Logical Theory of Automata," in *Cerebral Mechanisms of Behavior: The Hixon Symposium*, ed. Lloyd A. Jeffries (New York: John Wiley and Sons, 1951).

103. For reflection on the critique of finite-state languages as criticism of information theory and of engineers, see Noam Chomsky, "Of Minds and Language," *Biolinguistics* 1 (2007): 9–27, on 11.

104. Janet Martin-Nielsen, "'It Was All Connected': Computers and Linguistics in Early Cold War America," in *Cold War Social Science: Knowledge Production, Liberal Democracy, and Human Nature*, ed. Mark Solovey and Hamilton Cravens (New York: Palgrave Macmillan, 2012).

105. George A. Miller, *Occasional Paper #1: A Very Personal History* (Cam-

bridge, Mass.: MIT Cognitive Science Workshop, June 1, 1979), 7. Miller generously asserts that Chomsky did not seek to use his paper as an attack on the engineers and information theorists, but Chomsky asserts that the critique was intentional.

106. George A. Miller, "The Magical Number Seven, Plus or Minus Two: Some Limits on Our Capacity for Processing Information," *Psychological Review* 63 (1956): 81–97.

107. Claude Shannon, "A Mathematical Theory of Communication," *Bell Systems Technical Journal* 27 (1948): 379–423, 623–56.

108. See George A. Miller, "The Human Link in Communication Systems," in *The Psychology of Communication*. Originally published in *Proceedings of the National Electronics Conference* 12 (1956): 395–400.

109. Miller, *The Psychology of Communication*, vi.

110. "Information and Memory," in Miller, *The Psychology of Communication*, 8. Originally published as George A. Miller, "Information and Memory," *Scientific American*, August 1956: 42–46.

111. Miller, "The Human Link," 49.

112. This was also a departure from the research agenda that Miller himself had been pursuing for almost a decade. During that decade Miller and his collaborators had completed one study after another that explored human abilities to recognize, react to, or remember words that were statistical approximations of English. In this he followed the method of Claude Shannon. In this early work, Miller contended that people have a fixed channel capacity. See George A. Miller, *Language and Communication* (New York: McGraw-Hill, 1951), 84.

113. Herbert A. Simon, *Models of My Life* (New York: Basic Books, 1991), 205–7.

114. Margaret A. Boden, "Personal Politics and Ideology Machines," in *Artificial Intelligence and Natural Man* (New York: Basic Books, 1977); Boden, *Mind as Machine: A History of Cognitive Science*, 709–12.

115. Edwin G. Boring, "Mind and Mechanism," *American Journal of Psychology* 59, no. 2 (1946): 173–92; Saul Gorn, "On the Mechanical Stimulation of Habit-Forming and Learning," *Information and Control* 2 (1959): 226–59; Howard S. Hoffman, "The Analogue Lab: A New Kind of Teaching Device," *American Psychologist* 17, no. 10 (1962): 684–94.

116. Laurence D. Smith, "Metaphors of Knowledge and Behavior in the Behaviorist Tradition," 249–54.

117. Boring, "Mind and Mechanism," 178. For discussion of the exchange between Boring and Wiener, see Peter Galison, "The Ontology of the Enemy: Norbert Wiener and the Cybernetic Vision," 247, 251–52.

118. Boring, "Mind and Mechanism," 183–84.

119. Allen Newell, J. C. Shaw, and Herbert Simon, "Report on a General Problem-Solving Program," in *Proceedings of the International Conference on*

Information Processing (Paris: 1959). Cited in Miller, Galanter, and Pribram, *Plans*, 189.

120. Miller, Galanter, and Pribram, *Plans*, 41.

121. Lawrence D. Smith has noted the tendency of neo-behaviorists to see the mode of animal thinking described in their work as functioning in essentially the same fashion as their announced prescriptions (philosophy of science) for their own thinking. See *Behaviorism and Logical Positivism*.

122. A notable exception to this is Robert Abelson's simulation of Barry Goldwater's ideology. For discussion of this program see Boden, "Personal Politics and Ideology Machines."

123. Miller, "Thinking, Cognition, and Learning," 149–50.

Chapter Six

1. Angela N. H. Creager, Elizabeth Lunbeck, and M. Norton Wise, *Science without Laws: Model Systems, Cases, Exemplary Narratives, Science and Cultural Theory* (Durham, N.C.: Duke University Press, 2007); Angela N. H. Creager, *The Life of a Virus: Tobacco Mosaic Virus as an Experimental Model, 1930–1965* (Chicago: University of Chicago Press, 2002).

2. Howard Gardner, *The Mind's New Science*; M. Posner and G. L. Schulman, "Cognitive Science," in *The First Century of Experimental Psychology*, ed. E. Hearst (Hillsdale, N.J.: Lawrence Erlbaum, 1979).

3. Noam Chomsky, *Cartesian Linguistics* (New York and London: Harper and Row, 1966); Noam Chomsky, *Aspects of the Theory of Syntax* (Cambridge, Mass.: MIT Press, 1965).

4. See, for instance, Jerome S. Bruner, *Toward a Theory of Instruction* (Cambridge, Mass.: Belknap Press of Harvard University Press, 1966); Jerome S. Bruner, *The Process of Education* (Cambridge, Mass.: Harvard University Press, 1960); Jerome S. Bruner, "The Act of Discovery," *Harvard Educational Review* 31, no. 1 (1961): 21–32.

5. The psychologists included Roger Brown, Mary Henle, Ulric Neisser, Donald Norman, Bärbel Inhelder, Nancy Waugh, Patricia Greenfield, Susan Carey, Jacques Mehler, Doris Aaronson, Daniel Kahneman, John Carroll, David McNeill, David Olson, and Walter Reitman. Linguistics was represented by Noam Chomsky, Roman Jakobson, Jerry Fodor, and Jerrold Katz, biology by Ernst Mayer, mathematics by Benoit Mandelbrot and Seymour Papert, pediatrics by T. Berry Brazelton, history by H. Stuart Hughes, psychiatry and psychoanalysis by Jacques Lacan and Joseph Jaffe, and decision theory and industrial administration by Herbert Simon. The roster of the Center's students, visitors, and directors can be found in Jerome S. Bruner, "Founding the Center for Cognitive Studies," in *The Making of Cognitive Science: Essays in Honor of George A.*

Miller, ed. William Hirst (Cambridge and New York: Cambridge University Press, 1988); Harvard University Center for Cognitive Studies, *Annual Reports* (Cambridge, Mass.: 1961–69).

6. Jerome Bruner and George Miller to Nathan Pusey, January 4, 1962, Papers of the Dean of the Faculty of Arts and Sciences, letters to/from, 1961–62, Box 2, Folder: Cognitive Studies, Center for, UA III 5.55.26. Harvard University Archives.

7. NIH Grant Application, "Studies in Cognition," December 13, 1965, Administrative Papers of the Center for Cognitive Studies [hereafter CCS], Grant Requests, Proposals & Reports, Folder: NIH Proposal, UA V. 290.15, Harvard University Archives.

8. Following common usage by mid-twentieth-century psychologists, "tool" here refers to an individual physical instrument, an assemblage of the same, an experimental method, or an organized way of categorizing, seeing, or approaching the world.

9. Lecture notes from "The Materialist Tradition," Lecture 4, 5. Social Sciences 8: Psychological Conceptions of Man, Harvard University. Delivered on October 6, 1960. Personal papers of George A. Miller, Princeton University Department of Psychology.

10. For a more recent discussion of the linkage between new tools and new theories in cognitive psychology, see Gigerenzer, "Discovery in Cognitive Psychology: New Tools Inspire New Theories"; Gigerenzer, "From Tools to Theories: A Heuristic of Discovery in Cognitive Psychology."

11. For discussion of the role of the war on fostering task-oriented, interdisciplinary research, see Peter Galison, "The Americanization of Unity," *Daedalus* 127, no. 1 (1998): 45–71, on 59–65. For discussion of how this practice developed out of Warren Weaver's earlier style at the Rockefeller Foundation, see Mindell, *Between Human and Machine: Feedback, Control, and Computing before Cybernetics*, 189. On reading an earlier draft of this chapter, George Miller remarked that "both Jerry [Bruner] and I were forced by WWII to become interdisciplinary (although: re different disciplines), and that we found it exciting and wanted to revive it in 1960. The war had given both of us intellectual claustrophobia." Personal communication, December 29, 1997.

12. General discussion of Lincoln Laboratory may be found in Leslie, *The Cold War and American Science*.

13. Miller, Galanter, and Pribram, *Plans and the Structure of Behavior*.

14. Miller explained to his department chairman: "My plan is to spend this time at the Institute for Advanced Study in Princeton. I hope to learn more of the mathematics I need to develop some of my own theoretical notions. Since von Neumann has a group of men working on his theory of games, the Institute seems an ideal place to become familiar with this kind of theory construction." George A. Miller to Edwin B. Newman, December 15, 1949, Papers of the Dean

of the Faculty of Arts and Sciences, letters to/from, 1949–50, Phillips - S (Misc), Folder: Psychology, UA III 5.55.26, Harvard University Archives.

15. Bruner has reflected that "there have been times when I thought I would have been better off in the seventeenth century, when it was more usual to follow one's curiosity than the straighter arrow of specialized study. I am not a good 'discipline' man and do not like boundaries." Bruner, *In Search of Mind*, 8.

16. George A. Miller, "What Is Information Measurement?" *American Psychologist* 8 (1953): 3–11; George A. Miller and F. C. Frick, "Statistical Behavioristics and Sequences of Responses," *Psychological Review* 56 (1949): 311–24; George A. Miller and Jennifer A. Selfridge, "Verbal Context and the Recall of Meaningful Material," *American Journal of Psychology* 63 (1950): 176–85; George A. Miller, G. A. Heise, and W. Lichten, "The Intelligibility of Speech as a Function of the Context of the Test Materials," *Journal of Experimental Psychology* 41 (1951): 329–35; Miller, "The Magical Number Seven"; George A. Miller and E. A. Friedman, "The Reconstruction of Mutilated English Texts," *Information and Control* 1 (1957): 38–55; Noam Chomsky and George A. Miller, "Finite State Languages." For further discussion of Miller's postwar work, see Crowther-Heyck, "George A. Miller, Language, and the Computer Metaphor of Mind"; Edwards, *The Closed World: Computers and the Politics of Discourse in Cold War America*, chapter 7.

17. Jerome S. Bruner and Cecile C. Goodman, "Value and Need as Organizing Factors in Perception," *Journal of Abnormal and Social Psychology* 42, no. 1 (1947): 33–44; Bruner and Postman, "On the Perception of Incongruity"; Posner and Schulman, "Cognitive Science."

18. Bruner, Goodnow, and Austin, *A Study of Thinking*. Once the book was published, Bruner would tell Jean Piaget "if it had not been for you, for von Neumann and for Claude Shannon, it never would have occurred to me to take this approach." Jerome S. Bruner to Jean Piaget, May 17, 1956, Papers of Jerome Bruner, General Correspondence, 1965–71, Papousek–Ry (gen'l), Folder: Piaget, HUG 4242.5, Harvard University Archives.

19. Crowther-Heyck, "George A. Miller, Language, and the Computer Metaphor of Mind."

20. Jerome S. Bruner and George A. Miller to Nathan Pusey, January 4, 1962, Papers of the Dean of the Faculty of Arts and Sciences, letters to/from, 1961–62, Box 2, Folder: Cognitive Studies, Center for, UA III 5.55.26, Harvard University Archives.

21. Center for Cognitive Studies, *First Annual Report, 1960–1961*, 1–2.

22. Bruner initially presented the results of his research on the R and D group at a conference in 1958. For his paper in the conference proceedings, see Bruner, "The Conditions of Creativity."

23. Jerome Bruner and George Miller to John W. Gardner, April 7, 1960, Papers of the Dean of the Faculty of Arts and Sciences, letters to/from, 1959–60,

Box 6, Folder: Center for Cognitive Studies, UA III 5.55.26, Harvard University Archives.

24. "Weekly Log," 28 November, 1960, CCS, Correspondence, Budget and Equipment Papers, Harvard University Archives.

25. George Miller to Jerome Bruner, Dictaphone recording Friday, November 29, 1963, Bruner Papers, General Correspondence, 1965–71, Folder: Miller, George, HUG 4242.5, Harvard University Archives.

26. Chomsky and Miller, "Finitary Models of Language Users"; Chomsky and Miller, "Introduction to the Formal Analysis of Natural Languages."

27. Jerome S. Bruner to Mary Jane Aschner, June 28, 1961, Bruner Papers, General Correspondence, 1962–64, Box A, Folder: Aschner. HUG 4242.5. Harvard University Archives.

28. George A. Miller, "Language and Psychology," in *New Directions in the Study of Language*, ed. Eric H. Lenneberg (Cambridge, Mass.: MIT Press, 1964), 94.

29. See, for instance, CCS, *Third Annual Report, 1962–1962*, 5.

30. David French, quoted in CCS, *First Annual Report, 1960–1961*, 4.

31. See Rose R. Olver, "A Developmental Study of Cognitive Equivalence" (Ph.D. diss., Harvard University, 1962).

32. George A. Miller, "Project Grammarama," in *The Psychology of Communication* (New York: Basic Books, 1967), 125–87.

33. George A. Miller, "Computers, Communication, and Cognition," in *The Psychology of Communication*, 114–15.

34. For instance, see Chomsky, "Three Models for the Description of Language."

35. Chomsky, *Syntactic Structures*, 18–25. The original proof appears in Chomsky, "Three Models for the Description of Language."

36. George A. Miller, Lecture Notes for Psychology 165: Psychology of Speech and Communication. October 4–18, 1957, HUC 8957.272.165, Harvard University Archives.

37. On the significance of embedded sentences to the work of the Center, see *Third Annual Report, 1962–1963*, 39.

38. George A. Miller, "Computers, Communication, and Cognition."

39. A sampling of this line of Bruner's research follows. Bruner and Goodman, "Value and Need"; Bruner and Postman, "On the Perception of Incongruity"; Jerome S. Bruner and Mary C. Potter, "Interference in Visual Recognition," *Science* 144, no. 3617 (1964): 424–25; Jerome S. Bruner, "On Perceptual Readiness," *Psychological Review* 64 (1957): 123–52. The eye camera and associated research is discussed in J. F. Mackworth and N. H. Mackworth, "Eye Fixations Recorded on Changing Visual Scenes by the Television Eye-Marker," *Journal of the Optical Society of America* 48 (1958): 439–45; N. H. Mackworth, "A Stand Camera for Line-of-Sight Recording," *Perception and Psychophysics* 2 (1967):

119–27; N. H. Mackworth and E. L. Thomas, "A Head Mounted Eye Marker Camera," *Journal of the Optical Society of America* 52 (1962): 713–16.

40. CCS, *First Annual Report, 1960–1961,* 8.

41. CCS, *Second Annual Report, 1961–1962,* 5.

42. Ibid.

43. Donald A. Norman, interview, August 2, 1997. See also Donald A. Norman and Willem J. M. Levelt, "Life at the Center," in *The Making of Cognitive Science: Essays in Honor of George A. Miller,* ed. William Hirst (Cambridge and New York: Cambridge University Press, 1988).

44. Jerome S. Bruner, interview, May 7, 1997.

45. CCS, *Second Annual Report, 1961–1962,* 5.

46. Donald A. Norman, interview, August 2, 1997.

47. Bruner, "The Conditions of Creativity." This work was a part of a quickly growing research area concerned with understanding creativity and using group brainstorming to produce innovation among businessmen, engineers, scientists, and students. See, for instance, William J. J. Gordon, *Synectics* (New York: Harper and Brothers, 1961); A. F. Osborn, *Applied Imagination: Principles and Procedures of the Creative Process,* rev. ed. (New York: Scribner's, 1953).

48. NIH Grant Application, "Studies in Cognition," December 13, 1965. CCS, Grant Requests, Proposals and Reports, Folder: NIH Proposal, 4. UA V. 290.15.

49. George A. Miller to Jerome S. Bruner, August 21, 1963, Bruner Papers, General Correspondence, 1962–64, Box L-O, Folder: M, HUG 4242.5, Harvard University Archives.

50. CCS, *Annual Report,* 1965–66, 39–40.

51. For discussion of the role of behaviorism in establishing the disciplinary autonomy of psychology, see O'Donnell, *The Origins of Behaviorism: American Psychology, 1870–1920.*

52. Donald A. Norman, interview, August 8, 1997; Norman and Levelt, "Life at the Center."

53. On Postman, see Bruner Papers, General Correspondence, 1948–1951, Box Meikle-Postman, Folder: Postman, Harvard University Archives. On Mackworth, see CCS Papers, Correspondence, UA V.290.5, Harvard University Archives.

54. Jerome S. Bruner et al., *Studies in Cognitive Growth* (New York: John Wiley and Sons, 1966); Jerome S. Bruner, "The Growth and Structure of Skill," in *Motor Skills in Infancy,* ed. K. J. Connolly (London and New York: Academic Press, 1971); Chomsky, *Aspects of the Theory of Syntax*; Patricia M. Greenfield and Jerome S. Bruner, "Culture and Cognitive Growth," in *Handbook of Socialization Theory and Research,* ed. David A. Goslin (Chicago: Rand McNally, 1969); Miller, "Project Grammarama"; Mackworth, "A Stand Camera."

55. Randy Allen Harris, *The Linguistics Wars* (New York and Oxford: Oxford University Press, 1993), 68, 271.

56. Norman and Levelt, "Life at the Center," 101. Also CCS *Annual Reports* (especially before 1965); NIH Grant Application, "Studies in Cognition," December 13, 1965, CCS, Grant Requests, Proposals and Reports, Folder: NIH Proposal, 3–4. UA V. 290.15, Harvard University Archives; Donald A. Norman, interview, August 2, 1997; CCS, "Final Report 1965–1972," Carnegie Corporation Grant Files, Series 2, Box 606: Harvard University—Thought Processes, Columbia University Archives; Bruner, *In Search of Mind*; Bruner, "Founding the Center."

57. CCS, *Third Annual Report, 1962–1963*, 8.

58. NIH Grant Application, "Studies in Cognition," December 13, 1965, Harvard University Archives.

59. Interview, name withheld.

60. CCS, *First Annual Report, 1960–1961*, 10.

61. Jerome S. Bruner to Ross Stanger, February 9, 1962, Bruner Papers, General Correspondence, 1961–62, Box N-Z, Folder S, HUG 4242.5, Harvard University Archives. In the last couple of years before the move to William James Hall, the Center had an outpost on Garden Street. However, the split that ensued did not draw energy from the Center's Kirkland Street home base because the majority of the Center's activities remained in the original location.

62. CCS, *Fifth Annual Report, 1964–65*, 7–8.

63. Interview with Donald A. Norman, August 2, 1997.

64. One was George Miller. Interview, September 25, 1997.

65. Doris Aaronson, Personal Communication, August 23, 1997; interview with Mary Potter, October 8, 1997; Interview with George Miller, September 25, 1997.

66. Jerome Bruner to Hilde Himmelweit, March 22, 1971, Bruner Papers, General Correspondence, 1965–71, Head Start – I, Folder: Himmelweit, Hilde, HUG 4242.5, Harvard University Archives.

67. Nathan Pusey, handwritten memo to himself outlining a meeting with Bruner and Miller, January 12, 1962, Papers of the Dean of the Faculty of Arts and Sciences, letters to/from, 1961–62, Box 2, Folder: Cognitive Studies, Center for, UA III 5.55.26, Harvard University Archives.

68. Bruner to Bertie Woolf, October 26, 1970. Bruner Papers, Personal Correspondence and Papers, N-P, Folder: NICHHD 03049 Request for Renewal, HUG 4242.18, Harvard University Archives.

69. Bruner mentions this group briefly in "Founding the Center for Cognitive Studies." For details on Bread and Roses, see Nancy Hawley, "Dear Sisters," October 8, 1970, Annie Popkin Papers, Box 2, Folder 39, Schlesinger Library; Kathy McAfee and Myrna Wood, "Bread and Roses," in *Voices from Women's Liberation*, ed. Leslie Tanner (New York: Signet Books, 1970); Alice Echols, *Daring to Be Bad: Radical Feminism in America, 1967–1975* (Minneapolis: University of Minnesota Press, 1989); Ann Hunter Popkin, "Bread and Roses: An

Early Moment in the Development of Socialist-Feminism" (Ph.D. diss., Brandeis University, 1978).

70. Rose Olver, interview, September 8, 1998.

71. Among the numerous accounts these events have received are Diane Ravitch, *The Troubled Crusade: American Education 1945–1980* (New York: Basic Books, 1983); Richard Norton Smith, *The Harvard Century: The Making of a University to a Nation* (New York: Simon and Schuster, 1986). The full story remains clouded, however, as Harvard's administrative records on the event are sealed until at least 2029.

72. Jerome Bruner to Daniel Kahneman, March 22, 1971. Bruner Papers, General Correspondence, 1965–71, Kagan–Kennedy, Folder: Kahneman, Daniel. HUG 4242.5. Harvard University Archives.

73. Ulric Neisser, *Cognitive Psychology* (Englewood Cliffs, N.J.: Prentice-Hall, 1967), 5.

74. CCS, "Final Report 1965–1972," Carnegie Corporation Grant Files, Series 2, Box 606: Harvard University—Thought Processes, Columbia University Archives.

75. Among these are Willem Levelt's role in founding the Max Plank Institute in Nijmegen, Jacques Mehler's in the *Journal of Cognition*, and George Miller's in the program in cognitive science at Princeton. In addition, Howard Gardner shaped Project Zero at Harvard and Donald Norman shaped the program at UCSD along the lines they saw at the Center. Donald A. Norman, interview, August 2, 1997; Howard Gardner, *To Open Minds* (New York: Basic Books, 1989), 54–55.

76. Bruner, "Founding the Center," 97.

77. CCS, *Seventh Annual Report, 1966–1967* (Cambridge, Mass.), 1.

Chapter Seven

1. Charles Laird, *Through These Eyes* (National Film Board of Canada, 2004).

2. Peter B. Dow, *Schoolhouse Politics: Lessons from the Sputnik Era* (Cambridge Mass.: Harvard University Press, 1991), 164–65.

3. Analysis in this chapter centers on the ways MACOS grew from cognitive psychology and was embedded in science curriculum reform. For discussion of how MACOS fit within reform of social studies curricula, see Ronald W. Evans, *The Hope for American School Reform: The Cold War Pursuit of Inquiry Learning in Social Studies* (New York: Palgrave Macmillan, 2011); Ronald W. Evans, *The Tragedy of American School Reform: How Curriculum Politics and Entrenched Dilemmas Have Diverted Us from Democracy* (New York: Palgrave Macmillan, 2011)

4. Ellen Condliffe Lagemann, *An Elusive Science: The Troubling History of Education Research* (Chicago: University of Chicago Press, 2000), 47; Westbrook, *John Dewey and American Democracy*, 97, 101–2, 167–73.

5. Diane Ravitch, *The Troubled Crusade: American Education 1945–1980* (New York: Basic Books, 1983), 55, 67–68.

6. Andrew Hartman, *Education and the Cold War: The Battle for the American School*, 101–6; Michelle Nickerson, "Domestic Threats: Women, Gender and Conservatism in Los Angeles, 1945–1966" (Ph.D. diss., Yale University, 2003), 132, 147; Perrin Selcer, "Patterns of Science: Developing Knowledge for a World Community at UNESCO" (Ph.D. diss., University of Pennsylvania, 2011), chapter 2. Nickerson also provides general discussion of the role of educational activities as a critical part of the conservative movement, helping to pave the way for women to move from political inactivity into more nationally visible and male-dominated organizations, such as the John Birch Society.

7. Nickerson, "Domestic Threats," 174–75.

8. Kurt Lewin, "Group Decision and Social Change," in *Readings in Social Psychology*, ed. E. E. Macoby, T. M. Newcomb, and E. L. Hartley (New York: Henry Holt and Company, 1958); Ronald Lippett, *Training in Community Relations* (New York: Harper and Brothers, 1949).

9. The roots of conference planning in group dynamics are discussed in Mead and Byers, *The Small Conference*.

10. Selcer, "Patterns of Science," chapter 2.

11. Ibid.

12. Nickerson, "Domestic Threats," 150.

13. Hartman, *Education and the Cold War*, 104.

14. Langton T. Crane, *The National Science Foundation and Pre-College Science Education, 1950–1975: Report Prepared for the Subcommittee on Science, Research, and Technology of the Committee on Science and Technology, U.S. House of Representatives* (Washington: Science Policy Research Division, Congressional Research Service, Library of Congress, 1975), 51–55; J. Merton England, *A Patron for Pure Science: The National Science Foundation's Formative Years, 1945–57* (Washington, D.C.: National Science Foundation, 1983), 242–43.

15. Arthur Bestor, *Educational Wastelands: The Retreat from Learning in Our Public Schools* (Urbana: University of Illinois Press, 1953). On Bestor, see Ravitch, *The Troubled Crusade*, 75–76; Arthur Zilversmit, *Changing Schools: Progressive Education Theory and Practice, 1930–1960* (Chicago: University of Chicago Press, 1993), 109–10; Lawrence Arthur Cremin, *The Transformation of the School: Progressivism in American Education, 1876–1957* (New York: Vintage Books, 1961), 344–47.

16. John L. Rudolph, *Scientists in the Classroom: The Cold War Reconstruction of American Science Education* (New York: Palgrave, 2002), 115, 131, 144.

17. Ibid., 124-25; Jack S. Goldstein, *A Different Sort of Time: The Life of Jerrold Zacharias*, 164-65, 187.

18. Goldstein, *A Different Sort of Time*, 142.

19. James B. Conant, "The Superintendent Was the Target," *The New York Times*, April 29, 1951.

20. Zilversmit, *Changing Schools*, 216-17 n. 7.

21. Rudolph, *Scientists in the Classroom* 121, 235 n. 39.

22. Ibid., 120.

23. Proposal to the Ford Foundation for the Support of a Curriculum Development Program in the Humanities and Social Sciences by American Council of Learned Societies and Educational Services Incorporated, 9/1/62 to 9/30/63. Signed by Frederick Burkhardt, president, ACLS; James Killian, chairman of the Corporation, MIT; Jerrold Zacharias, ESI vice president; Carroll Newsom, president, ESI. Jerome S. Bruner, Correspondence: Misc. Organizations, ESI, PERD, Publishers, 1964-65, Box 1 of 8, Folder: ESI Pre-1965 Files, HUG 4242.9, Harvard University Archives.

24. Barbara Barksdale Clowse, *Brainpower for the Cold War: The Sputnik Crisis and National Defense Education Act of 1958* (Westport, Conn.: Greenwood Press, 1981); Ravitch, *The Troubled Crusade*, 229.

25. Wayne W. Welsch, "Twenty Years of Science Curriculum Development: A Look Back," *Review of Research in Education* 7 (1979): 282-306, on 285.

26. Bruner, *The Process of Education*, vii, ix.

27. On the New Math, see William Wooton, *SMSG: The Making of a Curriculum* (New Haven, Conn.: Yale University Press, 1965); Lagemann, *An Elusive Science: The Troubling History of Education Research*, 166-68; Christopher James Phillips, "The American Subject: The New Math and the Making of a Citizen" (Ph.D. diss., Harvard University, 2011).

28. Rudolph, *Scientists in the Classroom*, 121.

29. 1965 program description in Hillier Krieghbaum and Hugh Rawson, *An Investment in Knowledge: The First Dozen Years of the National Science Foundation's Summer Institutes to Improve Secondary School Science and Mathematics Teaching, 1954-1965* (New York: New York University Press, 1969), 9. The cost of the summer institutes and curriculum development is given in Lagemann, *An Elusive Science*, 169.

30. Welsch, "Twenty Years of Science Curriculum Development," 284.

31. Sales numbers from Peter B. Dow, "Innovation's Perils: An Account of the Origins, Development, Implementation, and Public Reaction to *Man: A Course of Study*" (Ed. D., Harvard University, 1979), 72.

32. Theodore W. Hipple, Thomas R. Giblin, and Jack Megenity, "Have Your Students Read . . .?" *Phi Delta Kappan* 53, no. 7 (1972): 441-42, on 441. The most widely assigned book was Charles E. Silberman, *Crisis in the Classroom: The Remaking of American Education* (New York: Random House, 1970).

33. Harold G. Shane, "Significant Writings That Have Influenced the Curriculum: 1906–81," *Phi Delta Kappan* 62, no. 5 (1981): 311–14, on 311.

34. Bruner, Goodnow, and Austin, *A Study of Thinking.*

35. Jerome S. Bruner, Rose R. Olver, and Patricia M. Greenfield, introduction to *Studies in Cognitive Growth: A Collaboration at the Center for Cognitive Studies* (New York: John Wiley and Sons, 1966); Bruner, *In Search of Mind: Essays in Autobiography.*

36. One case was Miller's discussion of how the distance covered by a falling object could be coded in a table with entries for distance covered at each moment or a single equation: distance = $gt^2/2$. George A. Miller, *Language and Communication* (New York: McGraw-Hill, 1951), 234.

37. Jerome S. Bruner, "Going Beyond the Information Given," in *Beyond the Information Given: Studies in the Psychology of Knowing* (New York: W. W. Norton, 1973), 226. Bruner's comment here is based on work originally reported in Bruner, Goodnow, and Austin, *A Study of Thinking.*

38. Bruner, "Going Beyond," 224.

39. For a related view that thinking is defined by filling in gaps between available information, see also Frederic C. Bartlett, *Thinking: An Experimental and Social Study* (New York: Basic Books, 1958).

40. Rudolph, *Scientists in the Classroom*, 115–18. Compare Dewey's advocacy of teaching where direct experience comes first and laws second. "The young begin with active occupations having a social origin and use, and proceed to a scientific insight in the materials and laws involved." John Dewey, *Democracy and Education: An Introduction to the Philosophy of Education* (New York: Macmillan, 1916), 227.

41. For a transcript of this teaching method in operation—psychologist Mary Henle teaching one child to identify nouns, verbs, and tenses—see "Instructional Interview #2," n.d. Box 5, Folder 1. Peter Dow, *Man: A Course of Study* Records, Monroe Gutman Library Special Collections. For discussion of the use of transformational grammars in the course, see Jerome S. Bruner, *Occasional Paper No. 3: Man: A Course of Study* (Cambridge, Mass.: Educational Services Incorporated, 1965), 8. One indication of the novelty of this material was that Bruner was still putting "kernel sentences" in quotes both in this paper and the reprint for a wider audience. See Jerome S. Bruner, "Man: A Course of Study," in *Toward a Theory of Instruction* (Cambridge, Mass.: Belknap Press of Harvard University Press, 1966), 79. Compare this mode of teaching to John Dewey's assertion that teaching language for its own sake is an absurdity. John Dewey, *School and Society* (Chicago: University of Chicago Press, 1899), 49.

42. Robert B. Westbrook, "John Dewey," *Prospects: The Quarterly Review of Comparative Education* 23, no. 1/2 (1993): 277–91.

43. Jerome S. Bruner, "After John Dewey, What?" in *On Knowing: Essays for the Left Hand* (Cambridge, Mass.: Belknap Press of Harvard University Press,

1962), 121. This chapter was originally published in *The Saturday Review*, Supplement, June 17, 1961.

44. Bruner, *The Process of Education*, 14. The claim that people and organisms had inherent interest in or were rewarded by learning itself was an antibehaviorist position. See, for instance Chomsky, "Review of B. F. Skinner, *Verbal Behavior.*"

45. Bruner, *The Process of Education*, 33.

46. Ibid., 44. In making this point Bruner was articulating in the pedagogical domain one of the findings of Jean Piaget and Bärbel Inhelder: that children developed topological before Euclidean geometry. At this time Bruner was in close communication with Piaget and Inhelder, and Inhelder herself had recently spent a year as a visitor at the Center for Cognitive Studies.

47. Bruner to J. B. Conant, 3/3/60, Jerome S. Bruner, General Correspondence, 1961–62, Box: G-M HUG, Folder: Harriet Gibney, Harvard University Archives.

48. Bruner, *The Process of Education*, 19.

49. "Report: Conference of Social Studies and Humanities Curriculum Program," June 9–23, 1962, 2. MACOS papers, Box 1, Folder 3.

50. Goldstein, *A Different Sort of Time.*

51. "Report: Conference of Social Studies and Humanities Curriculum Program," 3.

52. Ibid., 4.

53. Ibid.

54. Dow, *Schoolhouse Politics*, 42.

55. "Report: Conference of Social Studies and Humanities Curriculum Program," 18.

56. "Brief Report of a Two Week Conference held at Endicott House," June 9–23, 1962, 4. MACOS papers, Box 1, Folder 3.

57. Ibid., 5.

58. Ibid.

59. Dow, *Schoolhouse Politics*, 44–45.

60. The Endicott House report notes that Bruner articulated the pedagogical sequence for the social studies curricula as Action—Imagery—Notation. "A Brief Report," 6. Compare this to the discussion of "enactive," "ikonic," and "symbolic" modes of cognition discussed in Jerome S. Bruner, "On Cognitive Growth," in *Studies in Cognitive Growth: A Collaboration at the Center for Cognitive Studies*, ed. Jerome S. Bruner, Rose R. Olver, and Patricia M. Greenfield (New York: John Wiley and Sons, 1966).

61. "Report: Conference of Social Studies and Humanities Curriculum Program," 7; "Brief Report," 3, 5.

62. Educational Services Incorporated, "A Short History of the Social Studies Program" Spring 1965, 2. MACOS records, Box 1, Folder 15.

63. "Report: Conference of Social Studies and Humanities Curriculum Program," 24.

64. ESI Social Studies Curriculum Program, "A Narrative Report, 1962–1964," November 1964, 10. MACOS papers, Box 3, Folder 6.

65. Dow, *Schoolhouse Politics*, 56.

66. ESI Social Studies Curriculum Program, "A Narrative Report, 1962–1964," 11.

67. Ibid., 6.

68. The members of the IRG included Bruner, Courtney Cazden, Blythe Clinchy, Gloria Cooper, Margaret Donaldson, James Gallagher, Howard Gardner, Carol Gilbert, Jean Berko Gleason, Joseph Glick, Mary Henle, Richard Jones, Helen Kenney, Judy Krieger, David McNeill, Eli Maranda, Jess Reid, Anne Ryle, Roger Wales, Dan Washburn, Babbett Whipple, and Barbara White. The envisioned role of the IRG is discussed in a "Supplementary Statement" submitted with grant proposal to the NSF, September 1, 1964, 4–5. MACOS Papers, Box 35, Folder 2.

69. ESI Social Studies Curriculum Program, "A Narrative Report, 1962–1964," 20.

70. Dow, *Schoolhouse Politics*, 61.

71. Ibid., 66.

72. Ibid., 71, 77. There may have also been political reasons for Bruner's adoption of DeVore's tactics. It is possible that he saw Oliver's plan as too Eurocentric. Several years later, the ESI staff was concerned enough about charges of racism that they struck a section on the !Kung from the curriculum because "it had become politically unacceptable to use materials that showed partially naked, dark-skinned 'primitives' in a public school classroom" (Dow, *Schoolhouse Politics*, 122). Further discussion of this development can be found in Erika Milam, "Public Science of the Savage Mind: Contesting Cultural Anthropology in the Cold War Classroom," *Journal for the History of the Behavioral Sciences* 49, no. 3 (2013).

73. Gardner, *To Open Minds*, 52–55.

74. Dow, *Schoolhouse Politics*, 98–99.

75. Ibid., 98.

76. Bruner, "The Conditions of Creativity."

77. Babette Whipple, "The Think-In," December 1965. MACOS papers, Box 8, Folder 4.

78. Babette Whipple, *Occasional Paper No. 10: The Grouptalk* (Cambridge, Mass.: Educational Services Incorporated, 1967); *Man: A Course of Study. Evaluation Strategies*, (Cambridge, Mass.: EDC, 1970).

79. Bruner, *Occasional Paper No. 3: Man: A Course of Study*, 3.

80. *Man: A Course of Study. Seminars for Teachers* (Cambridge, Mass.: EDC, 1969, 1970).

81. On the anti-racist goals of MACOS, see the commentary of Asen Balikci in Charles Laird, *Through These Eyes*.

82. Bruner, *Occasional Paper No. 3: Man: A Course of Study*, 4.

83. A similar anti-biological argument for culture and not universal biological inheritance as the essential feature of human nature can be found in the contemporary works of the anthropologist Clifford Geertz, who had been trained in the Department of Social Relations. See especially his "The Growth of Culture and the Evolution of Mind" (1962) and "The Impact of Culture on the Concept of Man" (1966). These are reprinted in Clifford Geertz, *The Interpretation of Cultures* (New York: Basic Books, 1973).

84. Jerome S. Bruner, *A Time for Learning* (Association-Sterling Films, 1973).

85. Bruner, *Toward a Theory of Instruction*, 92.

86. Peter B. Dow, "Man: A Course of Study: A Continuing Exploration of Man's Humanness," in *Man: A Course of Study. Talks to Teachers* (Cambridge, Mass.: EDC, 1968), 16.

87. *Man: A Course of Study. Evaluation Strategies*, 11.

88. Bruner, *A Time for Learning*.

89. *Man: A Course of Study. The Observer's Handbook* (Cambridge, Mass.: EDC, 1970).

90. Mary Henle, "On Coping with Ambiguity," in *Man: A Course of Study. Seminars for Teachers*.

91. Dow, *Schoolhouse Politics*, 153–56.

92. *Man: A Course of Study. Evaluation Strategies*, 12.

Chapter Eight

1. On MACOS course adoption rates, see Dorothy Nelkin, *Science Textbook Controversies and the Politics of Equal Time* (Cambridge, Mass: MIT Press, 1977), 108; cited in Mark Solovey "Social Science Funding under Siege in Conservative Times: The Case of the U.S. National Science Foundation during the 1970s." Paper presented at the Ninth Annual Workshop on the History of Economics as History of Science, École Normale Supérieure de Cachan, June 16, 2012.

2. For discussion and defense of this methodology, see Kenneth Keniston, *The Uncommitted: Alienated Youth in American Society* (New York: Dell Publishing Co., 1970 [1965]).

3. Ibid., 9.

4. On the novelty of modern alienation, see ibid., 3.

5. On the role of Mills for the New Left, see James Miller, *Democracy Is in the Streets: From Port Huron to the Siege of Chicago* (Cambridge, Mass.:

Harvard University Press, 1994), 79-91. On the continuity of his thought with other social thinkers of the 1950s period, see Daniel Geary, *Radical Ambition: C. Wright Mills, the Left, and American Social Thought* (Berkeley and Los Angeles: University of California Press, 2009). Also see Kevin Mattson, *Intellectuals in Action: The Origins of the New Left and Radical Liberalism, 1945–1970* (University Park: Pennsylvania State University Press, 2002).

6. C. Wright Mills, "On Knowledge and Power," *Dissent*, July 1955, 207, 212.

7. Herbert Marcuse, *One Dimensional Man: Studies in the Ideology of Advanced Industrial Society* (Boston: Beacon Press, 1964).

8. Penina Migdal Glazer, "The New Left: A Style of Protest," *The Journal of Higher Education* 38, no. 3 (1967): 119–30.

9. "The Port Huron Statement," 331–34, 365. As reprinted in Miller, *Democracy Is in the Streets*, 329–74.

10. "The Port Huron Statement," 330–32, 335, 374. Several years later, Columbia University's SDS branch likewise noted the fragmented nature of American society. Columbia Students for a Democratic Society, "The Columbia Statement," in *The University Crisis Reader*, ed. Immanuel Maurice Wallerstein and Paul Starr (New York: Random House, 1971), 1:26.

11. Free Speech Movement, "We Want a University," in *The Berkeley Student Revolt: Facts and Interpretations*, ed. Seymour Martin Lipset and Sheldon S. Wolin (Garden City, N.Y.: Doubleday and Company, 1965), 209.

12. Columbia Students for a Democratic Society, "The Columbia Statement," 41–42.

13. "The Port Huron Statement," 330.

14. Mills, "Mass Society and Liberal Education," 357–58.

15. Martin L. King Jr., "The Role of the Behavioral Scientist in the Civil Rights Movement," *Journal of Social Issues* 24, no. 1 (1968): 1–12, on 10–11.

16. Steve Halliwell, "Columbia: An Explanation, or Applesauce, Epicycles, and the Joplin Proviso," in *The New Left: A Collection of Essays*, ed. Priscilla Long (Boston: Porter Sargent, 1969), 203–5; Evelyn Goldfield, Sue Munaker, and Naomi Weisstein, "A Woman Is a Sometime Thing: Cornering Capitalism by Removing 51% of Its Commodities," in *The New Left: A Collection of Essays*, 262–63.

17. Noam Chomsky, "Introduction," in *American Power and the New Mandarins* (New York: Pantheon Books, 1969), 9.

18. Talcott Parsons, Edward A. Shils, and with the assistance of James Olds, "Values, Motives, and Systems of Action," in *Toward a General Theory of Action*, ed. Talcott Parsons and Edward A. Shils (Cambridge, Mass.: Harvard University Press, 1951), 82, 85–86, 89–90. For discussion of the significance of this paper and Parsons's other work for modernization theory, see Gilman, *Mandarins of the Future*, 86–94. For other insightful analysis of modernization theory see David C. Engerman, ed., *Staging Growth: Modernization, Development,*

and the Global Cold War (Amherst: University of Massachusetts Press, 2003); Michael Latham, *Modernization as Ideology: American Social Science and "Nation Building " in the Kennedy Era* (Chapel Hill: University of North Carolina Press, 2000).

19. Chomsky, "Objectivity and Liberal Scholarship"; Noam Chomsky, "The Responsibility of Intellectuals," *The New York Review of Books*, February 23, 1967. On the administration's dishonesty, see also Howard Zinn, *Vietnam: The Logic of Withdrawal* (Boston: Beacon Press, 1967), 35, 36, 38, 59, 77, 78.

20. For instance, see the offhand remark about dishonesty in Allen Guttman, "Protest against the War in Vietnam," *Annals of the American Academy of Political and Social Science* 382 (1969): 53–63 n. 18. On Schlesinger Jr.'s lies, see *New York Times*, November 26, 1965. On the disconnect between administration rhetoric and policy, see Franz Schurmann, Peter Dale Scott, and Reginald Zelnik, *The Politics of Escalation in Vietnam* (Boston: Beacon Press, 1966).

21. Steven Brown, "Consistency and the Persistence of Ideology: Some Experimental Results," *The Public Opinion Quarterly* 34, no. 1 (1970): 60–68.

22. See the discussion on Project Camelot in Mark Solovey, "Project Camelot and the 1960s Epistemological Revolution: Rethinking the Politics-Patronage-Social Science Nexus," *Social Studies of Science* 31, no. 2 (2001): 171–206; Ron Robin, *The Making of the Cold War Enemy: Culture and Politics in the Military-Intellectual Complex* (Princeton, N.J.: Princeton University Press, 2001), 206–25; Herman, *Romance of American Psychology*, 153–73.

23. Richard M. Pfeffer, ed., *No More Vietnams? The War and the Future of American Policy* (New York, Evanston, and London: Harper and Row for the Adlai Stevenson Institute for International Affairs, 1968), 50.

24. Feminist critique on this point can be found in Goldfield, Munaker, and Weisstein, "A Woman Is a Sometime Thing," 262–63.

25. Chomsky, "The Responsibility of Intellectuals."

26. See, for instance, C. Wright Mills, "The New Left," *New Left Review*, September-October 1960. Reprinted in C. Wright Mills, *Power, Politics, and People: The Collected Essays of C. Wright Mills* (London and New York: Oxford University Press, 1967), 247–69; Christian Bay, "The Cheerful Science of Dismal Politics," in *The Dissenting Academy*, ed. Theodore Roszak (New York: Pantheon Books, 1968).

27. "Critiques of American Society: A Series of Twelve Talks and Discussions," n.d. [other information indicates 1968], Lucy Candib Papers, Box 1, Folder: Critiques; Schlesinger Library; a public lecture series that also counted for course credit at MIT and Harvard.

28. To members of the unity of science movement, being an engineer rather than a scientist was to be susceptible to anti-democratic authoritarian ideology. See Phillip Frank, "Science and Democracy."

29. Halliwell, "Columbia: An Explanation," 203–5.

30. James Jacobs, *S. M. Lipset, Social Scientist of the Smooth Society* (Boston: New England Free Press, n.d).

31. See Chomsky, "Introduction," to *American Power*, 16; Zinn, *Vietnam*, 7, 50, 95, 119.

32. Irving Lester Janis, *Victims of Groupthink: A Psychological Study of Foreign Policy Decisions and Fiascos* (Boston: Houghton Mifflin, 1972). On rigidity and intolerance of ambiguity in Rostow and Bundy, see David Grossman Armstrong, "The True Believer: Walt Whitman Rostow and the Path to Vietnam" (Ph.D. diss., University of Texas at Austin, 2000); Kai Bird, *The Color of Truth: McGeorge Bundy and William Bundy, Brothers in Arms: A Biography* (New York: Simon and Schuster, 1998); David Milne, *America's Rasputin: Walt Rostow and the Vietnam War* (New York: Hill and Wang, 2008).

33. Zinn, *Vietnam*, 90, 92. This comparison even appeared at a conference of leading foreign policy experts at the Adlai Stevenson Institute for International affairs. Attendees included Henry Kissinger, Daniel Ellsberg, Samuel Huntington, Ithiel de Sola Pool, Arthur Schlesinger Jr., and people with RAND and DOD ties such as Albert Wohlstetter and Adam Yarmolinsky. See the comments by Eqbal Ahmad as reprinted in the meeting transcript (Pfeffer, ed., *No More Vietnams?* 14–15).

34. On the inherent connection between the social sciences and democracy, see Lerner, "The Social Sciences: Whence and Whither?" 31.

35. Stephen Earl Bennett, "Modes of Resolution of a 'Belief Dilemma' in the Ideology of the John Birch Society," *The Journal of Politics* 33, no. 3 (1971): 735–72. Work on cognitive dissonance originated with Leon Festinger, *When Prophecy Fails: A Social and Psychological Study of a Modern Group That Predicted the Destruction of the World* (New York: Harper and Row, 1956).

36. Donald Granberg and Gail Corrigan, "Authoritarianism, Dogmatism and Orientations Toward the Vietnam War," *Sociometry* 35, no. 3 (1972): 468–76.

37. Sheldon G. Levy, "The Psychology of Political Activity," *Annals of the American Academy of Political and Social Science* 391 (1970): 83–96; Marc Pilisuk et al., "War Hawks and Peace Doves: Alternate Resolutions of Experimental Conflicts," *The Journal of Conflict Resolution* 9, no. 4 (1965): 491–508. This mode of analysis continued into the 1980s. See James Sidanius, "Cognitive Functioning and Sociopolitical Ideology Revisited," *Political Psychology* 6, no. 4 (1985): 637–61.

38. Keniston's views are analyzed in Daniel Geary, "Children of *The Lonely Crowd*: David Riesman, the Young Radicals, and the Splitting of Liberalism in the 1960s" (unpublished ms., 2012).

39. John Schultz, *The Chicago Conspiracy Trial*, rev. ed. (New York: Da Capo Press, 1993), 398.

40. Examples of this can be found in Richard Flacks, "The Liberated Gener-

ation: An Exploration of the Roots of the Student Protest," *Journal of Social Issues* 23, no. 3 (1967): 52–75; Pilisuk et al., "War Hawks and Peace Doves."

41. Rhoda Unger, personal communication; Naomi Weisstein, "Psychology Constructs the Female; or the Fantasy Life of the Male Psychologist (with Some Attention to the Fantasies of His Friends, the Male Biologist and the Male Anthropologist)," in *Radical Feminism*, ed. Anne Koedt, Ellen Levine, and Anita Rapone (New York: Quadrangle Books, 1973).

42. Rose Olver, interview; Rhoda Kesler Unger, *Resisting Gender: Twenty-Five Years of Feminist Psychology* (London and Thousand Oaks, Calif.: Sage Publications, 1998). On the use of psychology in the second wave, see Herman, *Romance of American Psychology*, 290–304.

43. Discussion of the relationship between feminism and Weisstein's science may be found in an unpublished job talk she gave at Brown University and in a draft of "All Mountains Moved In Fire." Radcliffe Archives. Weisstein's published critique of the science came after first, personal experiences, and second, politicization through discussion with the Westside Group. This chronology is inverted in her publications. In print, her technical critique of science predates her personal narrative. The account of her experience can be found in Naomi Weisstein, "'How Can a Little Girl Like You Teach a Great Big Class of Men?' The Chairman Said, and Other Adventures of a Woman in Science," in *Working It Out: 23 Women Writers, Artists, Scientists, and Scholars Talk About Their Lives and Work*, ed. Sara Ruddick and Pamela Daniels (New York: Pantheon, 1977). The critique of psychological methods is Naomi Weisstein, "'Kinder, Küche, Kirche' as Scientific Law: Psychology Constructs the Female," in *Sisterhood Is Powerful: An Anthology of Writings from the Women's Liberation Movement*, ed. Robin Morgan (New York: Random House, 1970). In subsequent versions Weisstein expanded her critique to include anthropologists. See Weisstein, "Psychology Constructs the Female."

44. Weisstein, "'How Can a Little Girl Like You Teach a Great Big Class of Men?'"

45. Naomi Weisstein, Virginia Blaisdell, and Jesse Lemisch, *The Godfathers* (New Haven, Conn.: Belladonna Publishing, 1975), 2.

46. Weisstein's comment about "social devices" is a quotation of Holton. Weisstein, Blaisdell, and Lemisch, *The Godfathers*, 16, 20. "Meritocratic elite" is from Zbigniew Brzezinski, "America in the Technotronic Age," *Encounter*, January 1968, 24.

47. Weisstein, Blaisdell, and Lemisch, *The Godfathers*, 20.

48. Helen S. Astin, *The Woman Doctorate in America: Origins, Career, and Family* (New York: Russell Sage Foundation, 1969); Linda S. Fidell, "Empirical Verification of Sex Discrimination in Hiring Practices in Psychology," *American Psychologist* 25 (1970): 1094–98; Sandra L. Bem and Daryl J. Bem, "Case

Study of a Nonconscious Ideology: Training the Woman to Know Her Place," in *Beliefs, Attitudes, and Human Affairs*, ed. Daryl J. Bem (Belmont, Calif.: Brooks/Cole Publishing Company, 1970), 90. In a more recent study, researchers have found that peer reviewers systematically evaluate men better than women to such an extent that women seeking a postdoctoral fellowship need more than twice the number of publications to receive the same scores as men seeking that fellowship. C. Wenneras and A. Wold, "Nepotism and Sexism in Peer Review," *Nature* 387 (1997): 341. The finding on evaluators' preferential ranking of men's CVs has been replicated. See R. E. Steinpreis, K. A. Anders, and D. Ritzke, "The Impact of Gender on the Review of the Curriculum Vitae of Job Applicants and Tenure Candidates: A National Empirical Study," *Sex Roles* 41 (1999): 509.

49. Philip Goldberg, "Are Women Prejudiced against Women?" *Trans-Action* 5 (1968): 28–30; Gail I. Pheterson, Sara B. Kiesler, and Philip A. Goldberg, "Evaluation of the Performance of Women as a Function of Their Sex, Achievement, and Personal History," *Journal of Personality and Social Psychology* 19, no. 1 (1971): 114–18. A more recent ethnography made a similar point. Though not focused on gender, it noted that readers evaluated the quality of scientific papers as much by who the author was as by the paper's content. Latour and Woolgar, *Laboratory Life*.

50. The study was Robert Rosenthal, *Pygmalion in the Classroom: Teacher Expectations and Pupils' Intellectual Development* (New York: Holt, Rinehart, and Winston, 1968).

51. Historical discussion of the problem of experimenter effects can be found in Jerry M. Suls and Ralph L. Rosnow, "Concerns About Artifacts in Psychological Experiments," in *The Rise of Experimentation in American Psychology*, ed. Jill G. Morawski (New Haven, Conn.: Yale University Press, 1988).

52. Naomi Weisstein, "Psychology Constructs the Female."

53. Arthur R. Jensen, "How Much Can We Boost IQ and Scholastic Achievement?" *Harvard Educational Review* 39 (1969): 1–123.

54. Richard Lewontin, "Race and Intelligence," *Bulletin of the Atomic Scientists*, March 1970, 2–8.

55. Richard J. Herrnstein, "IQ," *Atlantic Monthly*, September 1971.

56. A useful collection of essays on both sides is Ned Joel Block and Gerald Dworkin, *The I.Q. Controversy: Critical Readings* (New York: Pantheon Books, 1976). Historical examination of the debate can be found in Graham Richards, *Race, Racism, and Psychology: Towards a Reflexive History* (London and New York: Routledge, 1997), chap. 9.

57. See Richards, *Race, Racism, and Psychology*. On sociobiologist E. O. Wilson having water poured on his head during a public symposium on sociobiology, see David L. Hull, *Science as a Process: An Evolutionary Account of the*

Social and Conceptual Development of Science (Chicago: University of Chicago Press, 1988), 229.

58. William H. Tucker, "Fact and Fiction in the Discovery of Sir Cyril Burt's Flaws," *Journal of the History of the Behavioral Sciences* 30 (1994): 335–47, on 336.

59. David Layzer, "Science or Superstition? (A Physical Scientist Looks at the IQ Controversy)," *Cognition* 1, nos. 2–3 (1972): 265–99.

60. Skinner's *Beyond Freedom and Dignity* went through fifteen printings between 1971 and 1972. It was part of the Book-Of-The-Month-Club and was serialized in *Psychology Today* and *The New York Post*.

61. Noam Chomsky, "Psychology and Ideology," *Cognition* 1, no. 1 (1971): 11–46.

62. Harlow, "The Nature of Love."

63. Free Speech Movement, "We Want a University," 209–10, 213–15. Leaflet originally distributed January 4, 1965; University of California, Berkeley, Study Commission on University Governance, *The Culture of the University: Governance and Education* (Berkeley and Los Angeles: University of California Press, 1968), 12, 13.

64. On the desire for a truly liberal education as a motivation for protest on the part of some supporters of the Free Speech Movement, see David A. Hollinger, "A View from the Margins," in *The Free Speech Movement: Reflections on Berkeley in the 1960s*, ed. Robert Cohen and Reginald E. Zelnik (Berkeley and Los Angeles: University of California Press, 2002).

65. Free Speech Movement, "We Want a University," 214–15. The question about whether education at Berkeley was education for slaves also appears in Paul Goodman, "Thoughts on Berkeley," *New York Review of Books*, January 14, 1965.

66. Free Speech Movement, "We Want a University," 214; Bell, *The Reforming of General Education*; Kerr, *The Uses of the University*, 15, 27, 31.

67. "The Port Huron Statement," 334–35.

68. On the intelligentsia or students, rather than the industrial working class, as the source of social change, see "The Port Huron Statement," 373–74; Mills, "The New Left." For examples of the job-training functions of the university, see Bell, *The Reforming of General Education*, 106. On students' self-perception as a class, even as the "lumpen proletariat" or as a minority within a ghetto, see Kerr, *The Uses of the University*, 78, 101. Historical consideration of students as a class or as ghetto residents can be found in, *inter alia*, Douglas C. Rossinow, "Mario Savio and the Politics of Authenticity," in *The Free Speech Movement: Reflections on Berkeley in the 1960s*; Miller, *Democracy Is in the Streets*. On the connections between the revival of a truly liberal education as a motivation for some supporters of the Free Speech Movement, see Hollinger, "A View from the Margins."

69. On political action as a means of combating alienation from the self, see Douglas C. Rossinow, *The Politics of Authenticity: Liberalism, Christianity, and the New Left in America* (New York: Columbia University Press, 1998).

70. For a similar statement that emphasized the lack of political consciousness and social isolation, see Noam Chomsky, "Knowledge and Power: Intellectuals and the Welfare-Warfare State," in *The New Left: A Collection of Essays*, 190.

71. Columbia Students for a Democratic Society, "The Columbia Statement," 25–26.

72. Casey Hayden and Mary King, "Sex and Caste: A Kind of Memo," reprinted in *Takin' It to the Streets*, ed. Alexander Bloom and Wini Breines (New York: Oxford University Press, 1995); Alice Echols, *Daring to Be Bad: Radical Feminism in America, 1967–1975*, chap. 1; Sara M. Evans, *Personal Politics: The Roots of Women's Liberation in the Civil Rights Movement and the New Left* (New York: Knopf, 1979), chaps. 7–8.

73. Naomi Weisstein papers, draft of "All Mountains Moved In Fire: A Personal, Political, and Scientific Memoir of the Early Women's Liberation Movement," unpublished ms. by Weisstein and Candace Lyle Hogan, 14–20, Carton 14, Radcliffe Archives.

74. For discussions of the importance of community in the Boston and Chicago second-wave feminist movements, see Amy Kesselman et al., "Our Gang of Four: Friendship and Women's Liberation," in *The Feminist Memoir Project: Voices from Women's Liberation*, ed. Rachel Blau DuPlessis and Ann Snitow (New York: Three Rivers Press, 1998); Popkin, "Bread and Roses: An Early Moment in the Development of Socialist-Feminism." More generally on the movement's development out of the New Left, see Evans, *Personal Politics*; Echols, *Daring to Be Bad*.

75. Memories of the relationship between community and political activity in the feminist movement can be found in Popkin, "Bread and Roses"; Kesselman et al., "Our Gang of Four."

76. Margaret Mead, "Purposive Systems: Proceedings of the First Annual Symposium of the American Society for Cybernetics," 8–9. For general discussion of declining politeness and its political implications, see Kenneth Cmiel, "The Politics of Civility," in *The Sixties: From Memory to History*, ed. David Farber (Chapel Hill: University of North Carolina Press, 1994).

77. David S. Brown, *Richard Hofstadter: An Intellectual Biography*; Geary, "Children of *The Lonely Crowd*."

78. For discussion of this incident, see Irving Howe, *A Margin of Hope: An Intellectual Autobiography* (San Diego: Harcourt Brace Jovanovich, 1982), 293; Joseph Dorman, *Arguing the World: The New York Intellectuals in Their Own Words* (New York: Free Press, 2000).

79. John R. Searle, "A Foolproof Scenario for Student Revolts," in *The Uni-*

versity Crisis Reader, ed. Immanuel Maurice Wallerstein and Paul Starr (New York: Random House, 1971), 2:34; originally published in *The New York Times Magazine,* December 29, 1968.

80. Geary, "Children of *The Lonely Crowd.*"

81. George F. Kennan, "Rebels Without a Program," speech delivered at Swarthmore College, December 1967, in George Frost Kennan, *Democracy and the Student Left* (Boston: Little Brown, 1968), 4–6; originally in the *New York Times Magazine,* January 21, 1968.

82. Jerome Karabel, *The Chosen: The Hidden History of Admission and Exclusion at Harvard, Yale, and Princeton,* 71; Robert R. Weyeneth, "The Architecture of Racial Segregation: The Challenges of Preserving the Problematical Past," *The Public Historian* 27, no. 4 (2005): 11–44; Kathleen L. Wolgemuth, "Woodrow Wilson and Federal Segregation," *The Journal of Negro History* 44, no. 2 (1959): 158–73; Nancy J. Weiss, "The Negro and the New Freedom: Fighting Wilsonian Segregation," *Political Science Quarterly* 84, no. 1 (1969): 61–79.

83. On the connections between civil rights activism and the Free Speech Movement, see Rossinow, "Mario Savio and the Politics of Authenticity." It was not only at Berkeley where working for civil rights was the foundation for student activism. On Texas, see Rossinow, *The Politics of Authenticity.*

84. Lewis Samuel Feuer, *The Conflict of Generations: The Character and Significance of Student Movements* (New York: Basic Books, 1969).

85. Zbigniew Brzezinski, "Revolution and Counterrevolution (but Not Necessarily About Columbia)," *The New Republic,* June 1, 1968.

86. Daniel Bell, "Columbia and the New Left," *Public Interest* 13 (1968): 61–101; Nathan Glazer, "'Student Power' in Berkeley," *Public Interest* 13 (1968): 3–21, on 21.

87. Glazer, "'Student Power' in Berkeley."

88. Bell, "Columbia and the New Left," 92.

89. Gustav Le Bon, *The Crowd: A Study of the Popular Mind* (Atlanta: Cherokee Publishing Company, 1982 [1895]).

90. Harry F. Wolcott, "The Middlemen of MACOS," *Anthropology and Education Quarterly* 38, no. 2 (2007): 195–206, on 200; Dow, *Schoolhouse Politics,* 199; Nick Thimmesch, "The Grass-Roots Dollar Chase—Ready on the Right," *New York Magazine,* June 9, 1975.

91. Mike Royko, "Wife-Swapping Eskimos vs. TV Morals," *The Denver Post,* February 9, 1976. Reagan's attack on MACOS occurred on August 22, 1980, and is quoted on the back cover of Phyllis Schlafly, *Child Abuse in the Classroom: Excerpts from Official Transcript of Proceedings before the U.S. Department of Education* (Alton, Ill.: Pere Marquette Press, 1984). On the development of direct mail as a social and political technology for the right wing, see Bruce J. Schulman, *The Seventies: The Great Shift in American Culture, Society, and Politics* (New York: Free Press, 2001), 197–99.

92. John A. Steinbacher, *Robert Francis Kennedy, the Man, the Mysticism, the Murder* (Los Angeles: Impact Publishers, 1968).

93. William C. Martin, *With God on Our Side: The Rise of the Religious Right in America* (New York: Broadway Books, 1996), 113.

94. John A. Steinbacher, *The Conspirators: Men against God* (Whittier, Calif.: Orange Tree Press, 1972), preface. Boldface emphasis in the original.

95. Dow, *Schoolhouse Politics*, 197.

96. On the Gablers, see Dorothy Nelkin, *Science Textbook Controversies and the Politics of Equal Time* (Cambridge, Mass.: MIT Press, 1977), 47–48.

97. Nadine Winterhalter, President, Parents in Action to T. S. Hancock, Executive Director, Region IV Education Service Center, Houston, Texas, February 7, 1973, 2. MACOS archive.

98. Dow, *Schoolhouse Politics*, 198–99.

99. Ibid., 179.

100. Steinbacher, *The Conspirators*, 51, 53.

101. Pat Robertson, *The New World Order* (Dallas: World Publishing, 1991). The similarities between Robertson's analysis and that of MACOS's earliest critics could either come from Robertson's familiarity with the works of Steinbacher or from the reliance of both authors on common source texts. On Robertson's book, see Jacob Heilbrunn, "His Anti-Semitic Sources," *The New York Review of Books*, April 20, 1995.

102. Minutes of the Board of Trustees, Madison County School District no. 38, Phoenix Arizona, October 28, 1971, 134a. On the criticism of MACOS's attention to physiology, see the allegations against "MACOS: Man A Course of Study" by parents in Bellevue, Washington, n.d. (either late 1971 or early 1972). On the class charges of evolution in and the dismissal of this charge, see "Majority Report" of the Citizens Committee for Review of MAN: A Course of Study, Presented to William Webb, Superintendent of Public Schools, Collier County, Florida, November 3, 1975, 7–8, MACOS Archive. Perhaps the difference between these positions comes from a separation between the course materials themselves and the supplemental materials available to teachers. The supplemental materials included essays for the teachers that included material from Robert Trivers and Hans Guggenheim on species evolution. However, the same booklet opened with an essay from Peter Dow, a central figure in the MACOS project. He noted "much of the data of Man: A Course of Study leads toward the study of human evolution. The course does not pursue this fascinating path, because in developing the material we came to feel that a comparative rather than an evolutionary approach to the study of man was the most effective way to introduce the subject to young children." Dow, "Man: A Course of Study: A Continuing Exploration of Man's Humanness," 15, Box 36, Folder 3 for Phoenix; Box 37, Folder 1 for Bellevue; Box 37, Folder 6 for Collier County, MACOS Archive.

103. Peter Woolfson, "The Fight over 'MACOS': An Ideological Conflict in Vermont," *Council on Anthropology and Education Quarterly* 5, no. 3 (1974): 27–30, on 28.

104. Alan Jehlen, "Social Studies Course Assailed," *The Patriot Ledger*, October 8, 1975.

105. Nadine Winterhalter to T. S. Hancock, February 7, 1973, 1.

106. "MACOS: Man A Course of Study" by parents in Bellevue, Washington, n.d., 2.

107. Bruner, "Occasional Paper No. 3: Man: A Course of Study," 4.

108. "Majority Report" of the Citizens Committee for Review of MAN: A Course of Study, Collier County, Florida, November 3, 1975, MACOS archive; "Underlying the 'MACOS' Controversy," *The Phoenix Gazette*, September 24, 1971, 10.

109. For examples of this particular mode of analysis see James J. Kilpatrick, "Is Eskimo Sex Life a School Subject?" *Boston Evening Globe*, April 2, 1975; "MACOS and Moral Values," *The Wall Street Journal*, July 21, 1975. Given the consistency of this trope that conflated discussion with approval and the fact that conservative commentary and fliers were more readily accessible than the MACOS materials themselves, this specific reading of MACOS either indicates that many conservatives had not actually read the MACOS materials, or, if they did, then this is a case that indicates the power of readers rather than authors to determine the meaning of texts.

110. Phyllis Musselman, "We're All Annimals [*sic*]—Kids Are Taught Here," *Weekly American News*, September 22, 1971. William Merse, Deacon, Covenant Presbyterian Church and State Committeeman, American Party of Collier County to Roland Anderson, Chairman, Collier County School Board, William Webb, Collier County Superintendent, Tom Morris, Middle School Supervisor, and Robert Mung, Principal, Pine Ridge Middle School, n.d. MACOS archive, Box 37, Folder 6.

111. Minutes of the Board of Trustees, Madison County School District no. 38, Phoenix Arizona, October 28, 1971, 134a; "MACOS Social? Study?" *Weekly American News*, September 1, 1971.

112. See Susan M. Marshner, *Man: A Course of Study—Prototype for Federalized Textbooks?* (Washington, D.C.: The Heritage Foundation, 1975), 26.

113. Phyllis Musselman, "Shofstall Gives Some Advice on MACOS," *Weekly American News*, September 29, 1971.

114. Onallee McGraw, "Violation of Constitutional Rights," *Bethesda-Chevy Chase Tribune*, March 2, 1973.

115. Melvyn New, "Nineteen Eighty-Four," *American Speech* 51, no. 3/4 (1976): 278–79; Joseph J. Thorndike, "'The Sometimes Sordid Level of Race and Segregation': James J. Kilpatrick and the Virginia Campaign against *Brown*," in

The Moderate's Dilemma: Massive Resistance to School Desegregation in Virginia, ed. Matthew D. Lassiter and Andrew B. Lewis (Charlottesville: The University Press of Virginia, 1998).

116. Kilpatrick was particularly bothered by the bill's provision that it would be applied only to states with low voter registration rates and that the bill used statistical measures to judge where discrimination had occurred. He remarked that the one-party tradition was more responsible than discrimination for low rates of voter registration. James J. Kilpatrick, "Must We Repeal the Constitution to Give the Negro the Vote?" *The National Review*, April 20, 1965, 322.

117. "Majority Report" of the Citizens Committee for Review of MAN: A Course of Study, Collier County, Florida, November 3, 1975, MACOS archive; "Underlying the 'MACOS' Controversy."

118. District School Board of Collier County, "Man: A Course of Study. Position Paper," n.d. (internal evidence indicates fall of 1975), MACOS archive. Box 37, Folder 6.

119. "Parent Evaluation–Man A Course of Study," October 26, 1971; "Parents Comments–*Man A Course of Study*–Madison Park School, Phoenix Arizona, n.d. MACOS archive. Box 36, Folder 3.

120. Arthur Block "Phoenix Reactions to MACOS," n.d., MACOS archive. Box 17, Folder 7.

121. Harriet Macintosh, Freedom for Readers, Inc., to Peter Dow, February 17, 1973. MACOS archive. Box 17, Folder 7.

122. Arthur Block, "Advantages of MACOS," *Boston Globe*, June 19, 1975.

123. Marshner, *Man: A Course of Study—Prototype for Federalized Textbooks?* 11.

124. Jerrold R. Zacharias, "Research Scholars and Curriculum Development," *PMLA* 80, no. 2 (1965): 13–15, on 13.

125. Woolfson, "The Fight over 'MACOS,'" 29. For a call for memorization, an attack on discovery-based learning, and a false assertion that NSF-sponsored curricula including New Math and MACOS had not been tested before implementation, see James J. Kilpatrick, "War Flares over Scientific Funds," *Boston Globe*, January 20, 1976.

126. P. J. MacDonald to the School Board of Madison County, n.d. [1971], 3, MACOS archive. Box 36, Folder 3; "Minority Report of Joanne McAuley, Including Dissenting and Additional Views to Accompany the Report of the Science Curriculum Implementation Review Group to the Chairman, Committee on Science and Technology," U.S. House of Representatives, October 20, 1975, in "National Science Foundation Curriculum Development and Implementation for Pre-College Science Education," report prepared for the Committee on Science and Technology, U.S. House of Representative, Ninety-Fourth Congress, 1st sess., Serial Q, November 1975. (U.S. Government Printing Office: 1975), 18–19.

127. Marshner, *Man: A Course of Study—Prototype for Federalized Textbooks?* 38.

128. Ibid., 35, 43.

129. Jerome S. Bruner, Rose R. Olver, and Patricia M. Greenfield, *Studies in Cognitive Growth: A Collaboration at the Center for Cognitive Studies* (New York: John Wiley and Sons, 1966).

130. Bruner specified this particular end of MACOS in a film distributed to explain course goals and methods, *A Time for Learning.*

131. *Who Do We Eat?* Pamphlet circulated to encourage opposition to MACOS in Phoenix, n.d. [1971], MACOS archive. Box 36, Folder 3.

132. See for instance, Marshner, *Man: A Course of Study—Prototype for Federalized Textbooks?* 26; Onallee McGraw, *Secular Humanism and the Schools: The Issue Whose Time Has Come* (Washington, D.C.: The Heritage Foundation, 1976), 4.

133. "Minority Report of Joanne McAuley," 18.

134. This account of the congressional maneuvers around MACOS follows Dow, *Schoolhouse Politics,* 199–215.

135. Welsch, "Twenty Years of Science Curriculum Development."

136. For discussion of the role of creativity at RAND, see Sharon Ghamari-Tabrizi, "Simulating the Unthinkable: Gaming Future War in the 1950s and 1960s." Discussion of the movement of RAND intellectuals to DOD can be found in David R. Jardini, "Out of the Blue Yonder: The Transfer of Systems Thinking from the Pentagon to the Great Society, 1961–1965," in *Systems, Experts, and Computers: The Systems Approach in Management and Engineering, World War II and After,* ed. Agatha C. Hughes and Thomas P. Hughes (Cambridge, Mass.: MIT Press, 2000).

Conclusion

1. Thus, their activities were precisely the opposite of the boundary-reinforcing activities that sociologist Thomas Gieryn has discussed. Thomas Gieryn, "Boundary-Work and the Demarcation of Science from Non-Science: Strains and Interests in Professional Ideologies of Scientists," *American Sociological Review* 48 (1983): 781–95.

2. Leahey, "The Mythical Revolutions of American Psychology"; Gardner, *The Mind's New Science*; Baars, *The Cognitive Revolution in Psychology*; Mandler, "Origins of the Cognitive (R)Evolution"; John D. Greenwood, "Understanding the 'Cognitive Revolution' in Psychology," *Journal of the History of the Behavioral Sciences* 35, no. 1 (1999): 1–22; Boden, *Mind as Machine.*

3. Exemplary studies that track instruments in natural science include David Kaiser, *Drawing Theories Apart: The Dispersion of Feynman Diagrams in*

Postwar Physics (Chicago: University of Chicago Press, 2005); Adele Clarke and Joan H. Fujimura, *The Right Tools for the Job: At Work in Twentieth-Century Life Sciences* (Princeton, N.J.: Princeton University Press, 1992); Robert E. Kohler, *Lords of the Fly: Drosophila Genetics and the Experimental Life* (Chicago: University of Chicago Press, 1994); Galison, *Image and Logic*; Nicolas Rasmussen, *Picture Control: The Electron Microscope and the Transformation of Biology in America, 1940–1960* (Stanford, Calif.: Stanford University Press, 1997).

4. Daston and Galison, *Objectivity*.

5. On pluralist practice in contemporary procedures of academic evaluation, see Michèle Lamont, *How Professors Think: Inside the Curious World of Academic Judgment* (Cambridge, Mass.: Harvard University Press, 2009).

6. Christopher Lasch, *The Culture of Narcissism: American Life in an Age of Diminishing Expectations* (New York: Norton, 1978).

7. On the politics of these methodological debates, see Daniel T. Rodgers, *The Age of Fracture* (Cambridge, Mass.: Belknap Press of Harvard University Press, 2011).

8. Adoption of cognitive linguistics into politically charged literary theory appears in Stanley Fish, *Is There a Text in This Class? The Authority of Interpretive Communities* (Cambridge, Mass.: Harvard University Press, 1980).

9. Allan Bloom, *The Closing of the American Mind: How Higher Education Has Failed Democracy and Impoverished the Souls of Today's Students* (New York: Simon and Schuster, 1988), 183, 324.

10. Discussion of conservative Christian interest in phonics can be found in Mark B. Thogmartin, "The Prevalence of Phonics Instruction in Fundamentalist Christian Schools," *Journal of Research on Christian Education* 3, no. 1 (1994): 103–32.

11. On colleges as "indoctrination centers," see the commentary by Rick Santorum reported in *Inside Higher Education*, February 27, 2012; http://www.insidehighered.com/news/2012/02/27/santorums-views-higher-education-and-satan (accessed January 5, 2013). For rejection of critical thinking, see Andrew Rosenthal "No Comment Necessary: Texas GOP 2012 Platform Opposes Teaching 'Critical Thinking Skills,'" *New York Times*, June 29, 2012; http://takingnote.blogs.nytimes.com/2012/06/29/no-comment-necessary-texas-gops-2012-platform-opposes-teaching-critical-thinking-skills (accessed January 5, 2013).

12. Kevin Durrheim, "Theoretical Conundrum: The Politics and Science of Theorizing Authoritarian Cognition," *Political Psychology* 18, no. 3 (1997): 625–47; Samelson, "The Authoritarian Character."

13. Bob Altemeyer, *Right-Wing Authoritarianism* (Winnipeg: University of Manitoba Press, 1981); Bob Altemeyer, *Enemies of Freedom: Understanding Right-Wing Authoritarianism* (San Francisco: Jossey-Bass Publishers, 1988).

14. "Bush's Final Approval Rating: 22 Percent." CBS News, February 11, 2009; http://www.cbsnews.com/2100-500160_162-4728399.html (accessed May 17, 2013).

15. David S. Amodio et al., "Neurocognitive Correlates of Liberalism and Conservatism," *Nature Neuroscience* 10 (2007): 1246–47; John T. Jost et al., "Political Conservatism as Motivated Social Cognition," *Psychological Bulletin* 129, no. 3 (2003): 339–75; John W. Dean, *Conservatives without Conscience* (New York: Viking, 2006); Sean Wilentz, "Confounding Fathers: The Tea Party's Cold War Roots," *The New Yorker*, October 18, 2010; Andrew Sullivan, "Trying to Understand the Tea Party II," *The Atlantic*, June 15, 2010; Paul Krugman, "Paranoia Strikes Deep," *New York Times*, November 9, 2009. For an argument that diagnoses conservative mental frameworks and then adds that such views have been disproven by cognitive science, see George Lakoff, *Moral Politics: How Liberals and Conservatives Think*, 2d ed. (Chicago: University of Chicago Press, 2002).

16. Daniel Kahneman, *Thinking Fast and Slow* (New York: Farrar, Straus and Giroux, 2011).

17. The alignment of centrist politics with behavioral economics can be found in Richard H. Thaler and Cass R. Sunstein, *Nudge: Improving Decisions About Health, Wealth, and Happiness* (New Haven, Conn.: Yale University Press, 2008).

References

Unpublished Sources

AMERICAN PHILOSOPHICAL SOCIETY

Papers of Anne Roe
Papers of S. S. Wilks

CARNEGIE MELLON UNIVERSITY

Herbert Simon Papers

COLUMBIA UNIVERSITY ARCHIVES

Carnegie Corporation Grant Files
Carnegie Corporation Oral History Project

FORD FOUNDATION ARCHIVES

Ford Foundation Grant Files

HARVARD UNIVERSITY ARCHIVES

Papers of E. G. Boring
Papers of Jerome S. Bruner
Papers of Paul Herman Buck
Papers of the Center for Cognitive Studies
Papers and Minutes of the Committee on General Education in a Free Society
Papers of the Dean of the Faculty of Arts and Sciences
Papers of Clyde K. M. Kluckhohn
MACOS Archive
Papers of George A. Miller
Papers of the Department of Psychology
Papers of the Department of Psychology and Social Relations

Papers of Talcott Parsons
Papers of the Department of Social Relations
Harvard University Course Lecture Notes

LIBRARY OF CONGRESS MANUSCRIPT DIVISION

Papers of Vannevar Bush
Papers of J. Robert Oppenheimer

NATIONAL LIBRARY OF MEDICINE

Papers of Joshua Lederberg

RADCLIFFE INSTITUTE

Papers of Lucy Candib
Papers of Annie Popkin
Papers of Naomi Weisstein

ROCKEFELLER ARCHIVE CENTER

Papers of the Social Science Research Council

YALE UNIVERSITY ARCHIVES

Papers of Carl Hovland

Personal Interviews and Communications

Doris Aaronson, interview, August 21, 1997
Doris Aaronson, personal communications, August 15 and 23, 1997
Jerome S. Bruner, interview, May 7, 1997
Charles Gross, personal communication, 2002
George A. Miller, personal communication, December 29, 1997
George A. Miller, interviews, March 7 and September 25, 1997
Donald A. Norman, interview, August 2, 1997
Mary C. Potter, interview, October 8, 1997
Rose Olver, interview, September 8, 1998

Published Sources

"Academy Conferences." *Bulletin of the American Academy of Arts and Sciences* 10, no. 1 (1956): 2–4.
Ackerman, James S. *The Architecture of Michelangelo*. New York: Viking Press, 1961.
———. "On *Scientia*." In *Science and Culture: A Study of Cohesive and Disjunctive Forces*, edited by Gerald Holton, 14–23. Boston: Beacon Press, 1967.
Adler, Les K., and Thomas G. Paterson. "Red Fascism: The Merger of Nazi Ger-

many and Soviet Russia in the American Image of Totalitarianism, 1930s–1950s." *American Historical Review* 75 (1970): 1046–64.

Adler, Mortimer J. "God and the Professors." In *Science, Philosophy, and Religion: A Symposium*, 120–38. New York: Conference on Science, Philosophy and Religion in their Relation to the Democratic Way of Life, Inc., 1941.

——. "This Pre-War Generation." *Harpers*, October 1940, 524–34.

Adorno, Theodor W., Else Frenkel-Brunswik, Daniel J. Levinson, and R. Nevitt Sanford. *The Authoritarian Personality*. New York: Harper and Brothers, 1950.

Aikin, Wilford Merton. *The Story of the Eight-Year Study: With Conclusions and Recommendations*. New York: Harper and Brothers, 1942.

Allen, Raymond B. "Communists Should Not Teach in American Colleges." *Educational Forum* 13, no. 4 (1949): 433–40.

Allport, Gordon W. *Becoming: Basic Considerations for a Psychology of Personality*. New Haven, Conn.: Yale University Press, 1955.

——. *The Nature of Prejudice*. Cambridge, Mass.: Addison-Wesley Pub. Co., 1954.

——. "The Psychologist's Frame of Reference." *Psychological Bulletin* 37, no. 1 (1940): 1–28.

Alpers, Benjamin Leontief. *Dictators, Democracy, and American Public Culture: Envisioning the Totalitarian Enemy, 1920s–1950s*. Cultural Studies of the United States. Chapel Hill: University of North Carolina Press, 2003.

Alpert, Harry. *Emile Durkheim and His Sociology*. New York: Columbia University Press, 1939.

——. "The National Science Foundation and Social Science Research." *American Sociological Review* 19, no. 2 (1954): 208–11.

——. "The Social Sciences and the National Science Foundation: 1945–1955." *American Sociological Review* 20, no. 6 (1955): 653–61.

Altemeyer, Bob. *Enemies of Freedom: Understanding Right-Wing Authoritarianism*. Jossey-Bass Social and Behavioral Science Series. San Francisco: Jossey-Bass Publishers, 1988.

——. *Right-Wing Authoritarianism*. Winnipeg: University of Manitoba Press, 1981.

Amadae, S. M. *Rationalizing Capitalist Democracy: The Cold War Origins of Rational Choice Liberalism*. Chicago: University of Chicago Press, 2003.

Amodio, David S., John T. Jost, Sarah L. Master, and Cindy M. Yee. "Neurocognitive Correlates of Liberalism and Conservatism." *Nature Neuroscience* 10 (2007): 1246–47.

Anzieu, Didier. *Freud's Self-Analysis*. The International Psycho-Analytical Library, no. 118. London: Hogarth Press and the Institute of Psycho-analysis, 1986.

Arend, Lawrence E. "Spatial Differential and Integral Operations in Human Vision." *Psychological Review* 80, no. 5 (1973): 374–95.

Asch, S. E. "Effects of Group Pressure Upon the Modification and Distortion of Judgments." In *Groups, Leadership and Men: Research in Human Relations*, edited by Harold Guetzkow, 177–90. Pittsburgh, Pa.: Carnegie Press, 1951.

Ash, Mitchell G. "Gestalt Psychology: Origins in Germany and Reception in the United States." In *Points of View in the Modern History of Psychology*, edited by Claude E. Buxton, 295–344. Orlando: Academic Press, 1985.

Astin, Helen S. *The Woman Doctorate in America: Origins, Career, and Family.* New York: Russell Sage Foundation, 1969.

Baars, Bernard J. *The Cognitive Revolution in Psychology.* New York: Guilford Press, 1986.

Baker, Keith Michael. *Condorcet: From Natural Philosophy to Social Mathematics.* Chicago: University of Chicago Press, 1975.

Bales, Robert F. "How People Interact in Conferences." *Scientific American* 192 (1955): 31–55.

Barber, Bernard. *Science and the Social Order.* Glencoe, Ill.: The Free Press, 1952.

Barron, Frank. "Complexity-Simplicity as a Personality Dimension." *Journal of Abnormal and Social Psychology* 48, no. 2 (1953): 163–72.

———. *Creativity and Personal Freedom.* Princeton, N.J.: Van Nostrand, 1968.

———. *Creativity and Psychological Health: Origins of Personal Vitality and Creative Freedom.* Princeton, N.J.: Van Nostrand, 1963.

———. "The Needs for Order and for Disorder as Motives in Creative Activity." In *Scientific Creativity: Its Recognition and Development*, edited by Calvin W. Taylor and Frank Barron, 153–60. New York: John Wiley and Sons, 1963.

———. "Originality in Relation to Personality and Intellect." *Journal of Personality* 25 (1957): 730–42.

Bartlett, Frederic C. *Thinking: An Experimental and Social Study.* New York: Basic Books, 1958.

Bay, Christian. "The Cheerful Science of Dismal Politics." In *The Dissenting Academy*, edited by Theodore Roszak, 208–30. New York: Pantheon Books, 1968.

Beach, Frank A. "The Snark Was a Boojum." *American Psychologist* 5, no. 5 (1950): 115–24.

Bechtel, William, Adele Abrahamsen, and George Graham. "The Life of Cognitive Science." In *A Companion to Cognitive Science*, edited by William Bechtel and William Graebner, 1–104. Malden, Mass.: Blackwell, 1998.

The Behavioral Sciences at Harvard. Cambridge, Mass.: Harvard University, 1955.

Bell, Daniel. "America as a Mass Society: A Critique." In *The End of Ideol-*

ogy: On the Exhaustion of Political Ideas in the Fifties, 21–38. New York: The Free Press, 1962.

———. "Columbia and the New Left." *Public Interest* 13 (1968): 61–101.

———. *The Coming of Post-Industrial Society: A Venture in Social Forecasting.* New York: Basic Books, 1973.

———. "The Disjunction of Culture and Social Structure: Some Notes on the Meaning of Social Reality." *Daedalus* 94, no. 1 (1965): 208–22. Reprinted in *Science and Culture: A Study of Cohesive and Disjunctive Forces*, edited by Gerald Holton, 236–50. Boston: Beacon Press, 1967.

———. *The End of Ideology: On the Exhaustion of Political Ideas in the Fifties.* New York: The Free Press, 1960.

———. "Notes on the Post-Industrial Society." *The Public Interest*, no. 6 (1967): 24–35.

———. *The Reforming of General Education: The Columbia College Experience in Its National Setting.* New York: Columbia University Press, 1966.

———. "The Sensibility of the Sixties." In *The Cultural Contradictions of Capitalism*, 120–45. New York: Basic Books, 1976.

Bem, Sandra L., and Daryl J. Bem. "Case Study of a Nonconscious Ideology: Training the Woman to Know Her Place." In *Beliefs, Attitudes, and Human Affairs*, edited by Daryl J. Bem, 89–99. Belmont, Calif.: Brooks/Cole Publishing Company, 1970.

Benjamin, Ludy T. "The Psychological Roundtable: Revolution of 1936." *American Psychologist* 32, no. 7 (1977): 542–49.

Bennett, John W., and Kurt H. Wolff. "Toward Communication between Sociology and Anthropology." *Yearbook of Anthropology* (1955): 329–51.

Bergmann, Gustav, and Kenneth W. Spence. "Operationism and Theory in Psychology." *Psychological Review* 48, no. 1 (1941): 1–14.

Bestor, Arthur. *Educational Wastelands: The Retreat from Learning in Our Public Schools.* Urbana: University of Illinois Press, 1953.

Biagioli, Mario. *Galileo, Courtier: The Practice of Science in the Culture of Absolutism.* Chicago: University of Chicago Press, 1993.

Bird, Kai. *The Color of Truth: McGeorge Bundy and William Bundy, Brothers in Arms: A Biography.* New York: Simon and Schuster, 1998.

Birnbaum, Lucille C. "Behaviorism in the 1920s." *American Quarterly* 7, no. 1 (1955): 15–30.

Bissell, Richard M., Jr. "Comment on Kennan." In *The American Style: Essays in Value and Performance*, edited by Elting E. Morison, 334–52. New York: Harper and Brothers, 1958.

Bittel, L. R. "Brainstorming: Better Way to Solve Plant Problems." *Factory Management* 114, no. 5 (1956): 98–107.

Bitterman, M. E. "Toward Comparative Psychology of Learning." *American Psychologist* 15 (1960): 704–12.

Blackwell, Gordon W. "Multidisciplinary Team Research." *Social Forces* 33, no. 4 (1955): 367–74.

Block, Ned Joel, and Gerald Dworkin. *The I.Q. Controversy: Critical Readings.* New York: Pantheon Books, 1976.

Bloom, Allan. *The Closing of the American Mind: How Higher Education Has Failed Democracy and Impoverished the Souls of Today's Students.* New York: Simon and Schuster, 1988.

Boden, Margaret A. *Mind as Machine: A History of Cognitive Science.* Oxford and New York: Oxford University Press, 2006.

———. "Personal Politics and Ideology Machines." In *Artificial Intelligence and Natural Man*, 64–94. New York: Basic Books, 1977.

Boorstin, Daniel J. *The Genius of American Politics.* Chicago: University of Chicago Press, 1953.

Boring, Edwin G. "Mind and Mechanism." *American Journal of Psychology* 59, no. 2 (1946): 173–92.

———. *The Physical Dimensions of Consciousness.* New York: The Century Co., 1933.

Pierre Bourdieu, "The Production and Reproduction of Legitimate Language," in *Language and Symbolic Power: The Economy of Linguistic Exchanges*, 43–65. Cambridge, Mass: Harvard University Press, 1991.

Brick, Howard. "Displacement of Economy in an Age of Plenty." In *Transcending Capitalism: Visions of a New Society in Modern American Thought*, 152–85. Ithaca, N.Y.: Cornell University Press, 2006.

———. "Talcott Parsons's 'Shift Away from Economics,' 1937–1946." *Journal of American History* 87, no. 2 (2000): 490–514.

Brinkley, Alan. "The Illusion of Unity in Cold War Culture." In *Rethinking Cold War Culture*, edited by Peter J. Kuznick and James Gilbert, 61–73. Washington, D.C.: Smithsonian Institution Press, 2001.

———. "The Problem of American Conservatism." *American Historical Review* 99, no. 2 (1994): 409–29.

Brinton, Crane, and George Caspar Homans. *The Society of Fellows.* Cambridge, Mass.: Society of Fellows of Harvard University; distributed by Harvard University Press, 1959.

Bronowski, Jacob. *Science and Human Values.* New York: Julian Messner, 1956.

Brower, Daniel. "The Problem of Quantification in Psychological Science." *Psychological Review* 56, no. 6 (1949): 325–33.

Brown, David S. *Richard Hofstadter: An Intellectual Biography.* Chicago: University of Chicago Press, 2006.

Brown, G. "An Experiment in the Teaching of Creativity." *School Review* 72, no. 4 (1964): 437–50.

Brown, Steven. "Consistency and the Persistence of Ideology: Some Experimental Results." *The Public Opinion Quarterly* 34, no. 1 (1970): 60–68.

Bruner, Jerome S. "The Act of Discovery." *Harvard Educational Review* 31, no. 1 (1961): 21–32.

———. "After John Dewey, What?" In *On Knowing: Essays for the Left Hand*. Cambridge, Mass.: Belknap Press of Harvard University Press, 1979 [1962].

———. "The Conditions of Creativity." In *Contemporary Approaches to Creative Thinking*, edited by Howard E. Gruber, Glenn Terrell, and Michael Wertheimer, 1–30. New York: Atherton Press, 1962.

———. "Founding the Center for Cognitive Studies." In *The Making of Cognitive Science: Essays in Honor of George A. Miller*, edited by William Hirst, 90–99. Cambridge and New York: Cambridge University Press, 1988.

———. "Going Beyond the Information Given." In *Beyond the Information Given: Studies in the Psychology of Knowing*, 218–38. New York: W. W. Norton, 1973.

———. "The Growth and Structure of Skill." In *Motor Skills in Infancy*, edited by K. J. Connolly. London and New York: Academic Press, 1971.

———. *In Search of Mind: Essays in Autobiography*. New York: Harper and Row, 1983.

———. "Man: A Course of Study." In *Toward a Theory of Instruction*, 73–101. Cambridge, Mass.: Belknap Press of Harvard University Press, 1966.

———. "Mechanism Riding High: Review of Kenneth W. Spence, *Behavior Theory and Conditioning*." *Contemporary Psychology* 2, no. 6 (1957): 155–57.

———. "Neural Mechanisms in Perception." *Psychological Review* 64 (1957): 340–58.

———. *Occasional Paper No. 3: Man: A Course of Study*. Cambridge, Mass.: Educational Services Incorporated, 1965.

———. "On Cognitive Growth." In *Studies in Cognitive Growth: A Collaboration at the Center for Cognitive Studies*, edited by Jerome S. Bruner, Rose R. Olver, and Patricia M. Greenfield. New York: John Wiley and Sons, 1966.

———. "On Learning Mathematics." In *On Knowing: Essays for the Left Hand*, 97–111. Cambridge, Mass.: Harvard University Press, 1979 [1962].

———. "On Perceptual Readiness." *Psychological Review* 64 (1957): 123–52.

———. *The Process of Education*. Cambridge, Mass.: Harvard University Press, 1960.

———. *A Time for Learning*. University-at-Large, released by Association-Sterling Films, 1973. Filmstrip.

———. *Toward a Theory of Instruction*. Cambridge, Mass.: Belknap Press of Harvard University Press, 1966.

Bruner, Jerome S., and Gordon W. Allport. "Fifty Years of Change in American Psychology." *Psychological Bulletin* 37, no. 10 (1940): 757–77.

Bruner, Jerome S., and Cecile C. Goodman. "Value and Need as Organizing Factors in Perception." *Journal of Abnormal and Social Psychology* 42, no. 1 (1947): 33–44.

Bruner, Jerome S., Jacqueline J. Goodnow, and George A. Austin. *A Study of Thinking*. London: John Wiley and Sons, 1956.

Bruner, Jerome S., Rose R. Olver, and Patricia M. Greenfield. *Studies in Cognitive Growth: A Collaboration at the Center for Cognitive Studies*. New York: John Wiley and Sons, 1966.

Bruner, Jerome S., Rose R. Olver, Patricia M. Greenfield, Joan Rigney Hornsby, Helen J. Kenney, Michael Maccoby, Nancy Modiano, Frederic A. Mosher, David R. Olson, Mary C. Potter, Lee C. Reich, and Anne McKinnon Sonstroem. *Studies in Cognitive Growth*. New York: John Wiley and Sons, 1966.

Bruner, Jerome S., and Leo J. Postman. "On the Perception of Incongruity: A Paradigm." *Journal of Personality* 18 (1949): 206–23.

Bruner, Jerome S., and Mary C. Potter. "Interference in Visual Recognition." *Science* 144, no. 3617 (1964): 424–25.

Bryson, Lyman. *Science and Freedom*. New York: Columbia University Press, 1948.

Bryson, Lyman, Louis Finkelstein, and Robert M. MacIver, eds. *Approaches to National Unity: Seventh Symposium, Conference on Science, Philosophy and Religion*. New York: Harper and Brothers, 1945.

Brzezinski, Zbigniew. "America in the Technotronic Age." *Encounter*, January 1968, 16–26.

———. "Revolution and Counterrevolution (but Not Necessarily About Columbia)." *The New Republic*, June 1, 1968.

Buchner, Edward Franklin. "Psychological Progress in 1910." *Psychological Bulletin* 8, no. 1 (1911): 1–10.

Buck, Peter. "Adjusting to Military Life: The Social Sciences Go to War, 1941–1950." In *Military Enterprise and Technological Change: Perspectives on the American Experience*, edited by Merritt Roe Smith, 203–52. Cambridge, Mass.: MIT Press, 1985.

Buckley, Kerry W. *Mechanical Man: John Broadus Watson and the Beginnings of Behaviorism*. New York and London: Guilford Press, 1989.

Buckley, William F. *God and Man at Yale: The Superstitions of Academic Freedom*. Chicago: H. Regnery Co., 1951.

Buckley, William F., and Brent Bozell. *McCarthy and His Enemies: The Record and Its Meaning*. Chicago: H. Regnery Co., 1954.

Bundy, McGeorge. "The Attack on Yale." *The Atlantic*, November 1951, 50–52.

———. "McGeorge Bundy Replies." *The Atlantic*, December 1951, 82, 84.

Burt, C. "Eugenics: Intelligence and Genius." *British Medical Bulletin* 6 (1949): 78–79.

Camic, Charles. "Introduction: Talcott Parsons before *the Structure of Social Action*." In *Talcott Parsons: The Early Essays*, edited by Charles Camic. Chicago: University of Chicago Press, 1991.

Campbell, Donald T. "Ethnocentrism of Disciplines and the Fish-Scale Model of Omniscience." In *Interdisciplinary Relationships in the Social Sciences*, edited by Muzafer Sherif and Carolyn Sherif, 328–48. Chicago: Aldine, 1969.

———. "A Tribal Model of the Social System Vehicle Carrying Scientific Knowledge." *Knowledge* 1, no. 2 (1979): 181–201.

Cantril, Hadley. "An Inquiry Concerning the Characteristics of Man." *Journal of Abnormal and Social Psychology* 45, no. 3 (1950): 490–503.

———. "Toward a Scientific Morality." *The Journal of Psychology* 27 (1949): 363–76.

Capes, Mary. *Communication or Conflict: Conferences: Their Nature, Dynamics, and Planning.* New York: Association Press, 1960.

Capshew, James H. *Psychologists on the March: Science, Practice, and Professional Identity in America, 1929–1969.* Cambridge and New York: Cambridge University Press, 1999.

Carnegie Corporation of New York. *Annual Report*, 1946.

Carson, John. *The Measure of Merit: Talents, Intelligence, and Inequality in the French and American Republics, 1750–1940.* Princeton, N.J.: Princeton University Press, 2007.

Cattell, Raymond B. "The Personality and Motivation of the Researcher from Measurements of Contemporaries and from Biography." In *Scientific Creativity: Its Recognition and Development*, 119–31. New York: John Wiley and Sons, 1963.

Chomsky, Noam. *Aspects of the Theory of Syntax.* Cambridge, Mass.: MIT Press, 1965.

———. *Cartesian Linguistics.* New York and London: Harper and Row, 1966.

———. "Introduction." In *American Power and the New Mandarins*, 3–22. New York: Pantheon Books, 1969.

———. "Knowledge and Power: Intellectuals and the Welfare-Warfare State." In *The New Left: A Collection of Essays*, edited by Priscilla Long, 172–99. Boston: Porter Sargent, 1969.

———. "Objectivity and Liberal Scholarship." In *American Power and the New Mandarins*, 23–158. New York: Pantheon Books, 1969.

———. "Of Minds and Language." *Biolinguistics* 1 (2007): 9–27.

———. "Psychology and Ideology." *Cognition* 1, no. 1 (1971): 11–46.

———. "The Responsibility of Intellectuals." *The New York Review of Books*, February 23, 1967.

———. "Review of B. F. Skinner, *Verbal Behavior.*" *Language* 35, no. 1 (1959): 26–58. Reprinted in *The Structure of Language*, edited by J. A. Fodor and J. J. Katz. Englewood Cliffs, N.J.: Prentice-Hall, 1959.

———. *Syntactic Structures.* The Hague: Mouton Publishers, 1957.

———. "Three Models for the Description of Language." *IRE Transactions on Information Theory*, IT-2, no. 3 (1956): 113–24.

Chomsky, Noam, and George A. Miller. "Finitary Models of Language Users." In *Handbook of Mathematical Psychology*, vol. 2, edited by R. D. Luce, R. Bush, and E. Galanter, 419–91. New York: John Wiley and Sons, 1963.

———. "Finite State Languages." *Information and Control* 1 (1958): 91–112.

———. "Introduction to the Formal Analysis of Natural Languages." In *Handbook of Mathematical Psychology*, vol. 2, edited by R. D. Luce, R. Bush, and E. Galanter, 269–322. New York: John Wiley and Sons, 1963.

Christie, Richard. "Authoritarianism Reexamined." In *Studies in the Scope and Method of the Authoritarian Personality*, edited by Richard Christie and Marie Jahoda, 123–96. Glencoe, Ill.: The Free Press, 1954.

———. "Some Experimental Approaches to Authoritarianism: I. A Retrospective Perspective on the *Einstellung* (Rigidity?) Paradigm." In *Strength and Weakness: The Authoritarian Personality Today*, edited by William F. Stone, Gerda Lederer, and Richard Christie, 70–98. New York: Springer-Verlag, 1993.

Clark, B. "You Can Learn to Think Creatively." *Reader's Digest* 77 (1960): 66–70.

Clarke, Adele, and Joan H. Fujimura. *The Right Tools for the Job: At Work in Twentieth-Century Life Sciences*. Princeton, N.J.: Princeton University Press, 1992.

Clowse, Barbara Barksdale. *Brainpower for the Cold War: The Sputnik Crisis and National Defense Education Act of 1958*. Westport, Conn.: Greenwood Press, 1981.

Cmiel, Kenneth. "The Politics of Civility." In *The Sixties: From Memory to History*, edited by David Farber, 263–90. Chapel Hill: University of North Carolina Press, 1994.

Cofer, Charles N., and George Horsley Smith. "A Simple Inexpensive Classroom Demonstration of Concept Formation." *American Psychologist* 2, no. 11 (1947): 521–22.

Cohen, Albert Kircidel. *Delinquent Boys: The Culture of the Gang*. Glencoe, Ill.: Free Press, 1955.

Cohen, D., J. W. Whitmyre, and W. H. Funk. "Effect of Group Cohesiveness and Training on Creative Thinking." *Journal of Applied Psychology* 44 (1960): 319–22.

Cohen-Cole, Jamie. "Thinking About Thinking in Cold War America." Ph.D. diss., Princeton University, 2003.

Colodny, Robert Garland, ed. *Mind and Cosmos: Essays in Contemporary Science and Philosophy*. University of Pittsburgh Series in the Philosophy of Science, vol. 3. Pittsburgh, Pa.: University of Pittsburgh Press, 1966.

Columbia Students for a Democratic Society. "The Columbia Statement." In *The University Crisis Reader*, vol. 1, edited by Immanuel Maurice Wallerstein and Paul Starr, 23–47. New York: Random House, 1971.

Commager, Henry Steele. *The American Mind: An Interpretation of American*

Thought and Character since the 1880s. New Haven, Conn.: Yale University Press, 1950.

Conant, James B. *Education in a Divided World: The Function of the Public Schools in Our Unique Society.* Cambridge, Mass.: Harvard University Press, 1948.

———. "Foreword." In *The Copernican Revolution,* xiii–xviii. Cambridge, Mass.: Harvard University Press, 1957.

———. *Modern Science and Modern Man.* New York: Doubleday, 1952.

———. *My Several Lives: Memoirs of a Social Inventor.* New York: Harper and Row, 1970.

———. *On Understanding Science.* New York: New American Library, 1951.

———. *Science and Common Sense.* New Haven, Conn.: Yale University Press, 1951.

———. "Some Aspects of Modern Harvard." *Journal of General Education* 4, no. 3 (1950): 175–83.

———. "The Superintendent Was the Target." *The New York Times,* April 29, 1951.

———. "Unity and Diversity in Secondary Education." *Vital Speeches* 18, no. 15 (1952): 463–65.

Cottrell, Leonard S., Jr. "Strengthening the Social Sciences in the Universities." In *The Social Sciences at Mid-Century,* 21–36. Freeport, N.Y.: Books For Libraries Press, 1968.

Craik, Kenneth James Williams. *The Nature of Explanation.* Cambridge: University Press, 1943.

Crane, Langton T. *The National Science Foundation and Pre-College Science Education, 1950–1975: Report Prepared for the Subcommittee on Science, Research, and Technology of the Committee on Science and Technology, U.S. House of Representatives.* Washington: Science Policy Research Division, Congressional Research Service, Library of Congress, 1975.

Creager, Angela N. H. *The Life of a Virus: Tobacco Mosaic Virus as an Experimental Model, 1930–1965.* Chicago: University of Chicago Press, 2002.

Creager, Angela N. H., Elizabeth Lunbeck, and M. Norton Wise. *Science without Laws: Model Systems, Cases, Exemplary Narratives, Science and Cultural Theory.* Durham, N.C.: Duke University Press, 2007.

Cremin, Lawrence Arthur. *The Transformation of the School: Progressivism in American Education, 1876–1957.* New York: Vintage Books, 1961.

Crowther-Heyck, Hunter. "George A. Miller, Language, and the Computer Metaphor of Mind." *History of Psychology* 2 (1999): 37–64.

———. *Herbert A. Simon: The Bounds of Reason in Modern America.* Baltimore, Md.: Johns Hopkins University Press, 2005.

———. "Herbert Simon and the GSIA: Building an Interdisciplinary Community." *Journal of the History of the Behavioral Sciences* 42, no. 4 (2006): 311–34.

——. "Patrons of the Revolution: Ideals and Institutions in Postwar Behavioral Science." *Isis* 97, no. 3 (2006): 420–46.

Crutchfield, Richard S. "Conformity and Character." *American Psychologist* 10, no. 5 (1955): 191–98.

——. "Conformity and Creative Thinking." In *Contemporary Approaches to Creative Thinking*, edited by Howard E. Gruber, Glenn Terrell, and Michael Wertheimer, 120–40. New York: Atherton Press, 1962.

Curcher, B. G. "A Loudness Scale for Industrial Noise Measurements." *Journal of the Acoustical Society of America* 6 (1935): 216–26.

Darwin, Charles. *The Expression of the Emotions in Man and Animals. With Photographic and Other Illustrations. With a Preface by Margaret Mead.* New York: Philosophical Library, 1955.

Daston, Lorraine, and Peter Galison. *Objectivity.* New York: Zone Press, 2007.

Davis, Mike. *City of Quartz: Excavating the Future in Los Angeles.* New York: Vintage Books, 1992.

de Sola Pool, Ithiel, and Manfred Kochen. "Contacts and Influence." *Social Networks* 1, no. 1 (1978): 5–51.

Dean, John W. *Conservatives without Conscience.* New York: Viking, 2006.

Demos, Raphael. "Philosophical Aspects of the Recent Harvard Report on Education." *Philosophy and Phenomenological Research* 7, no. 2 (1946): 187–213.

Dewey, John. *Democracy and Education: An Introduction to the Philosophy of Education.* New York: Macmillan, 1916.

——. *School and Society.* Chicago: University of Chicago Press, 1899.

——. "Unity of Science as a Social Problem." In *International Encyclopedia of Unified Science*, edited by Otto Neurath, Rudolf Carnap, and Charles Morris. Chicago: University of Chicago Press, 1938.

——. "The Higher Learning in America" *Social Frontier* 3 no. 24 (1937), 167–69.

Di Palma, Giuseppe, and Herbert McClosky. "Personality and Conformity: The Learning of Political Attitudes." *The American Political Science Review* 64, no. 4 (1970): 1054–73.

Dollard, Charles. "Strategy for Advancing the Social Sciences." In *The Social Sciences at Mid-Century*, 12–20. Freeport, N.Y.: Books For Libraries Press, 1968.

Dollard, John, and Neal E. Miller. *Personality and Psychotherapy: An Analysis in Terms of Learning, Thinking, and Culture.* New York: McGraw-Hill, 1950.

Dorman, Joseph. *Arguing the World: The New York Intellectuals in Their Own Words.* New York: Free Press, 2000.

Doss, Erika. "The Art of Cultural Politics: From Regionalism to Abstract Expressionism." In *Recasting America: Culture and Politics in the Age of Cold War*, edited by Larry May, 195–220. Chicago: University of Chicago Press, 1989.

Dow, Peter B. "Innovation's Perils: An Account of the Origins, Development, Implementation, and Public Reaction to *Man: A Course of Study*." Ed. D. diss., Harvard University, 1979.

———. "Man: A Course of Study: A Continuing Exploration of Man's Humanness." In *Man: A Course of Study. Talks to Teachers.* Cambridge, Mass.: EDC, 1968.

———. *Schoolhouse Politics: Lessons from the Sputnik Era.* Cambridge, Mass.: Harvard University Press, 1991.

Dukes, William F. "The Snark Revisited." *American Psychologist* 15 (1960): 157.

Dunlop, J. T., M. P. Gilmore, C. K. Kluckhohn, Talcott Parsons, and O. H. Taylor. "Toward a Common Language for the Area of the Social Sciences." Unpublished white paper. Cambridge, Mass.: Harvard University, 1941. http://hollis.harvard.edu/?itemid=|library/m/aleph|004641298.

DuPlessis, Rachel Blau, and Ann Barr Snitow. *The Feminist Memoir Project: Voices from Women's Liberation.* New York: Three Rivers Press, 1998.

Dupuy, Jean-Pierre. *The Mechanization of the Mind: On the Origins of Cognitive Science.* Princeton, N.J.: Princeton University Press, 2000.

Durrheim, Kevin. "Theoretical Conundrum: The Politics and Science of Theorizing Authoritarian Cognition." *Political Psychology* 18, no. 3 (1997): 625–47.

Eagly, Alice H. "Sex Differences in Influenceability." *Psychological Bulletin* 85, no. 1 (1978): 86–116.

Echols, Alice. *Daring to Be Bad: Radical Feminism in America, 1967–1975.* Minneapolis: University of Minnesota Press, 1989.

Educational Policies Commission. *American Education and International Tensions.* Washington, D.C.: Educational Policies Commission, 1949.

———. *Education for All American Youth.* Washington, D.C.: Educational Policies Commission, National Education Association of the United States, and the American Association of School Administrators, 1944.

———. *The Education of Free Men in American Democracy.* Washington, D.C.: Educational Policies Commission, National Education Association of the United States, and the American Association of School Administrators, 1941.

Edwards, Allen L. "The Signs of Incipient Fascism." *Journal of Abnormal and Social Psychology* 39, no. 3 (1944): 301–16.

———. "Unlabled Fascist Attitudes." *Journal of Abnormal and Social Psychology* 36 (1941): 575–82.

Edwards, Paul N. *The Closed World: Computers and the Politics of Discourse in Cold War America.* Cambridge, Mass.: MIT Press, 1996.

Einstein, Albert, and L. Infeld. *The Evolution of Physics.* 2d ed. New York: Simon and Schuster, 1942.

England, J. Merton. *A Patron for Pure Science: The National Science Foundation's Formative Years, 1945–57.* Washington, D.C.: National Science Foundation, 1983.

Ericson, Sanford C. "Unity in Psychology: A Survey of Some Opinions." *Psychological Review* 48, no. 1 (1941): 73–82.

Eurich, Alvin C. "A Renewed Emphasis Upon General Education." In *General Education in the American College: Thirty-Eighth Yearbook of the National Society for the Study of Education,* edited by Guy Montrose Whipple. Bloomington, Ill.: Public School Publishing Co., 1939.

Evans, Ronald W. *The Hope for American School Reform: The Cold War Pursuit of Inquiry Learning in Social Studies.* New York: Palgrave Macmillan, 2011.

———. *The Tragedy of American School Reform: How Curriculum Politics and Entrenched Dilemmas Have Diverted Us from Democracy.* New York: Palgrave Macmillan, 2011.

Evans, Sara M. *Personal Politics: The Roots of Women's Liberation in the Civil Rights Movement and the New Left.* New York: Knopf, 1979.

Farley, John. *Brock Chisholm, the World Health Organization, and the Cold War.* Vancouver, B.C.: UBC Press, 2008.

Feigl, Herbert. "The Philosophical Embarrassments of Psychology." *American Psychologist* 14, no. 3 (1959): 115–28.

Feuer, Lewis Samuel. *The Conflict of Generations: The Character and Significance of Student Movements.* New York: Basic Books, 1969.

Fidell, Linda S. "Empirical Verification of Sex Discrimination in Hiring Practices in Psychology." *American Psychologist* 25 (1970): 1094–98.

Fish, Stanley. *Is There a Text in This Class? The Authority of Interpretive Communities.* Cambridge, Mass.: Harvard University Press, 1980.

Fisher, Donald. *Fundamental Development of the Social Sciences: Rockefeller Philanthropy and the United States Social Science Research Council.* Ann Arbor: University of Michigan Press, 1993.

Fisher, Jerome. "The Memory Process and Certain Psychosocial Attitudes, with Special Reference to the Law of Prägnanz." *Journal of Personality* 19, no. 4 (1951): 406–20.

Flanagan, John C. "The Definition and Measurement of Ingenuity." In *Scientific Creativity: Its Recognition and Development,* edited by Calvin W. Taylor and Frank Barron, 89–98. New York: John Wiley and Sons, 1963.

Foran, John. "Discursive Subversions: *Time* Magazine, the CIA Overthrow of Mussaddiq, and the Installation of the Shah." In *Cold War Constructions: The Political Culture of United States Imperialism,* edited by Christian G. Appy, 157–82. Amherst: University of Massachusetts Press, 2000.

Forman, Paul. "Behind Quantum Electronics: National Security as a Basis for Physical Research in the United States, 1940–1960." *Historical Studies in the Physical and Biological Sciences* 18, no. 1 (1987): 149–229.

Francotte, Xavier. *La génie et la folie*. Brussels: Société belge de librairie, 1890.

Frank, Lawrence K. *Nature and Human Nature: Man's New Image of Himself*. New Brunswick, N.J.: Rutgers University Press, 1951.

———. "Psychology and Social Order." In *The Human Meaning of the Social Sciences*, edited by Daniel Lerner, 214–41. New York: Meridian Books, 1959.

Frank, Phillip. "Science and Democracy." In *Science, Philosophy, and Religion: A Symposium*, 215–28. New York: Conference on Science, Philosophy and Religion in their Relation to the Democratic Way of Life, Inc., 1941.

Free Speech Movement. "We Want a University." In *The Berkeley Student Revolt: Facts and Interpretations*, edited by Seymour Martin Lipset and Sheldon S. Wolin, 208–16. Garden City, N.Y.: Doubleday and Company, 1965.

Fremont-Smith, Frank. "The Josiah Macy, Jr. Foundation Conference Program." In *Cybernetics: Circular, Causal, and Feedback Mechanisms in Biological and Social Systems, 10th Annual Meeting*, edited by Heinz Von Foerster, 11–13. New York: Josiah Macy, Jr. Foundation, 1955.

Frenkel-Brunswik, Else. "Comprehensive Scores and Summary of Interview Results." In *The Authoritarian Personality*, edited by Theodor W. Adorno et al., 468–86. New York: Harper and Brothers, 1950.

———. "Dynamic and Cognitive Personality Organization as Seen through the Interviews." In *The Authoritarian Personality*, 442–67. New York: Harper and Brothers, 1950.

———. "Further Explorations by a Contributor to *The Authoritarian Personality*." In *Studies in the Scope and Method of the Authoritarian Personality*, edited by Richard Christie and Marie Jahoda, 226–75. Glencoe, Ill.: The Free Press, 1954.

———. "Intolerance of Ambiguity as an Emotional and Perceptual Personality Variable." *Journal of Personality* 18, no. 1 (1949): 108–43.

———. "Sex, People and Self as Seen through the Interviews." In *The Authoritarian Personality*, edited by Theodor W. Adorno et al., 390–441. New York: Harper and Brothers, 1950.

———. "Tolerance toward Ambiguity as a Personality Variable." *American Psychologist* 3 (1948): 268.

Frodin, Reuben. "Editorial Comment." *Journal of General Education* 3, no. 3 (1949): 161–62.

Fromm, Erich. *Escape from Freedom*. New York: Farrar and Rinehart, 1941.

Fuller, Steve. *Thomas Kuhn: A Philosophical History for Our Times*. Chicago: University of Chicago Press, 2000.

Funatoli, P. *Il Genio e La Follia*. Siena: Tip. dell'Ancora, 1885.

Furumoto, Laurel. "Shared Knowledge: The Experimentalists, 1904–1929."

In *The Rise of Experimentation in American Psychology*, edited by Jill G. Morawski, 94–113. New Haven, Conn.: Yale University Press, 1988.

Gabriel, H. W. "On Teaching Creative Engineering." *Journal of Engineering Education* 45 (1955): 794–801.

Galanter, Eugene, and Murray Gerstenhaber. "On Thought: The Extrinsic Theory." *Psychological Review* 63, no. 4 (1956): 218–27.

Galison, Peter L. "The Americanization of Unity." *Daedalus* 127, no. 1 (1998): 45–71.

———. *Image and Logic: A Material Culture of Microphysics*. Chicago: University of Chicago Press, 1997.

———. "The Ontology of the Enemy: Norbert Wiener and the Cybernetic Vision." *Critical Inquiry* 21 (1994): 228–68.

Galison, Peter L., and Bruce Hevly, eds. *Big Science: The Growth of Large-Scale Research*. Stanford, Calif.: Stanford University Press, 1992.

Galison, Peter L., Bruce Hevly, and Rebecca Lowen. "Controlling the Monster: Stanford and the Growth of Physics Research, 1935–1962." In *Big Science: The Growth of Large Scale Research*, edited by Peter Galison and Bruce Hevly, 46–77. Stanford, Calif.: Stanford University Press, 1992.

Gammon, Howard. "Some Practical Techniques for Increasing Creativity in Engineering." *Plant Maintenance and Engineering* 21 (1960): 25–27.

Gardner, Howard. *The Mind's New Science: A History of the Cognitive Revolution*. New York: Basic Books, 1985.

———. *To Open Minds*. New York: Basic Books, 1989.

Gardner, John W. *Excellence: Can We Be Equal and Excellent Too?* New York: Harper and Row, 1961.

———. "International Relations and Sociology: Discussion." *American Sociological Review* 13, no. 3 (1948): 273–75.

———. *Self-Renewal: The Individual and the Innovative Society*. New York: Harper and Row, 1963.

Geary, Daniel. "Children of *The Lonely Crowd*: David Riesman, the Young Radicals, and the Splitting of Liberalism in the 1960s." Unpublished manuscript. 2012.

———. *Radical Ambition: C. Wright Mills, the Left, and American Social Thought*. Berkeley and Los Angeles: University of California Press, 2009.

Geertz, Clifford. *The Interpretation of Cultures*. New York: Basic Books, 1973.

Geiger, Roger L. "American Foundations and Academic Social Science, 1945–1960." *Minerva* 26 (1988): 315–41.

———. *Research and Relevant Knowledge: American Research Universities since World War II*. New York and Oxford: Oxford University Press, 1993.

General Education in a Free Society. Cambridge, Mass.: Harvard University Press, 1945.

Ghamari-Tabrizi, Sharon. "Simulating the Unthinkable: Gaming Future War in the 1950s and 1960s." *Social Studies of Science* 30, no. 2 (2000): 163–223.

———. *The Worlds of Herman Kahn: The Intuitive Science of Thermonuclear War*. Cambridge, Mass.: Harvard University Press, 2005.

Gideonse, Harry D. *The Higher Learning in a Democracy: A Reply to President Hutchins' Critique of the American University*. New York and Toronto: Farrar and Rinehart, 1937.

Gieryn, Thomas. "Boundary-Work and the Demarcation of Science from Non-Science: Strains and Interests in Professional Ideologies of Scientists." *American Sociological Review* 48 (1983): 781–95.

———. "The U.S. Congress Demarcates Natural Science and Social Science (Twice)." In *Cultural Boundaries of Science: Credibility on the Line*, 65–114. Chicago: University of Chicago Press, 1999.

Gigerenzer, Gerd. "Discovery in Cognitive Psychology: New Tools Inspire New Theories." *Science in Context* 5, no. 2 (1992): 329–50.

———. "From Tools to Theories: A Heuristic of Discovery in Cognitive Psychology." *Psychological Review* 98, no. 2 (1991): 254–67.

———. "Mind as Computer: The Social Origins of a Metaphor." In *Adaptive Thinking: Rationality in the Real World*, 26–43. New York: Oxford University Press, 2000.

Gillin, John. "Grounds for a Science of Social Man." In *For a Science of Social Man*, edited by John Gillin, 3–13. New York: Macmillan, 1954.

———, ed. *For a Science of Social Man: Convergences in Anthropology, Psychology, and Sociology*. New York: Macmillan, 1954.

Gilman, Nils. *Mandarins of the Future: Modernization Theory in Cold War America*. Baltimore, Md.: Johns Hopkins University Press, 2003.

Giuilhot, Nicholas. "Cyborg Pantocrator: International Relations Theory from Decisionism to Rational Choice." *Journal of the History of the Behavioral Sciences* 47, no. 3 (2011): 279–301.

Glazer, Nathan. "'Student Power' in Berkeley." *Public Interest*, no. 13 (1968): 3–21.

Glazer, Penina Migdal. "The New Left: A Style of Protest." *The Journal of Higher Education* 38, no. 3 (1967): 119–30.

Goldberg, Philip. "Are Women Prejudiced against Women?" *Trans-Action* 5 (1968): 28–30.

Goldfield, Evelyn, Sue Munaker, and Naomi Weisstein. "A Woman Is a Sometime Thing: Cornering Capitalism by Removing 51% of Its Commodities." In *The New Left: A Collection of Essays*, edited by Priscilla Long, 236–71. Boston: Porter Sargent, 1969.

Goldstein, Jack S. *A Different Sort of Time: The Life of Jerrold Zacharias*. Cambridge, Mass.: MIT Press, 1992.

Golvin, N. E. "The Creative Person in Science." In *Scientific Creativity: Its Recognition and Development*, edited by Calvin W. Taylor and Frank Barron. New York: John Wiley and Sons, 1963.

Goodman, Paul. *Growing up Absurd: Problems of Youth in the Organized System*. New York: Random House, 1960.

———. "Thoughts on Berkeley." *New York Review of Books*, January 14, 1965.

Gordon, William J. J. *Synectics*. New York: Harper and Brothers, 1961.

Gorn, Saul. "On the Mechanical Stimulation of Habit-Forming and Learning." *Information and Control* 2 (1959): 226–59.

Green, Christopher D. "Of Immortal Mythological Beasts: Operationism in Psychology." *Theory and Psychology* 2, no. 3 (1992): 291–320.

Greenfield, Patricia M., and Jerome S. Bruner. "Culture and Cognitive Growth." In *Handbook of Socialization Theory and Research*, edited by David A. Goslin, 633–54. Chicago: Rand McNally, 1969.

Greenwalt, C. H. *The Uncommon Man: The Individual in the Organization*. New York: McGraw-Hill, 1959.

Greenwood, John D. *The Disappearance of the Social in American Social Psychology*. Cambridge and New York: Cambridge University Press, 2004.

———. "Understanding the 'Cognitive Revolution' in Psychology." *Journal of the History of the Behavioral Sciences* 35, no. 1 (1999): 1–22.

Grinker, Roy R., ed. *Toward a Unified Theory of Human Behavior*. New York: Basic Books, 1956.

Gross, Charles G. "Psychological Round Table in the 1960s." *American Psychologist* 32, no. 12 (1977): 1120–21.

Gruber, Howard E., Kenneth R. Hammond, and Richard Jessor. "Foreword." In *Contemporary Approaches to Cognition*, v–vi. Cambridge, Mass.: Harvard University Press, 1957.

Guilbaut, Serge. *How New York Stole the Idea of Modern Art: Abstract Expressionism, Freedom, and the Cold War*. Chicago: University of Chicago Press, 1983.

Gutman, Herbert. "The Biological Roots of Creativity." In *Explorations in Creativity*, edited by Ross L. Mooney and Taher A. Razik, 3–32. New York: Harper and Row, 1967.

———. "The Biological Roots of Creativity." *Genetic Psychology Monographs* (1961): 419–58.

Gutmann, Amy, and Dennis F. Thompson. *Democracy and Disagreement*. Cambridge, Mass.: Belknap Press of Harvard University Press, 1996.

Guttman, Allen. "Protest against the War in Vietnam." *Annals of the American Academy of Political and Social Science* 382 (1969): 53–63.

Habermas, Jürgen. *Communication and the Evolution of Society*. Boston: Beacon Press, 1979.

Halberstam, David. *The Best and the Brightest.* New York: Random House, 1972.

Halliwell, Steve. "Columbia: An Explanation or Applesauce, Epicycles, and the Joplin Proviso." In *The New Left: A Collection of Essays,* edited by Priscilla Long, 200–215. Boston: Porter Sargent, 1969.

Hardcastle, Gary L. "The Cult of Experiment: The Psychological Round Table, 1936–1941." *History of Psychology* 3, no. 4 (2000): 334–70.

———. "S. S. Stevens and the Origins of Operationism." *Philosophy of Science* 62 (1995): 404–24.

Harlow, Harry F. "Current and Future Advances in Physiological and Comparative Psychology." *American Psychologist* 11 (1956): 273–77.

———. "Mice, Monkeys, Men, and Motives." *Psychological Review* 60, no. 1 (1953): 23–31.

———. "The Nature of Love." *American Psychologist* 13, no. 12 (1958): 673–85.

Harman, Harry H. *The Psychologist in Interdisciplinary Research.* Santa Monica, Calif.: RAND Corporation, 1955.

Harmon, Lindsay R. "The Development of a Criterion of Scientific Competence." In *Scientific Creativity: Its Recognition and Development,* edited by Calvin W. Taylor and Frank Barron, 44–52. New York: John Wiley and Sons, 1963.

Harris, Randy Allen. "Chomsky's Other Revolution." In *Chomskyan (R)Evolutions,* edited by Douglas A. Kibbee, 237–64. Philadelphia: John Benjamins Publishing Company, 2010.

———. *The Linguistics Wars.* New York and Oxford: Oxford University Press, 1993.

Hartman, Andrew. *Education and the Cold War: The Battle for the American School.* New York: Palgrave Macmillan, 2008.

Harvard University Center for Cognitive Studies. *Annual Reports.* Cambridge, Mass., 1961–69.

Haskell, Thomas L. *The Emergence of Professional Social Science: The American Social Science Association and the Nineteenth Century Crisis of Authority.* Urbana: University of Illinois Press, 1977.

Hauptmann, Emily. "The Constitution of Behavioralism: The Influence of the Ford Foundation's Behavioral Science's Program on Political Science." Paper presented at the Sixth Annual Workshop on the History of Economics as History of Science, École Normale Supérieure de Cachan, June 19, 2009.

Hayles, N. Katherine. *How We Became Posthuman: Virtual Bodies in Cybernetics, Literature, and Informatics.* Chicago: University of Chicago Press, 1999.

Hebb, D. O. "The American Revolution." *American Psychologist* 15 (1960): 735–45.

———. *The Organization of Behavior: A Neuropsychological Theory.* New York: John Wiley and Sons, 1949.

———. "The Problem of Consciousness and Introspection." In *Brain Mechanisms and Consciousness*, edited by E. D. Adrian, 402–17. Oxford: Blackwell, 1954.

Heil, Alan L. *Voice of America: A History*. New York: Columbia University Press, 2003.

Heilbrunn, Jacob. "His Anti-Semitic Sources." *The New York Review of Books*, April 20, 1995.

Hegarty, Peter. *Gentlemen's Disagreement: Alfred Kinsey, Lewis Terman, and the Sexual Politics of Smart Men*. Chicago: University of Chicago Press, 2013.

Heims, Steve J. *The Cybernetics Group: Constructing a Social Science for Postwar America*. Cambridge, Mass.: MIT Press, 1991.

Helson, Ravenna. "Creativity in Women." In *The Psychology of Women: Future Directions in Research*, edited by Julie A. Sherman and Florence L. Denmark, 553–604. New York: Psychological Dimensions, 1978.

Henle, Mary. "On Coping with Ambiguity." In *Man: A Course of Study. Seminars for Teachers*, 11–12. Cambridge, Mass.: EDC, 1969.

Herman, Ellen. *The Romance of American Psychology: Political Culture in an Age of Experts*. Berkeley and Los Angeles: University of California Press, 1995.

Herrnstein, Richard J. "IQ." *Atlantic Monthly*, September 1971.

Heymann, C. David. *The Georgetown Ladies' Social Club: Power, Passion, and Politics in the Nation's Capital*. New York: Atria Books, 2003.

Hilgard, Ernest R. *Theories of Learning*. New York: Appleton-Century-Crofts, 1948.

Hilgard, Ernest R., and Gordon H. Bower. *Theories of Learning*. 3d ed. New York: Appleton-Century-Crofts, 1966.

Hipple, Theodore W., Thomas R. Giblin, and Jack Megenity. "Have Your Students Read . . .?" *Phi Delta Kappan* 53, no. 7 (1972): 441–42.

Hoffman, Howard S. "The Analogue Lab: A New Kind of Teaching Device." *American Psychologist* 17, no. 10 (1962): 684–94.

Hofstadter, Richard, and Wilson Smith, eds. *American Higher Education: A Documentary History*. Chicago: University of Chicago Press, 1961.

Hollinger, David A. "Jewish Intellectuals and the De-Christianization of American Public Culture in the Twentieth Century." In Hollinger, *Science, Jews, and Secular Culture*, 17–41.

———. "Science as a Weapon in *Kulturkämpfe* in the United States during and after World War II." *Isis* 86 (1995): 440–54.

———. *Science, Jews, and Secular Culture: Studies in Mid-Twentieth Century Intellectual History*. Princeton, N.J.: Princeton University Press, 1996.

———. "Two NYUs and the 'Obligation of Universities to the Social Order' in the Great Depression." In Hollinger, *Science, Jews, and Secular Culture*, 60–79.

———. "The Unity of Knowledge and the Diversity of Knowers: Science as an Agent of Cultural Integration in the United States between the Two World Wars." *Pacific Historical Review* 80, no. 2 (2011): 211–30.

———. "A View from the Margins." In *The Free Speech Movement: Reflections on Berkeley in the 1960s*, edited by Robert Cohen and Reginald E. Zelnik, 178–84. Berkeley and Los Angeles: University of California Press, 2002.

Holton, Gerald. "On the Educational Philosophy of the Project Physics Course." In *The Scientific Imagination: Case Studies*, 284–98. Cambridge: Cambridge University Press, 1978.

———. "The Thematic Imagination in Science." In *Science and Culture: A Study of Cohesive and Disjunctive Forces*, edited by Gerald Holton, 88–108. Boston: Beacon Press, 1967.

———. *Thematic Origins of Scientific Thought*. 2d ed. Cambridge, Mass.: Harvard University Press, 1988.

Homans, George C. *Coming to My Senses: The Autobiography of a Sociologist*. New Brunswick, N.J.: Transaction Books, 1984.

Hook, Sidney. "Democracy and Education: Introduction." In *The Authoritarian Attempt to Capture Education: Papers from the Second Conference on the Scientific Spirit and Democratic Faith*, 10–12. New York: King's Crown Press, 1945.

———. *Reason, Social Myths and Democracy*. Buffalo, N.Y.: Prometheus Books, 1991.

Horowitz, Daniel. *Betty Friedan and the Making of the Feminine Mystique: The American Left, the Cold War, and Modern Feminism*. Amherst: University of Massachusetts Press, 1998.

Hounshell, David. "The Cold War, RAND, and the Generation of Knowledge." *Historical Studies in the Physical Sciences* 27 (1997): 236–67.

Howe, Irving. *A Margin of Hope: An Intellectual Autobiography*. San Diego: Harcourt Brace Jovanovich, 1982.

Hull, Clark L. "Mind, Mechanism and Adaptive Behavior." *Psychological Review* 44 (1937): 1–32.

Hull, David L. *Science as a Process: An Evolutionary Account of the Social and Conceptual Development of Science*. Chicago: University of Chicago Press, 1988.

Hunt, William A. *The Clinical Psychologist*. Springfield, Ill.: Charles C. Thomas, 1956.

Hutchins, Robert M. "The Confusion in Higher Education." *Harpers*, October 1936, 449–58.

———. *The Higher Learning in America*. New Haven, Conn.: Yale University Press, 1936.

———. "What Is a General Education?" *Harpers*, November 1936, 602–9.

Hyman, Herbert H., and Paul B. Sheatsley. "*The Authoritarian Personality*—a

Methodological Critique." In *Studies in the Scope and Method of the Author-itarian Personality*, edited by Richard Christie and Marie Jahoda, 50–122. Glencoe, Ill.: The Free Press, 1954.

Igo, Sarah E. *The Averaged American: Surveys, Citizens, and the Making of a Mass Public*. Cambridge, Mass.: Harvard University Press, 2007.

Inkeles, Alex, and David Horton Smith. *Becoming Modern: Individual Change in Six Developing Countries*. Cambridge, Mass.: Harvard University Press, 1974.

Irving, Allan. *Brock Chisholm: Doctor to the World*. Markham, Ont.: Fitzhenry and Whiteside, 1998.

Isaac, Joel. "The Harvard Pareto Circle Revisited." Paper presented at the Cambridge Seminars in Political Thought and Intellectual History, October 27, 2008.

——. "Tangled Loops: Theory, History and the Human Sciences in Modern America." *Modern Intellectual History* 6 (2009): 397–424.

——. "Theorist at Work: Talcott Parsons and the Carnegie Project on Theory, 1949–1951." *Journal of the History of Ideas* 71, no. 2 (2010): 287–311.

——. "Tool Shock: Technique and Epistemology in the Postwar Social Sciences." *History of Political Economy* 42 (Annual Supplement 2010): 133–64.

——. *Working Knowledge: Making the Human Sciences from Parsons to Kuhn*. Cambridge, Mass.: Harvard University Press, 2012.

Jackson, John P. *Science for Segregation: Race, Law, and the Case against Brown v. Board of Education*. New York: New York University Press, 2005.

——. *Social Scientists for Social Justice: Making the Case against Segregation*. New York: New York University Press, 2001.

Jacobs, James. *S. M. Lipset: Social Scientist of the Smooth Society*. Boston: New England Free Press, n.d.

Jacobson, Frank N., and Salomon Rettig. "Authoritarianism and Intelligence." *Journal of Social Psychology* 50, no. 2 (1959): 213–19.

Jardini, David R. "Out of the Blue Yonder: The Transfer of Systems Thinking from the Pentagon to the Great Society, 1961–1965." In *Systems, Experts, and Computers: The Systems Approach in Management and Engineering, World War II and After*, edited by Agatha C. Hughes and Thomas P. Hughes, 311–57. Cambridge, Mass.: MIT Press, 2000.

Jastrow, Joseph. "Psychology." In *Encyclopaedia of the Social Sciences*, edited by Edwin R. Seligman, 588–96. New York: The Macmillan Company, 1934.

Jay, Martin. *The Dialectical Imagination: A History of the Frankfurt School and the Institute of Social Research, 1923–1950*. Boston: Little, Brown and Company, 1973.

Jehlen, Alan. "Social Studies Course Assailed." *The Patriot Ledger*, October 8, 1975, 9.

Jensen, Arthur R. "How Much Can We Boost IQ and Scholastic Achievement?" *Harvard Educational Review* 39 (1969): 1–123.

Johnson, B. Lamar. "General Education Changes the College." *Journal of Higher Education* 9 (1938): 18–22.

Johnson, E. Parker. "On Readmitting the Mind." *American Psychologist* 11 (1956): 712–14.

Johnson-Laird, Philip N. *The Computer and the Mind: An Introduction to Cognitive Science.* Cambridge, Mass.: Harvard University Press, 1988.

Johnston, Barry V. "The Contemporary Crisis and the Social Relations Department at Harvard: A Case Study in Hegemony and Disintegration." *The American Sociologist* 29, no. 3 (1998): 26–42.

Jost, John T., Jack Glasser, Arie W. Kruglanski, and Frank J. Sulloway. "Political Conservatism as Motivated Social Cognition." *Psychological Bulletin* 129, no. 3 (2003): 339–75.

Jumonville, Neil. *Critical Crossings: The New York Intellectuals in Postwar America.* Berkeley and Los Angeles: University of California Press, 1991.

Kahn, Theodore C. "Clinically and Statistically Oriented Psychologists Split Our Profession." *American Psychologist* 10, no. 4 (1955): 171–72.

Kahneman, Daniel. *Thinking Fast and Slow.* New York: Farrar, Straus and Giroux, 2011.

Kaiser, David. "Cold War Requisitions, Scientific Manpower, and the Production of American Physicists after World War II." *Historical Studies in the Physical and Biological Sciences* 33, no. 1 (2002): 131–59.

———. *Drawing Theories Apart: The Dispersion of Feynman Diagrams in Postwar Physics.* Chicago: University of Chicago Press, 2005.

———. "The Postwar Suburbanization of American Physics." *American Quarterly* 56 (2004): 851–88.

Kaplan, Sidney J. "An Appraisal of an Interdisciplinary Social Science Course." *Journal of Educational Sociology* 34, no. 2 (1960): 70–77.

Karabel, Jerome. *The Chosen: The Hidden History of Admission and Exclusion at Harvard, Yale, and Princeton.* Boston: Houghton Mifflin, 2005.

Kay, Lily E. "From Logical Neurons to Poetic Embodiments of Mind: Warren S. McCulloch's Project in Neuroscience." *Science in Context* 4, no. 4 (2001): 591–614.

Keller, Morton, and Phyllis Keller. *Making Harvard Modern: The Rise of America's University.* New York: Oxford University Press, 2001.

Kelley, George A. "I Itch Too." *American Psychologist* 10, no. 4 (1955): 172–73.

Keniston, Kenneth. *The Uncommitted: Alienated Youth in American Society.* New York: Laurel, Dell Pub. Co., 1970; originally pub. 1965.

Kennan, George F. "America's Administrative Response to Its World Problems." In *The American Style: Essays in Value and Performance*, edited by Elting E. Morison, 124–44. New York: Harper and Brothers, 1958.

———. *Democracy and the Student Left*. Boston: Little Brown, 1968.

Kerr, Clark. *The Uses of the University*. 4th ed. Cambridge, Mass.: Harvard University Press, 1995.

Kesselman, Amy, Heather Booth, Vivian Rothstein, and Naomi Weisstein. "Our Gang of Four: Friendship and Women's Liberation." In *The Feminist Memoir Project: Voices from Women's Liberation*, edited by Rachel Blau DuPlessis and Ann Snitow, 25–53. New York: Three Rivers Press, 1998.

Kevles, Daniel J. *The Physicists: The History of a Scientific Community in Modern America*. Cambridge, Mass.: Harvard University Press, 1995.

Kilpatrick, James J. "Must We Repeal the Constitution to Give the Negro the Vote?" *The National Review*, April 20, 1965, 319–22.

———. "War Flares over Scientific Funds." *Boston Globe*, January 20, 1976.

King, Martin L., Jr. "The Role of the Behavioral Scientist in the Civil Rights Movement." *Journal of Social Issues* 24, no. 1 (1968): 1–12.

Kluckhohn, Clyde. *Mirror for Man: The Relation of Anthropology to Modern Life*. New York: McGraw-Hill, 1949.

———. "Shifts in American Values: Review of Max Lerner, *America as a Civilization: Life and Thought in the United States*." *World Politics* 11, no. 2 (1959): 251–61.

———. "Universal Values and Anthropological Relativism." In *Modern Education and Human Values*, 87–112. Pittsburgh, Pa.: University of Pittsburgh Press, 1952.

Koch, Sigmund. "Epilogue: Some Trends of Study I." In *Psychology: A Study of a Science*, vol. 3: *Formulations of the Person and the Social Context*, edited by Sigmund Koch, 729–88. New York: McGraw-Hill, 1959.

———. "Introduction to Volume 3." In *Psychology: A Study of a Science*, vol. 3: *Formulations of the Person and the Social Context*, edited by Sigmund Koch, 1–6. New York: McGraw-Hill, 1959.

Koestler, Arthur. *The Act of Creation*. New York: Macmillan, 1964.

———. *Darkness at Noon*. New York: Macmillan, 1941.

———. *The Sleepwalkers: A History of Man's Changing Vision of the Universe*. London: Hutchinson, 1959.

Kohler, Robert E. *Lords of the Fly: Drosophila Genetics and the Experimental Life*. Chicago: University of Chicago Press, 1994.

———. "Warren Weaver and the Rockefeller Foundation Program in Molecular Biology, a Case Study in the Management of Science." In *Sciences in the American Context: New Perspectives*, edited by Nathan Reingold, 249–93. Washington, D.C.: Smithsonian Institution Press, 1979.

Krantz, D. L., and L. Wiggins. "Personal and Impersonal Channels of Recruitment in the Growth of Theory." *Human Development* 16 (1973): 133–56.

Krech, David, Richard S. Crutchfield, and Edgerton L. Ballachey. *Individual in Society*. New York: McGraw-Hill, 1962.

Krechevesky, I. "The Genesis of 'Hypotheses' in Rats." *Psychological Review* 45 (1932): 107–33.

———. "'Hypothesis' vs. 'Chance' in the Pre-Solution Period in Sensory Discrimination-Learning." *University of California Publications in Psychology* 6 (1932): 27–44.

Kretschmer, E. "The Breeding of the Mental Endowments of Genius." *Psychiatric Quarterly* 4 (1930): 74–80.

Krieghbaum, Hillier, and Hugh Rawson. *An Investment in Knowledge: The First Dozen Years of the National Science Foundation's Summer Institutes to Improve Secondary School Science and Mathematics Teaching, 1954–1965*. New York: New York University Press, 1969.

Krugler, David F. *The Voice of America and the Domestic Propaganda Battles, 1945–1953*. Columbia: University of Missouri Press, 2000.

Krugman, Paul. "Paranoia Strikes Deep." *New York Times*, November 9, 2009.

Kubie, Lawrence S. "Blocks to Creativity." In *Explorations in Creativity*, edited by Ross L. Mooney and Taher A. Razik, 69–78. New York: Harper and Row, 1967.

———. *Neurotic Distortions of the Creative Processes*. Lawrence: University of Kansas Press, 1958.

Kuhn, Thomas S. *The Structure of Scientific Revolutions*. Chicago: University of Chicago Press, 1962.

Kuklick, Bruce. *Blind Oracles: Intellectuals and War from Kennan to Kissinger*. Princeton, N.J.: Princeton University Press, 2006.

———. *The Rise of American Philosophy: Cambridge, Massachusetts, 1860–1930*. New Haven, Conn.: Yale University Press, 1977.

Kuklick, Henrika. "A 'Scientific Revolution': Sociological Theory in the United States, 1930–1945." *Sociological Inquiry* 43 (1973): 3–22.

Kusch, Martin. "Recluse, Interlocutor, Interrogator: Natural and Social Order in Turn-of-the-Century Psychological Research Schools." *Isis* 86 (1995): 419–39.

Lagemann, Ellen Condliffe. *An Elusive Science: The Troubling History of Education Research*. Chicago: University of Chicago Press, 2000.

———. *The Politics of Knowledge: The Carnegie Corporation, Philanthropy, and Public Policy*. Middletown, Conn.: Wesleyan University Press, 1989.

Laird, Charles. *Through These Eyes*. National Film Board of Canada, 2004. Film.

Lakoff, George. *Moral Politics: How Liberals and Conservatives Think*. 2d ed. Chicago: University of Chicago Press, 2002.

Lamont, Michèle. *How Professors Think: Inside the Curious World of Academic Judgment*. Cambridge, Mass.: Harvard University Press, 2009.

Landau, Martin, Harold Proshansky, and William Ittelson. "The Interdisciplinary Approach and the Concept of Behavioral Sciences." In *Decisions, Values and Groups*, edited by Norman Washburne. Oxford: Pergamon, 1962.

Lasch, Christopher. *The Culture of Narcissism: American Life in an Age of Diminishing Expectations*. New York: Norton, 1978.

Lasswell, H. D. *Psychopathology and Politics*. Chicago: University of Chicago Press, 1930.

Latour, Bruno, and Steven Woolgar. *Laboratory Life: The Construction of Scientific Facts*. Princeton, N.J.: Princeton University Press, 1986.

Layzer, David. "Science or Superstition? (A Physical Scientist Looks at the IQ Controversy)." *Cognition* 1, no. 2–3 (1972): 265–99.

Le Bon, Gustav. *The Crowd: A Study of the Popular Mind*. Atlanta: Cherokee Publishing Company, 1982 [1895].

Leahey, Thomas H. "The Mythical Revolutions of American Psychology." *American Psychologist* 47 (1992): 308–18.

Lears, T. J. Jackson. "A Matter of Taste: Corporate Cultural Hegemony in a Mass-Consumption Society." In *Recasting America: Culture and Politics in the Age of Cold War*, edited by Lary May, 38–57. Chicago: University of Chicago Press, 1989.

Lemann, Nicholas. *The Big Test: The Secret History of the American Meritocracy*. New York: Farrar, Straus and Giroux, 1999.

Lemov, Rebecca M. *World as Laboratory: Experiments with Mice, Mazes, and Men*. New York: Hill and Wang, 2005.

Lentz, Theo F. "The Attitudes of World Citizenship." *Journal of Social Psychology* 32 (1950): 207–14.

Lerner, Daniel. "*The American Soldier* and the Public." In *Continuities in Social Research: Studies in the Scope and Method of the American Soldier*, edited by Robert K. Merton and Paul F. Lazarsfeld, 212–51. Glencoe, Ill.: The Free Press, 1950.

———. "The Social Sciences: Whence and Whither?" In *The Human Meaning of the Social Sciences*, edited by Daniel Lerner, 13–39. New York: Meridian Books, 1959.

Lerner, Max. *America as a Civilization: Life and Thought in the United States Today*. New York: Simon and Schuster, 1957.

Leslie, Stuart W. *The Cold War and American Science: The Military-Industrial-Academic Complex at M.I.T. and Stanford*. New York: Columbia University Press, 1993.

Levinson, Daniel J. "The Study of Ethnocentric Ideology." In *The Authoritarian Personality*, edited by Theodor W. Adorno et al., 102–50. New York: Harper and Brothers, 1950.

Lewin, Kurt. "Group Decision and Social Change." In *Readings in Social Psy-*

chology, edited by E. E. Macoby, T. M. Newcomb and E. L. Hartley, 197–211. New York: Henry Holt and Company, 1958.

Lewontin, R. C. "The Cold War and the Transformation of the Academy." In *The Cold War and the University*, edited by Noam Chomsky, Ira Katznelson, R. C. Lewontin, David Montgomery, Laura Nader, Richard Ohmann, Ray Siever, Immanuel Wallerstein, and Howard Zinn, 1–34. New York: The New Press, 1997.

Lilienthal, David E. "The Unification of Specialized Knowledge in Practical Affairs." In *Science, Philosophy, and Religion: Third Symposium*, 237–44. New York: Conference on Science, Philosophy and Religion in their Relation to the Democratic Way of Life, Inc., 1943.

———. *TVA: Democracy on the March*. Chicago: Quadrangle Books, 1966.

Lippett, Ronald. *Training in Community Relations*. New York: Harper and Brothers, 1949.

Lipset, Seymour Martin. *Political Man: The Social Bases of Politics*. Garden City, N.Y.: Doubleday, 1960.

———. "The Sources of the 'Radical Right.'" In *The Radical Right*, edited by Daniel Bell, 307–71. Garden City, N.Y.: Doubleday and Company, 1964.

———. "Three Decades of the Radical Right: Coughlinites, McCarthyites, and Birchers." In *The Radical Right*, edited by Daniel Bell, 373–446. Garden City, N.Y.: Doubleday and Company, 1964.

Lipset, Seymour Martin, and Leo Lowenthal. "Preface." In *Culture and Social Character: The Work of David Riesman Reviewed*, edited by Seymour Martin Lipset and Leo Lowenthal, v–xi. Glencoe, Ill.: The Free Press, 1961.

Locke, Alain. "Pluralism and Intellectual Democracy." In *Science, Philosophy and Religion*, 196–209. New York: Conference on Science, Philosophy and Religion, 2nd Symposium, 1942.

Logue, Alexandra W. "The Growth of Behaviorism: Controversy and Diversity." In *Points of View in the Modern History of Psychology*, edited by Claude E. Buxton, 169–96. Orlando, Fla.: Academic Press, 1985.

Lowen, Rebecca S. *Creating the Cold War University: The Transformation of Stanford*. Berkeley and Los Angeles: University of California Press, 1997.

Lubell, Samuel. "That 'Generation Gap.'" *Public Interest* no. 13 (1968): 52–60.

Lundberg, George A. "Quantitative Methods in Social Psychology." *American Sociological Review* 1 (1936): 38–54.

———. "The Social Sciences in the Post-War Era." *Sociometry* 8, no. 2 (1945): 137–49.

Luszki, Margaret B. *Interdisciplinary Team Research: Methods and Problems*. Washington, D.C.: National Training Laboratories, 1958.

MacCorquodale, Kenneth, and Paul E. Meehl. "On a Distinction between Hypothetical Constructs and Intervening Variables." *Psychological Review* 55 (1948): 95–107.

MacKinnon, Donald W. "IPAR's Contribution to the Conceptualization and Study of Creativity." In *Perspectives in Creativity*, edited by Irving A. Taylor and Jacob W. Getzels, 60–89. Chicago: Aldine Publishing Company, 1975.

———. "The Nature and Nurture of Creative Talent." *American Psychologist* 17, no. 7 (1962): 484–95.

Mackworth, J. F., and N. H. Mackworth. "Eye Fixations Recorded on Changing Visual Scenes by the Television Eye-Marker." *Journal of the Optical Society of America* 48 (1958): 439–45.

Mackworth, N. H. "A Stand Camera for Line-of-Sight Recording." *Perception and Psychophysics* 2 (1967): 119–27.

Mackworth, N. H., and E. L. Thomas. "A Head Mounted Eye Marker Camera." *Journal of the Optical Society of America* 52 (1962): 713–16.

"MACOS Social? Study?" *Weekly American News*, September 1, 1971.

Man: A Course of Study. Evaluation Strategies. Cambridge, Mass.: EDC, 1970

Man: A Course of Study. The Observer's Handbook. Cambridge, Mass.: EDC, 1970.

Man: A Course of Study. Seminars for Teachers. Cambridge, Mass.: EDC, 1969, 1970.

Mandler, George. "Origins of the Cognitive (R)Evolution." *Journal of the History of the Behavioral Sciences* 38, no. 4 (2002): 339–53.

Marcuse, Herbert. *One Dimensional Man: Studies in the Ideology of Advanced Industrial Society.* Boston: Beacon Press, 1964.

Marks, Jeannette. *Genius and Disaster: Studies in Drugs and Genius.* New York: Adelphi, 1925.

Marquis, D. G., Harold Guetzkow, and R. W. Heyns. "A Social Psychological Study of the Decision-Making Conference." In *Groups, Leadership and Men: Research in Human Relations*, edited by Harold Guetzkow, 55–67. Pittsburgh, Pa.: Carnegie Press, 1951.

Marshner, Susan M. *Man: A Course of Study—Prototype for Federalized Textbooks?* Washington, D.C.: The Heritage Foundation, 1975.

Martin, William C. *With God on Our Side: The Rise of the Religious Right in America.* New York: Broadway Books, 1996.

Martin-Nielsen, Janet. "'It Was All Connected': Computers and Linguistics in Early Cold War America." In *Cold War Social Science: Knowledge Production, Liberal Democracy, and Human Nature*, edited by Mark Solovey and Hamilton Cravens, 63–78. New York: Palgrave Macmillan, 2012.

Maslow, Abraham H. "The Authoritarian Character Structure." *Journal of Social Psychology* 18 (1943): 401–11.

———. "Cognition of the Particular and of the Generic." *Psychological Review* 55 (1948): 22–39.

———. "Creativity in Self-Actualizing People." In *Creativity and Its Cultivation*, edited by Harold H. Anderson, 83–95. New York: Harper and Row, 1959.

———. "The Expressive Component of Behavior." *Psychological Review* 56, no. 5 (1949): 261–72.

———. "Problem-Centering Versus Means-Centering in Science." *Philosophy of Science* 13 (1946): 326–31.

Mattson, Kevin. *Intellectuals in Action: The Origins of the New Left and Radical Liberalism, 1945–1970.* University Park: Pennsylvania State University Press, 2002.

Mayr, Ernest. "Concepts of Classification and Nomenclature in Higher Organisms and Microorganisms." *Annals of the New York Academy of Science* 56 (1952): 391–97.

McAfee, Kathy, and Myrna Wood. "Bread and Roses." In *Voices from Women's Liberation*, edited by Leslie Tanner, 413–33. New York: Signet Books, 1970.

McClelland, David C. "The Psychology of Mental Content Reconsidered." *Psychological Review* 62, no. 4 (1955): 297–302.

McGirr, Lisa. *Suburban Warriors: The Origins of the New American Right.* Princeton, N.J.: Princeton University Press, 2001.

McGrath, Earl J. "The General Education Movement." *Journal of General Education* 1, no. 1 (1946): 3–8.

McGrath, Joseph E. "Small Group Research, That Once and Future Field: An Interpretation of the Past with an Eye to the Future." *Group Dynamics: Theory, Research, and Practice* 1, no. 1 (1997): 7–27.

McGraw, Onallee. *Secular Humanism and the Schools: The Issue Whose Time Has Come.* Washington, D.C.: The Heritage Foundation, 1976.

———. "Violation of Constitutional Rights." *Bethesda-Chevy Chase Tribune*, March 2, 1973.

McGreevy, John T. "Thinking on One's Own: Catholicism in the American Intellectual Imagination, 1928–1960." *Journal of American History* 84, no. 1 (1997): 97–131.

McGuire, Frederick L. "On the Issue 'What Is Science?'" *American Psychologist* 11, no. 3 (1956): 152–53.

McKellar, P. *Imagination and Thinking: A Psychological Analysis.* New York: Basic Books, 1957.

Mead, Margaret. *And Keep Your Powder Dry: An Anthropologist Looks at America.* New York: William Morrow, 1942.

———. *Coming of Age in Samoa: A Psychological Study of Primitive Youth for Western Civilisation.* 1st Perennial Classics ed. New York: HarperCollins, 2001.

———. "Cybernetics of Cybernetics." In *Purposive Systems: Proceedings of the First Annual Symposium of the American Society for Cybernetics*, edited by Heinz von Foerster, John D. White, Larry J. Peterson and John K. Russell, 1–11. New York and Washington, D.C.: Spartan Books, 1968.

Mead, Margaret, and Paul Byers. *The Small Conference: An Innovation in Communication*. The Hague: Mouton, 1968.

Meloen, Jos D. "The F Scale as a Predictor of Fascism: An Overview of 40 Years of Authoritarianism." In *Strength and Weakness: The Authoritarian Personality Today*, edited by William F. Stone, Gerda Lederer, and Richard Christie, 47–69. New York: Springer-Verlag, 1993.

Melton, Arthur W. "Present Accomplishments and Future Trends in Problem-Solving and Learning Theory." *American Psychologist* 11 (1956): 278–81.

Merriam, Charles E. "Annual Report of the American Council of Learned Societies." *The American Political Science Review* 20, no. 1 (1926): 189–92.

———. "Annual Report of the Social Science Research Council." *The American Political Science Review* 21, no. 1 (1927): 159–61.

———. "Annual Report of the Social Science Research Council." *The American Political Science Review* 20, no. 1 (1926): 185–89.

Merton, Robert K. "The Mosaic of the Behavioral Sciences." In *The Behavioral Sciences Today*, edited by Bernard Berelson, 247–72. New York: Basic Books, 1963.

Milam, Erika. "Public Science of the Savage Mind: Contesting Cultural Anthropology in the Cold War Classroom." *Journal for the History of the Behavioral Sciences* 49, no. 3 (2013).

Miller, George A. "Computers, Communication, and Cognition." *Advancement of Science* (1965): 417–30; reprinted in *The Psychology of Communication*, 93–124. New York: Basic Books, 1967.

———. "The Human Link in Communication Systems." *Proceedings of the National Electronics Conference* 12 (1956): 395–400.

———. "Information and Memory." *Scientific American*, August 1956: 42–46.

———. *Language and Communication*. New York: McGraw-Hill, 1951.

———. "Language and Psychology." In *New Directions in the Study of Language*, edited by Eric H. Lenneberg, 89–107. Cambridge, Mass.: MIT Press, 1964.

———. "The Magical Number Seven, Plus or Minus Two: Some Limits on Our Capacity for Processing Information." *Psychological Review* 63 (1956): 81–97.

———. *Occasional Paper #1: A Very Personal History*. Cambridge, Mass.: MIT Cognitive Science Workshop, June 1, 1979.

———. "Project Grammarama." In *The Psychology of Communication*, 125–87. New York: Basic Books, 1967.

———. "Thinking, Cognition, and Learning." In *The Behavioral Sciences Today*, edited by Bernard Berelson, 139–50. New York: Basic Books, 1963.

———. "What Is Information Measurement?" *American Psychologist* 8 (1953): 3–11.

Miller, George A., and Noam Chomsky. "Finitary Models of Language Users." In *Handbook of Mathematical Psychology*, edited by R. D. Luce, R. R. Bush, and Eugene Galanter, 2:419–91. New York: Wiley, 1963.

Miller, George A., and F. C. Frick. "Statistical Behavioristics and Sequences of Responses." *Psychological Review* 56 (1949): 311–24.

Miller, George A., and E. A Friedman. "The Reconstruction of Mutilated English Texts." *Information and Control* 1 (1957): 38–55.

Miller, George A., Eugene Galanter, and Karl H. Pribram. *Plans and the Structure of Behavior.* New York: Henry Holt and Company, 1960.

Miller, George A., G. A. Heise, and W. Lichten. "The Intelligibility of Speech as a Function of the Context of the Test Materials." *Journal of Experimental Psychology* 41 (1951): 329–35.

Miller, George A., and Jennifer A. Selfridge. "Verbal Context and the Recall of Meaningful Material." *American Journal of Psychology* 63 (1950): 176–85.

Miller, James. *Democracy Is in the Streets: From Port Huron to the Siege of Chicago.* Cambridge, Mass.: Harvard University Press, 1994.

Millikan, Max. "Inquiry and Policy: The Relation of Knowledge to Action." In *The Human Meaning of the Social Sciences*, edited by Daniel Lerner, 158–80. New York: Meridian Books, 1959.

Mills, C. Wright. "Mass Society and Liberal Education." In *Power, Politics, and People: The Collected Essays of C. Wright Mills*, edited by Irving Louis Horowitz, 353–73. London and New York: Oxford University Press, 1967 [1954].

———. "The New Left." *New Left Review*, September–October 1960.

———. "On Knowledge and Power." *Dissent*, July 1955, 202–12.

———. *Power, Politics, and People: The Collected Essays of C. Wright Mills.* London: Oxford University Press, 1967.

Mindell, David A. *Between Human and Machine: Feedback, Control, and Computing before Cybernetics.* Baltimore, Md.: Johns Hopkins University Press, 2002.

Mirowski, Philip. *Machine Dreams: Economics Becomes a Cyborg Science.* Cambridge: Cambridge University Press, 2002.

Mooney, Ross L. "Groundwork for Creative Research." *American Psychologist* 9, no. 9 (1954): 544–48.

Mooney, Ross L., and Taher A. Razik. "Preface." In *Explorations in Creativity*, edited by Ross L. Mooney and Taher A. Razik, ix–x. New York: Harper and Row, 1967.

Morawski, Jill G. "Impossible Experiments and Practical Constructions: The Social Bases of Psychologists' Work." In *The Rise of Experimentation in American Psychology*, edited by Jill G. Morawski, 72–93. New Haven, Conn.: Yale University Press, 1988.

———. "Organizing Knowledge and Human Behavior at Yale's Institute of Human Relations." *Isis* 77 (1986): 219–42.

Morison, Elting E. "The Course of the Discussion." In *The American Style: Essays in Value and Performance*, edited by Elting E. Morison, 395–416. New York: Harper and Brothers, 1958.

Murdock, George Peter. "The Conceptual Basis of Area Research." *World Politics* 2, no. 4 (1950): 571–78.

Musselman, Phyllis. "Shofstall Gives Some Advice on MACOS." *Weekly American News*, September 29, 1971.

———. "We're All Annimals [*sic*]—Kids Are Taught Here." *Weekly American News*, September 22, 1971.

Myrdal, Gunnar, Richard Mauritz, Edvard Sterner, and Arnold Rose. *An American Dilemma: The Negro Problem and Modern Democracy.* 2 vols. New York: Harper and Brothers, 1944.

Nedell, Allan A. "Project Troy and the Cold War Annexation of the Social Sciences." In *Universities and Empire: Money and Politics in the Social Sciences During the Cold War*, edited by Christopher Simpson, 3–38. New York: The New Press, 1998.

———. "'Truth Is Our Weapon': Project Troy, Political Warfare, and Government-Academic Relations in the National Security State." *Diplomatic History* 17 no. 3 (1993): 399–420.

Neisser, Ulric. *Cognitive Psychology*. Englewood Cliffs, N.J.: Prentice-Hall, 1967.

Nelkin, Dorothy. *Science Textbook Controversies and the Politics of Equal Time.* Cambridge, Mass.: MIT Press, 1977.

New, Melvyn. "Nineteen Eighty-Four." *American Speech* 51, no. 3/4 (1976): 278–79.

Newell, Allen, J. C. Shaw, and Herbert A. Simon. "Elements of a Theory of Human Problem Solving." *Psychological Review* 65 (1958): 151–66.

———. "The Processes of Creative Thinking." In *Contemporary Approaches to Creative Thinking*, edited by Howard E. Gruber, Glenn Terrell, and Michael Wertheimer, 63–119. New York: Atherton Press, 1962.

———. "Report on a General Problem-Solving Program." In *Proceedings of the International Conference on Information Processing*. Paris, 1959.

Newell, Allen, and Herbert Simon. *The Simulation of Human Thought.* Santa Monica, Calif.: The RAND Corporation, 1959.

Nicholson, Ian A. M. "Gordon Allport, Character, and the 'Culture of Personality,' 1897–1937." *History of Psychology* 1, no. 1 (1998): 52–68.

Nickerson, Michelle. "Domestic Threats: Women, Gender and Conservatism in Los Angeles, 1945–1966." Ph.D. diss., Yale University, 2003.

Nisbet, J. F. *The Insanity of Genius.* New York: Scribner's, 1912.

Norman, Donald A., and Willem J. M. Levelt. "Life at the Center." In *The Making of Cognitive Science: Essays in Honor of George A. Miller*, edited by William Hirst, 100–109. Cambridge: Cambridge University Press, 1988.

Novick, Peter. *That Noble Dream: The "Objectivity Question" and the American Historical Profession.* Cambridge: Cambridge University Press, 1988.

"NSC-68: United States Objectives and Programs for National Security." 1950.

O'Connell, Charles Thomas. "Social Structure and Science: Soviet Studies at Harvard." Ph.D. diss., UCLA, 1990.

O'Donnell, John M. *The Origins of Behaviorism: American Psychology, 1870–1920.* New York: New York University Press, 1985.

Olver, Rose R. "A Developmental Study of Cognitive Equivalence." Ph.D. diss., Harvard University, 1962.

Oppenheimer, J. Robert. "Analogy in Science." *American Psychologist* 11, no. 3 (1956): 127–35.

———. "The Growth of Science and the Structure of Culture: Comments on Dr. Frank's Paper." In *Science and the Modern Mind*, edited by Gerald Holton, 63–73. Cambridge, Mass.: Harvard University Press, 1958.

———. "Theory Versus Practice in American Values and Performance." In *The American Style: Essays in Value and Performance*, edited by Elting E. Morison, 111–23. New York: Harper and Brothers, 1958.

Osborn, A. F. *Applied Imagination: Principles and Procedures of the Creative Process.* Rev. ed. New York: Scribner's, 1953.

OSS Assessment Staff. *Assessment of Men: Selection of Personnel for the Office of Strategic Services.* New York: Reinhart, 1948.

Parker, Richard. *John Kenneth Galbraith: His Life, His Politics, His Economics.* New York: Farrar, Straus and Giroux, 2005.

Parsons, Talcott. "Clyde Kluckhohn and the Integration of the Social Sciences." In *Culture and Life: Essays in Memory of Clyde Kluckhohn*, edited by Walter W. Taylor, John L. Fischer, and Evon Z. Vogt. Carbondale: Southern Illinois University Press, 1973.

———. *Essays in Sociological Theory.* Glencoe, Ill.: The Free Press, 1949.

———. "Evolutionary Universals in Society." *American Sociological Review* 29, no. 3 (1964): 339–57.

———. *The Structure of Social Action: A Study in Social Theory with Special Reference to a Group of Recent European Writers.* New York: McGraw-Hill, 1937.

Parsons, Talcott, and Edward Shils, eds. *Toward a General Theory of Action.* Cambridge, Mass.: Harvard University Press, 1951.

Pells, Richard. *The Liberal Mind in a Conservative Age: American Intellectuals in the 1940s and 1950s.* New York: Harper and Row, 1985.

Pfeffer, Richard M., ed. *No More Vietnams? The War and the Future of American Policy.* New York, Evanston, and London: Harper and Row for the Adlai Stevenson Institute for International Affairs, 1968.

Pheterson, Gail I., Sara B. Kiesler, and Philip A. Goldberg. "Evaluation of the Performance of Women as a Function of Their Sex, Achievement, and Personal History." *Journal of Personality and Social Psychology* 19, no. 1 (1971): 114–18.

Phillips, Christopher James. "The American Subject: The New Math and the Making of a Citizen." Ph.D. diss., Harvard University, 2011.

Phillips, Henry B. "Smoker in the Lounge." *Bulletin of the American Academy of Arts and Sciences* 6, no. 8 (1953): 1.

Platt, Jennifer. *A History of Sociological Research Methods in America: 1920–1960.* Ideas in Context, no. 40. New York: Cambridge University Press, 1996.

Poincaré, Henri. *Science and Method.* Translated by F. Maitland. New York: Dover, 1952.

Polya, G. *How to Solve It.* Princeton, N.J.: Princeton University Press, 1945.

———. *Mathematics and Plausible Reasoning.* Princeton, N.J.: Princeton University Press, 1954.

Polyani, Michael. *Personal Knowledge.* Chicago: University of Chicago Press, 1958.

Pooley, Jefferson, and Mark Solovey. "Marginal to the Revolution: The Curious Relationship between Economics and the Behavioral Sciences Movement in Mid-Twentieth-Century America." *History of Political Economy* 42, annual supplement (2010): 199–233.

Popkin, Ann Hunter. "Bread and Roses: An Early Moment in the Development of Socialist-Feminism." Ph.D. diss., Brandeis University, 1978.

Porter, Theodore M. *Trust in Numbers: The Pursuit of Objectivity in Science and Public Life.* Princeton, N.J.: Princeton University Press, 1995.

Posner, M., and G. L. Schulman. "Cognitive Science." In *The First Century of Experimental Psychology,* edited by E. Hearst, 371–406. Hillsdale, N.J.: Lawrence Erlbaum, 1979.

Powers, Marshall K. "Area Studies." *Journal of Higher Education* 26, no. 2 (1955): 82–89, 113.

Pratt, C. C. *The Logic of Modern Psychology.* New York: Macmillan, 1939.

"Preface to the Issue." *Daedalus* 94, no. 1 (1965): iii–iv.

Prentice, W. C. H. "Operationism and Psychological Theory: A Note." *Psychological Review* 53, no. 4 (1946): 247–49.

Radick, Gregory. *The Simian Tongue: The Long Debate About Animal Language.* Chicago: University of Chicago Press, 2007.

Rand, Christopher. *Cambridge, U.S.A.: Hub of a New World.* New York: Oxford University Press, 1964.

———. "Center of a New World," part 3. *The New Yorker,* April 25, 1964, 55–129.

Rasmussen, Nicolas. *Picture Control: The Electron Microscope and the Transformation of Biology in America, 1940–1960.* Stanford, Calif.: Stanford University Press, 1997.

Ravitch, Diane. *The Troubled Crusade: American Education 1945–1980.* New York: Basic Books, 1983.

Razik, Taher A. *Bibliography of Creativity Studies and Related Areas.* Buffalo: State University of New York, 1965.

Reed, Edward. *From Soul to Mind: The Emergence of Psychology from Erasmus Darwin to William James*. New Haven, Conn.: Yale University Press, 1997.

Reisch, George A. *How the Cold War Transformed Philosophy of Science: To the Icy Slopes of Logic*. Cambridge: Cambridge University Press, 2005.

Richards, Graham. "Of What Is History of Psychology a History?" *British Journal for the History of Science* 20 (1987): 201–11.

——. *Race, Racism, and Psychology: Towards a Reflexive History*. London and New York: Routledge, 1997.

Richards, Joan L. "Geometry in the Age of Reason: Euclid and the Enlightenment." Paper presented at the History of Science Society, Kansas City, 1998.

Riesman, David. *The Lonely Crowd: A Study of the Changing American Character*. New Haven, Conn.: Yale University Press, 1950.

Robertson, Pat. *The New World Order*. Dallas: World Publishing Co., 1991.

Robin, Ron. *The Making of the Cold War Enemy: Culture and Politics in the Military-Intellectual Complex*. Princeton, N.J.: Princeton University Press, 2001.

Robins, Richard W., Samuel D. Gosling, and Kenneth H. Craik. "An Empirical Analysis of Trends in Psychology." *American Psychologist* 54, no. 2 (1999): 117–28.

Rodgers, Daniel T. *The Age of Fracture*. Cambridge, Mass.: Belknap Press of Harvard University Press, 2011.

Roe, Anne. *The Making of a Scientist*. New York: Dodd, Mead, 1953.

——. "A Psychological Study of Physical Scientists." *Psychological Monographs* 43 (1951): 121–239.

Rogers, Carl R. "Persons or Science? A Philosophical Question." *American Psychologist* 10, no. 7 (1955): 267–78.

Rogin, Michael Paul. *The Intellectuals and McCarthy: The Radical Specter*. Cambridge, Mass.: MIT Press, 1967.

Rokeach, Milton. "'Narrow-Mindedness' and Personality." *Journal of Personality* 30, no. 2 (1951): 234–51.

——. *The Open and Closed Mind: Investigations into the Nature of Belief Systems and Personality Systems*. New York: Basic Books, 1960.

Roose, Kenneth. "Observations on Interdisciplinary Work in the Social Sciences." In *Interdisciplinary Relationships in the Social Sciences*, edited by Muzafer Sherif and Carolyn Sherif, 323–27. Chicago: Aldine, 1969.

Rosenthal, Robert. *Pygmalion in the Classroom; Teacher Expectations and Pupils' Intellectual Development*. New York: Holt, Rinehart, and Winston, 1968.

Ross, Dorothy. *The Origins of American Social Science*. Cambridge and New York: Cambridge University Press, 1991.

Rossinow, Douglas C. "Mario Savio and the Politics of Authenticity." In *The Free Speech Movement: Reflections on Berkeley in the 1960s*, edited by Robert Cohen and Reginald E. Zelnik, 533–51. Berkeley and Los Angeles: University of California Press, 2002.

———. *The Politics of Authenticity: Liberalism, Christianity, and the New Left in America*. New York: Columbia University Press, 1998.

Rostow, W. W. "The National Style." In *The American Style: Essays in Value and Performance*, edited by Elting E. Morison, 246–313. New York: Harper and Brothers, 1958.

Royko, Mike. "Wife-Swapping Eskimos vs. TV Morals." *The Denver Post*, February 9, 1976, 14.

Rudolph, Fredrick. *Curriculum: A History of the American Undergraduate Course of Study since 1636*. San Francisco: Jossey-Bass, 1977.

Rudolph, John L. *Scientists in the Classroom: The Cold War Reconstruction of American Science Education*. New York: Palgrave, 2002.

Samelson, Franz. "The Authoritarian Character from Berlin to Berkeley and Beyond: The Odyssey of a Problem." In *Strength and Weakness: The Authoritarian Personality Today*, edited by William F. Stone, Gerda Lederer, and Richard Christie, 22–43. New York: Springer-Verlag, 1993.

———. "Authoritarianism from Berlin to Berkeley: On Social Psychology and History." *Journal of Social Issues* 42, no. 1 (1986): 191–208.

———. "Organizing for the Kingdom of Behavior: Academic Battles and Organizational Policies in the Twenties." *Journal of the History of the Behavioral Sciences* 21 (1985): 33–47.

Sanford, R. Nevitt. "A Personal Account of the Study of Authoritarianism: Comment on Samelson." *Journal of Social Issues* 42 (1986): 209–14.

Sanford, S. H. "Annual Report of the Executive Secretary." *American Psychologist* 7 (1952): 686–96.

Saunders, Frances Stonor. *Who Paid the Piper? The CIA and the Cultural Cold War*. London: Granta Books, 1999.

Schelling, Thomas C. *The Strategy of Conflict*. Cambridge, Mass.: Harvard University, 1980.

Schlafly, Phyllis. *Child Abuse in the Classroom: Excerpts from Official Transcript of Proceedings before the U.S. Department of Education*. Alton, Ill.: Pere Marquette Press, 1984.

Schlamm, William S. "150 Drawings—but out of This World." *National Review*, May 23, 1956, 18.

———. "The Self-Importance of Picasso." *National Review*, July 13, 1957, 65–66.

Schlesinger, Arthur M., Jr. *The Vital Center: The Politics of Freedom*. Boston: Houghton-Mifflin, 1949.

Schrank, Sarah. "The Art of the City: Modernism, Censorship, and the Emergence of Los Angeles's Postwar Art Scene." *American Quarterly* 56, no. 3 (2004): 663–91.

Schrecker, Ellen. *The Lost Soul of Higher Education: Corporatization, the As-*

sault on Academic Freedom, and the End of the American University. New York: The New Press, 2010.

——. *No Ivory Tower: McCarthyism and the Universities*. New York: Oxford University Press, 1986.

Schrum, Ethan. "Administering American Modernity: The Instrumental University in the Postwar United States." Ph.D. diss., University of Pennsylvania, 2009.

Schulman, Bruce J. *The Seventies: The Great Shift in American Culture, Society, and Politics*. New York: The Free Press, 2001.

Schurmann, Franz, Peter Dale Scott, and Reginald Zelnik. *The Politics of Escalation in Vietnam*. Boston: Beacon Press, 1966.

Searle, John R. "A Foolproof Scenario for Student Revolts." *New York Times Magazine*, December 29, 1968. Reprinted in *The University Crisis Reader*, edited by Immanuel Maurice Wallerstein and Paul Starr, 2:31–41. New York: Random House, 1971.

Seidel, Robert. "The Origins of Lawrence Berkeley Laboratory." In *Big Science: The Growth of Large Scale Research*, edited by Peter Galison and Bruce Hevly, 21–45. Stanford, Calif.: Stanford University Press, 1992

Selcer, Perrin. "Patterns of Science: Developing Knowledge for a World Community at UNESCO." Ph.D. diss., University of Pennsylvania, 2011.

——. "The View from Everywhere: Disciplining Diversity in Post-World War II International Social Science." *Journal of the History of the Behavioral Sciences* 45, no. 4 (2009): 309–29.

Self-Study Committee. *A Report on the Behavioral Sciences at the University of Chicago*. Chicago: University of Chicago, 1954.

Sewell, William. "Some Reflections on the Golden Age of Interdisciplinary Social Psychology." *Annual Review of Sociology* 15 (1989): 1–16.

Shane, Harold G. "Significant Writings That Have Influenced the Curriculum: 1906–81." *Phi Delta Kappan* 62, no. 5 (1981): 311–14.

Shannon, Claude. "A Mathematical Theory of Communication." *Bell Systems Technical Journal* 27 (1948): 379–423, 623–56.

Shapin, Steven. "'The Mind Is Its Own Place': Science and Solitude in 17th-Century England." *Science in Context* 4 (1991): 191–218.

——. *The Scientific Life: A Moral History of a Late Modern Vocation*. Chicago: University of Chicago Press, 2008.

——. *A Social History of Truth: Civility and Science in Seventeenth-Century England*. Chicago: University of Chicago Press, 1994.

Shils, Edward. "Authoritarianism: 'Right' and 'Left.'" In *Studies in the Scope and Method of the Authoritarian Personality*, edited by Richard Christie and Marie Jahoda, 24–49. Glencoe, Ill.: The Free Press, 1954.

——. "Transition, Ecology, and Institution in the History of Sociology." *Daedalus* 4 (1970): 760–825.

Schorske, Carl E. *Fin-De-Siècle Vienna: Politics and Culture*. New York: Vintage Books, 1981.

———. "A Life of Learning." In *Recasting America: Culture and Politics in the Age of Cold War*, edited by Lary May, 93–103. Chicago: University of Chicago Press, 1989.

Shulman, Holly Cowan. *The Voice of America: Propaganda and Democracy, 1941–1945*. Madison: University of Wisconsin Press, 1990.

Silberman, Charles E. *Crisis in the Classroom: The Remaking of American Education*. New York: Random House, 1970.

Simon, Herbert A. "Cognitive Science: The Newest Science of the Artificial." *Cognitive Science* 4 (1980): 33–46.

———. *Models of My Life*. New York: Basic Books, 1991.

———. "Scientific Discovery and the Psychology of Problem Solving." In *Mind and Cosmos*, edited by Robert G. Colodny, 22–40. Pittsburgh, Pa.: University of Pittsburgh Press, 1966.

———. "Thinking by Computers." In *Mind and Cosmos*, edited by Robert G. Colodny, 3–21. Pittsburgh, Pa.: University of Pittsburgh Press, 1966.

———. "Understanding Creativity." In *Creativity: Its Educational Implications*, edited by John Curtis Gowan, George D. Demos, and E. Paul Torrence, 43–53. New York: John Wiley and Sons, 1967.

Simon, Herbert, and Allen Newell. "Computer Simulation of Human Thinking and Problem Solving." In *Management and the Computer of the Future*, edited by Martin Greenberger, 94–133. New York: MIT Press and Wiley, 1962.

Skaggs, E. B. "Personalistic Psychology as Science." *Psychological Review* 52 (1945): 234–48.

Skinner, B. F. "Are Theories of Learning Necessary?" *Psychological Review* 57, no. 4 (1950): 193–216.

———. "A Case History in Scientific Method." *American Psychologist* 11, no. 5 (1956): 221–33.

———. "Why I Am Not a Cognitive Psychologist." In *Reflections on Behaviorism and Society*, 97–112. Englewood Cliffs, N.J.: Prentice Hall, 1978.

Smith, Laurence. *Behaviorism and Logical Positivism: A Reassessment of the Alliance*. Stanford, Calif.: Stanford University Press, 1986.

———. "Metaphors of Knowledge and Behavior in the Behaviorist Tradition." In *Metaphors in the History of Psychology*, edited by David E. Leary, 237–66. Cambridge: Cambridge University Press, 1990.

Smith, M. Brewster. "*The Authoritarian Personality*: A Re-Review 46 Years Later." *Political Psychology* 18 (1997): 159–63.

———. "Review of T. W. Adorno, E. Frenkel-Brunswik, D. J. Levinson and R. Nevitt Sanford, *The Authoritarian Personality*." *Journal of Abnormal and Social Psychology* 45 (1950): 775–79.

Snow, C. P. *The Two Cultures and the Scientific Revolution*. The Rede Lecture, 1959. Cambridge: Cambridge University Press, 1959.

Sokal, Michael M. "The Gestalt Psychologists in Behaviorist America." *American Historical Review* 89, no. 5 (1984): 1240–63.

——, ed. *Psychological Testing and American Society, 1890–1930*. New Brunswick, N.J.: Rutgers University Press, 1987.

Solomon, Lawrence N. "The Paradox of the Experimental Clinician." *American Psychologist* 10, no. 4 (1955): 170–71.

Solovey, Mark. "Project Camelot and the 1960s Epistemological Revolution: Rethinking the Politics-Patronage-Social Science Nexus." *Social Studies of Science* 31, no. 2 (2001): 171–206.

——. "Riding the Natural Scientists' Coattails onto the Endless Frontier: The SSRC and the Quest for Scientific Legitimacy." *Journal of the History of the Behavioral Sciences* 40, no. 4 (2004): 393–422.

——. "Social Science Funding under Siege in Conservative Times: The Case of the U.S. National Science Foundation during the 1970s." Paper presented at the Ninth Annual Workshop on the History of Economics as History of Science, École Normale Supérieure de Cachan, June 16, 2012.

Solovey, Mark, and Jefferson D. Pooley. "The Price of Success: Sociologist Harry Alpert, the NSF's First Social Science Policy Architect." *Annals of Science* 68, no. 2 (2011): 229–60.

Spearman, Charles Edward. *Creative Mind*. New York: D. Appleton, 1931.

Spence, Kenneth W. *Behavior Theory and Conditioning*. New Haven, Conn.: Yale University Press, 1956.

——. "Clark Leonard Hull: 1884–1952." *American Journal of Psychology* 65 (1952): 639–46.

——. "The Empirical Basis and the Theoretical Structure of Psychology." *Philosophy of Science* 24, no. 2 (1957): 97–108.

Sperry, Roger W. "On the Neural Basis of the Conditioned Response." *British Journal of Animal Behavior* 3 (1955): 41–44.

Stanger, Ross. "Fascist Attitudes: An Exploratory Study." *Journal of Social Psychology* 7, no. 3 (1936): 309–19.

——. "Fascist Attitudes: Their Determining Conditions." *Journal of Social Psychology* 7, no. 4 (1936): 438–54.

Steinbacher, John A. *The Conspirators: Men against God*. Whittier, Calif.: Orange Tree Press, 1972.

——. *Robert Francis Kennedy, the Man, the Mysticism, the Murder*. Los Angeles: Impact Publishers, 1968.

Steinpreis, R. E., K. A. Anders, and D. Ritzke. "The Impact of Gender on the Review of the Curriculum Vitae of Job Applicants and Tenure Candidates: A National Empirical Study." *Sex Roles* 41 (1999): 509.

Stevens, S. S. "The Attributes of Tones." *Proceedings of the National Academy of Sciences* 20 (1934): 457–59.

——. "The Operational Basis of Psychology." *American Journal of Psychology* 43 (1935): 323–30.

——. "The Operational Definition of Psychological Concepts." *Psychological Review* 42 (1935): 512–27.

——. "Psychology and the Science of Science." *Psychological Bulletin* 36, no. 4 (1939): 221–63.

——. "Psychology, the Propaedeutic Science." *Philosophy of Science* 3 (1936): 90–103.

——. "The Relation of Pitch to Intensity." *Journal of the Acoustical Society of America* 6 (1935): 150–54.

——. "A Scale for the Measurement of a Psychological Magnitude." *Psychological Review* 43, no. 5 (1936): 405–16.

——. "Tonal Density." *Journal of Experimental Psychology* 17 (1934): 585–91.

——, ed. *Handbook of Experimental Psychology*. New York: Wiley, 1951.

Stone, William F., Gerda Lederer, and Richard Christie. "The Status of Authoritarianism." In *Strength and Weakness: The Authoritarian Personality Today*, edited by William F. Stone, Gerda Lederer, and Richard Christie, 229–45. New York: Springer-Verlag, 1993.

Stouffer, Samuel A. *Communism, Conformity, and Civil Liberties: A Cross Section of the Nation Speaks Its Mind*. New York: Doubleday, 1955.

——. "Research Techniques." In *The Social Sciences at Mid-Century*, 62–69. Freeport, N.Y.: Books For Libraries Press, 1968.

Stouffer, Samuel A., Edward A. Suchman, Leland DeVinney, Shirley A. Star, and Robin M. Williams Jr. *The American Soldier*. Princeton, N.J.: Princeton University Press, 1949.

Strupp, Hans H., George F. Castore, Richard A. Lake, Reed M. Merrill, and Leopold Bellak. "Comments on Rogers' 'Persons or Science.'" *American Psychologist* 11, no. 3 (1956): 153–57.

Sullivan, Andrew. "Trying to Understand the Tea Party II." *The Atlantic*, June 15, 2010.

"Summary of the Discussion at the Meeting on February 13." *Bulletin of the American Academy of Arts and Sciences* 5, no. 6 (1952): 2.

Summerfield, Jack D., Lorlyn Thatcher, and with commentary by Lyman Bryson, eds. *The Creative Mind and Method: Exploring the Nature of Creativeness in American Arts, Sciences, and Professions*. New York: Russell and Russell Inc., 1964.

Sundberg, Norman D. "Basic Readings in Psychology." *American Psychologist* 15 (1960): 343–45.

Taylor, Donald W. *Thinking*. New Haven, Conn.: Yale University Department of Psychology, 1962.

Taylor, Harold. "The Philosophical Foundations of General Education." In *Fifty First Year Book of the National Society for the Study of Education*, edited by Nelson B. Henry, 20–45. Chicago: National Society for the Study of Education, 1952.

Terman, Lewis M., ed. *Genetic Studies of Genius*. 4 vols. Stanford, Calif.: Stanford University Press, 1925, 1926, 1930, 1957.

Thaler, Richard H., and Cass R. Sunstein. *Nudge: Improving Decisions About Health, Wealth, and Happiness*. New Haven, Conn.: Yale University Press, 2008.

Thimmesch, Nick. "The Grass-Roots Dollar Chase—Ready on the Right." *New York Magazine*, June 9, 1975, 58–63.

Thogmartin, Mark B. "The Prevalence of Phonics Instruction in Fundamentalist Christian Schools." *Journal of Research on Christian Education* 3, no. 1 (1994): 103–32.

Thompson, Dennis F. *The Democratic Citizen: Social Science and Democratic Theory in the Twentieth Century*. Cambridge: Cambridge University Press, 1970.

Thorndike, Joseph J. "'The Sometimes Sordid Level of Race and Segregation': James J. Kilpatrick and the Virginia Campaign against *Brown*." In *The Moderate's Dilemma: Massive Resistance to School Desegregation in Virginia*, edited by Matthew D. Lassiter and Andrew B. Lewis, 51–71. Charlottesville: The University Press of Virginia, 1998.

Thorne, Frederick C. "Psychologists, Heal Thyselves!" *American Psychologist* 11, no. 3 (1956): 152.

Thorpe, Charles. *Oppenheimer: The Tragic Intellect*. Chicago: University of Chicago Press, 2006.

Toews, John E. "Historicizing Psychoanalysis: Freud in His Time and for Our Time." *Journal of Modern History* 63 (1991): 504–45.

Tolman, Edward C. "Cognitive Maps in Rats and Men." *Psychological Review* 55, no. 4 (1948): 189–208.

Trilling, Lionel. *The Liberal Imagination: Essays on Literature and Society*. New York: Viking Press, 1950.

Triplet, Rodney G. "Henry A. Murray and the Harvard Psychological Clinic, 1926–1938: A Struggle to Expand the Disciplinary Boundaries of Academic Psychology." Ph.D. diss., University of New Hampshire, 1983.

Tucker, William H. "Fact and Fiction in the Discovery of Sir Cyril Burt's Flaws." *Journal of the History of the Behavioral Sciences* 30 (1994): 335–47.

UNESCO. *The Technique of International Conferences: A Progress Report on Problems and Methods*. Paris: UNESCO/SS/3, 1951.

Unger, Rhoda Kesler. *Resisting Gender: Twenty-Five Years of Feminist Psychology, Gender and Psychology*. London and Thousand Oaks, Calif.: Sage Publications, 1998.

University of California–Berkeley, Study Commission on University Governance. *The Culture of the University: Governance and Education*. Berkeley: University of California, 1968.

University of North Carolina. *Survey of the Behavioral Sciences, 1953–4: Report of the Visiting Committee*. 1954.

Verba, Sidney. *Small Groups and Political Behavior: A Study of Leadership*. Princeton, N.J.: Princeton University Press, 1961.

Verplanck, William S. "A Glossary of Some Terms Used in the Objective Science of Behavior." *Psychological Review* 64, no. 6, part 2 (1957): i–vi, 1–42.

Veysey, Laurence R. *The Emergence of the American University*. Chicago: University of Chicago Press, 1965.

Von Eschen, Penny M. *Satchmo Blows Up the World: Jazz Ambassadors Play the Cold War*. Cambridge, Mass.: Harvard University Press, 2006.

von Neumann, John. "The General and Logical Theory of Automata." In *Cerebral Mechanisms of Behavior: The Hixon Symposium*, edited by Lloyd A. Jeffries, 1–31. New York: John Wiley and Sons, 1951.

Wall, Wendy. *Inventing the American Way: The Politics of Consensus from the New Deal to the Civil Rights Movement*. Oxford and New York: Oxford University Press, 2008.

Wallace, H. R. "Creative Thinking: A Factor in Sales Productivity." *Vocational Guidance Quarterly* 9 (1961): 223–26.

Wallis, W. Allen. *The 1953–54 Program of University Surveys of the Behavioral Sciences*. Behavioral Sciences Division, Ford Foundation, 1955.

Watson, John B. *Behaviorism*. New York: W. W. Norton and Company, 1925.

———. "Psychology as the Behaviorist Views It." *Psychological Review* 20 (1913): 158–77.

Weiss, Nancy J. "The Negro and the New Freedom: Fighting Wilsonian Segregation." *Political Science Quarterly* 84, no. 1 (1969): 61–79.

Weisskopf, Victor Frederick. *The Joy of Insight: Passions of a Physicist*. New York: Basic Books, 1991.

Weisstein, Naomi. "'How Can a Little Girl Like You Teach a Great Big Class of Men?' The Chairman Said, and Other Adventures of a Woman in Science." In *Working It Out: 23 Women Writers, Artists, Scientists, and Scholars Talk About Their Lives and Work*, edited by Sara Ruddick and Pamela Daniels, 241–50. New York: Pantheon, 1977.

———. "'Kinder, Küche, Kirche' as Scientific Law: Psychology Constructs the Female." In *Sisterhood Is Powerful: An Anthology of Writings from the Women's Liberation Movement*, edited by Robin Morgan, 205–20. New York: Random House, 1970.

———. "Psychology Constructs the Female: or, The Fantasy Life of the Male Psychologist (with Some Attention to the Fantasies of His Friends, the Male Biologist and the Male Anthropologist)." In *Radical Feminism*, edited by Anne

Koedt, Ellen Levine, and Anita Rapone, 178–97. New York: Quadrangle Books, 1973.

Weisstein, Naomi, Virginia Blaisdell, and Jesse Lemisch. *The Godfathers*. New Haven, Conn.: Belladonna Publishing, 1975.

Welsch, Wayne W. "Twenty Years of Science Curriculum Development: A Look Back." *Review of Research in Education* 7 (1979): 282–306.

Wenneras, C., and A. Wold. "Nepotism and Sexism in Peer Review." *Nature* 387 (1997): 341.

Westbrook, Robert B. "John Dewey." *Prospects: The Quarterly Review of Comparative Education* 23, no. 1/2 (1993): 277–91.

———. *John Dewey and American Democracy*. Ithaca, N.Y.: Cornell University Press, 1991.

Weyeneth, Robert R. "The Architecture of Racial Segregation: The Challenges of Preserving the Problematical Past." *The Public Historian* 27, no. 4 (2005): 11–44.

Whipple, Babette. *Occasional Paper No. 10: The Grouptalk*. Cambridge, Mass.: Educational Services Incorporated, 1967.

White, Leonard D. "Introduction." In *The State of the Social Sciences*, edited by Leonard D. White, v–xi. Chicago: University of Chicago Press, 1956.

White, R. K. "Note on the Psychopathology of Genius." *Journal of Social Psychology* 1 (1930): 311–15.

Whitehead, Alfred N. *Science in the Modern World*. New York: The Macmillan Company, 1925.

Whyte, William H., Jr. *The Organization Man*. New York: Simon and Schuster, 1956.

Wilentz, Sean. "Confounding Fathers: The Tea Party's Cold War Roots." *The New Yorker*, October 18, 2010.

Wilson, Sloan. *The Man in the Gray Flannel Suit*. New York: Simon and Schuster, 1955.

Winston, Andrew S. "'As His Name Indicates': R. S. Woodworth's Letters of Reference and Employment for Jewish Psychologists in the 1930s." *Journal of the History of the Behavioral Sciences* 32, no. 1 (1996): 30–43.

———. "The Defects of His Race: E. G. Boring and Antisemitism in American Psychology." *History of Psychology* 1, no. 1 (1998): 27–51.

Winston, Andrew S., and Daniel J. Blais. "What Counts as an Experiment? A Transdisciplinary Analysis of Textbooks, 1930–1970." *American Journal of Psychology* 109, no. 4 (1996): 559–616.

Wirth, Louis, ed. *Eleven Twenty-Six: A Decade of Social Science Research*. Chicago: University of Chicago Press, 1940.

Wise, M. Norton, ed. *The Values of Precision*. Princeton, N.J.: Princeton University Press, 1995.

Wohl, R. Richard. "Some Observations on the Social Organization of Interdisciplinary Social Science Research." *Social Forces* 33, no. 4 (1955): 374–83.

Wolcott, Harry F. "The Middlemen of MACOS." *Anthropology and Education Quarterly* 38, no. 2 (2007): 195–206.

Wolgemuth, Kathleen L. "Woodrow Wilson and Federal Segregation." *The Journal of Negro History* 44, no. 2 (1959): 158–73.

Woolfson, Peter. "The Fight over 'MACOS': An Ideological Conflict in Vermont." *Council on Anthropology and Education Quarterly* 5, no. 3 (1974): 27–30.

Wooton, William. *SMSG: The Making of a Curriculum*. New Haven, Conn.: Yale University Press, 1965.

Wrong, Dennis. "The United States in Comparative Perspective: Max Lerner's America as a Civilization." *The American Journal of Sociology* 65, no. 5 (1960): 499–604.

X [George F. Kennan]. "The Sources of Soviet Conduct." *Foreign Affairs* 25, no. 4 (July 1947): 566–82.

Yates, Richard. *Revolutionary Road*. Boston: Little Brown, 1961.

Zacharias, Jerrold R. "Research Scholars and Curriculum Development." *PMLA* 80, no. 2 (1965): 13–15.

Zenderland, Leila. *Measuring Minds: Henry Herbert Goddard and the Origins of American Intelligence Testing*. Cambridge and New York: Cambridge University Press, 1998.

Zilversmit, Arthur. *Changing Schools: Progressive Education Theory and Practice, 1930–1960*. Chicago: University of Chicago Press, 1993.

Zinn, Howard. *Vietnam: The Logic of Withdrawal*. Boston: Beacon Press, 1967.

Index